RESEARCH
FRONTIERS IN
MAGNETOCHEMISTRY

RESEARCH FRONTIERS IN MAGNETOCHEMISTRY

Edited by

Charles J. O'Connor

Research Professor
Department of Chemistry
University of New Orleans

World Scientific
Singapore • New Jersey • London • Hong Kong

o 5501118
CHEMISTRY

Published by

World Scientific Publishing Co. Pte. Ltd.
P O Box 128, Farrer Road, Singapore 9128
USA office: Suite 1B, 1060 Main Street, River Edge, NJ 07661
UK office: 73 Lynton Mead, Totteridge, London N20 8DH

ISBN 981-02-1246-1

Printed in Singapore by Utopia Press.

FOREWORD

The investigation and characterization of magnetic phenomena is a prominent and fruitful area of chemical and materials research. The fruits of magnetochemistry research reveal information that promotes the understanding and control of the physical properties of materials. The reviews that make up this volume are focused on the overlap between the study of magnetism and the study of chemistry. The study of magnetism as it is applied to chemistry encompasses many common areas of interest including solid state chemistry, materials science, coordination chemistry, organic chemistry, and biochemistry.

Over the past 25 years, there have been many advances in the understanding of magnetic phenomena in molecular systems. For example, a variety of low dimensional materials, many new ferromagnetic, antiferromagnetic, and ferrimagnetic systems have been synthesized and analyzed, metal cluster compounds that exhibit magnetic exchange have been examined, new orbital overlap theories have been proposed to explain magneto-structural correlations in exchange coupled systems, and efforts directed toward the preparation of an organic ferromagnetic material have produced new and interesting compounds. There have also been many advances in the use of magnetism as a probe of inorganic biomolecules.

This volume is conceived as an attempt to bring together reviews of current research in magnetochemistry that are written by the world's leading researchers in the fields of chemistry, physics, materials science, and magnetism. "Research Frontiers in Magnetochemistry" contains comprehensive and in-depth reviews that describe some of the current activities of these scientists. These review articles detail the research that has been accomplished by these experts and lay the foundation for future research endeavors.

This volume is intended to be used by graduate students as well as senior researchers in all fields of magnetochemistry. There are several excellent text books that introduce elementary concepts of magnetochemistry and the theoretical models that may be used to explain magnetic behavior. This book is not intended to compete with these texts; indeed, some familiarity with magnetochemical literature is expected. This volume extends the discussions of magnetochemistry texts to the more intricate details of research that is currently being conducted at the frontiers of magnetochemical research.

Charles J. O'Connor
New Orleans, January 1993

CONTENTS

NOVEL ONE-DIMENSIONAL COPPER(II) MAGNETIC SYSTEMS

William E. Hatfield and Kathleen L. Trojan

*Department of Chemistry, University of North Carolina,
Chapel Hill, North Carolina 27599-3290, USA.*

INTRODUCTION

This review is devoted to a discussion of exchange interactions in selected low-dimensional copper(II) chain compounds, especially chains that have unusual structures. Compounds with properties that illustrate an important facet of exchange have been selected for discussion. The practical importance of low-dimensional systems has been demonstrated, for example, by the recent discovery of high-temperature superconducting materials, which is thought to be propagated through low-dimensional networks. Chain systems may have other important properties and find uses in devices. A better understanding of the fundamental properties of low-dimensional systems will permit the design of substances with prescribed properties.

Copper(II) chain compounds are especially useful in the study of low-dimensional systems since a large body of theoretical work exists,[1] and analysis of data for complicated structures may be readily obtained[2] for use in magneto-structural correlations. The low spin of the metal ion eliminates the complications from features such as zero-field splittings and resultant anisotropy, and of course, systems containing ions with low spins are more reflective of a quantum system than those with large, approaching classical, spins.

Space does not permit an exhaustive survey, but important systems to be discussed include the linear chain that has received much attention, that being catena-di-μ-chloro(bis-pyridine)copper(II) as well as the best example of a Heisenberg linear chain yet characterized, that being catena-μ-N,N'-pyrazine-bis(nitrato)copper(II). Other compounds to be discussed include aqua[N-(salicylaldiminato)glycinato]copper(II), a chain that exhibits progressive dimerization, and related alternating chain compounds; the next-nearest neighbor chain [Cu(hydrazinium)Cl$_3$], a compound that has spin frustration and undergoes a phase transition to relieve the frustration, and related compounds; the alternating next-nearest neighbor chain, copper(II) phthalate; chains composed of trimeric units such as Ca$_3$Cu$_3$(PO$_4$)$_4$; compounds with large copper-bridge-copper distances; and mixed-valence chains.

THEORETICAL DEVELOPMENTS AND FOUNDATIONS

Considerable effort has been expended recently on the calculation of electronic structures and magnetic properties of clusters of spins with novel and unusual structures.[1,3] Calculations have been carried out within parameter space bound frequently by a number of chemical systems, and closed-form expressions have been obtained for a number of these. Systems that will receive attention in this review exhibit exchange coupling by a variety of superexchange mechanisms. A vector coupling scheme may be used to model the quantitative behavior of systems with exchange-coupled spins with the Heisenberg-Dirac-Van Vleck (HDVV) Hamiltonian[4,5]

$$\mathcal{H} = -2J_{ij}S_i \cdot S_j \tag{1.1}$$

where the subscripts i and j correspond to the two different spins and J_{ij} is the exchange coupling constant. When J is positive, the exchange interaction is ferromagnetic, and negative J designates antiferromagnetic exchange interactions. It is often necessary to take into account anisotropy in the exchange interaction. For the N identical spin system, the HDVV Hamiltonian may be rewritten as

$$\mathcal{H} = -2J\Sigma(\alpha S_i^x \cdot S_j^x + \beta S_i^y \cdot S_j^y + \gamma S_i^z \cdot S_j^z) \tag{1.2}$$

For $\alpha = \beta = \gamma$, the isotropic Heisenberg case arises. When $\alpha = \beta = 0$, $\gamma = 1$, the Ising case arises, and when $\alpha = \beta = 1$, $\gamma = 0$, the XY case arises. Since Heisenberg exchange predominates in copper chain compounds, this review will be devoted to such systems.

The quantum mechanical problem can not be solved exactly except at 0 K. It is known that the ground state is a band or continuum of degenerate states.[6,7] The seminal work in the field was carried out by Bonner and Fisher.[8] Magnetic susceptibilities of clusters with increasing numbers of spins were calculated with the aim toward extrapolating to the infinite limit so that properties of infinite systems could be described. Subsequent work has been patterned after this pioneering effort, although some approaches, such as Monte-Carlo calculations, have been reported. Following the concept put forth by Bonner and Fisher, it has been possible in some cases to generate reasonably compact empirical equations from the calculated data sets. These equations can then be used to extract magnetic parameters from magnetic susceptibility data for compounds with novel chain structures by using desk-top computers.[3] Systems with parameters outside the range studied still require calculations on large computers.

CALCULATIONAL APPROACH

The basis functions which are used are characterized by their m_s values. For a system consisting of N paramagnetic ions of spin S, a total spin quantum number S_T may be defined which is composed of component energy levels S_z having spin values NS, NS-1, NS-2, $\cdots$, 0, or 1/2. In addition, the degeneracy of each S_z level is given by $(S_z) = \Omega(S_z) - \Omega(S_{z+1})$, where $\Omega(S_z)$ is the sum of the coefficients of the S_z^{th} order terms in the expansion $(x^{Sz} + x^{Sz-1} + \cdots x^{-Sz+1} + x^{-Sz})$. Determination of the zero-field energies of each of the degenerate levels and the effect of the magnetic field on these levels will then allow the calculation of the magnetic susceptibility. For the N ion case, it is necessary to calculate the matrix elements $<S \mid H \mid S'>$, where the N ion spin wave function is $<S \mid = <S_1, S_2, \cdots, S_{N-1}, S_N \mid$. For each spin state $\mid S>$, the total spin S_T is represented by the sum of the S_z components such that $S_T = \Sigma_i M_s^i$. The Hamiltonian, including the Zeeman interaction, will have the form $\mathcal{H} = -2J \Sigma S_i \cdot S_j + g\beta \Sigma H \cdot S_i^z$, where states with different S_T will not mix with respect to the Zeeman term since $<S \mid S'> = \delta$, and each spin function is an exact eigenfunction of S_z. The zero field energies take the form

$$-2J<S_iS_j \mid S_i \cdot S_j \mid S_iS_j> \cdot <S_1, S_2, \cdots, S_{i-1}, S_{i+1}, \cdots, S_{j-1}, S_{j+1},$$
$$\cdots, S_{N-1}, S_N \mid S'_1, S'_2, \cdots, S'_{i-1}, S'_{i+1}, \cdots, S'_{j-1}, S'_{j+1}, \cdots, S'_{N-1}, S'_N>$$

where the first term is a simple "two-site" term, and the multiplicative factor is $<S_R \mid S'_R>$. If S_T is not equal to S'_T, then at least one of the above terms will be zero because of the orthogonality of the spin basis functions. There are only two cases where the second term and therefore the zero-field matrix element is non-zero. First, if $S_iS_j = S'_iS'_j$, there will be a non-zero term if $<S_R \mid S'_R> = 1$. Since $M_s^i + M_s^j = M_s^i + M_s^j$, then $S_T = S'_T$. The second possibility results from the ladder operators in the expansion of $S_i \cdot S_j$. These conditions have been taken advantage of in order to calculate the zero field energies and sizes of the Hamiltonian matrices since the entire Hamiltonian matrix can be block factored according to S_T, and each submatrix solved separately.

Next, for the calculation of magnetic properties the derivative $-\delta E_i / \delta H$ must be determined. The Hellmann-Feynman theorem[9] provides the means for doing this operation. It can be shown that $\delta E_i / \delta H = <\delta \mathcal{H} / \delta H>$. The only field dependence in the Hamiltonian occurs in the Zeeman term, $g\beta \Sigma H \cdot S_i^z$. Since the eigenvalues and eigenvectors from the solution of the determinantal equation are known, then the expectation value $<g\beta S_z>$ for each m_s state can be calculated by linear algebraic techniques, where

$$\delta E_i / \delta H = [i^{th} \text{ row eigenvector}] < g\beta S_z > \left| \begin{array}{l} i^{th} \\ \text{column} \\ \text{eigenvector} \end{array} \right|$$

Expressions from calculations on finite ring clusters have been used successfully for the analysis of magnetic susceptibility data for a relatively large number of compounds with novel and diverse structures and superexchange pathways, and the results have been found to be in agreement with magneto-chemical-structural correlations, or have been used as the basis for new magneto-chemical-structural correlations. Calculations within the Heisenberg approximation have been carried out for $S = 1/2$ chains in the parameter space $J = \pm 1.0$ to ± 20.0 cm^{-1} in units of 1 cm^{-1} and $\alpha = \pm 0.1$ to ± 1.0 in units of 0.1. In general, the expressions that have been generated do not hold outside the range of parameters for which the calculations were carried out.[10]

TYPES OF MAGNETIC CHAINS

Copper(II) forms a wide variety of chains with novel structures. This is largely a result of the ability of copper(II) to adopt a number of coordination numbers and coordination geometries. While coordination number three is rare, coordination numbers four, five, and six are common and numerous. Geometries of compounds and chains with coordination number four vary from nearly tetrahedral through a series of distorted structures to square planar. Geometries of compounds and chains with coordination number five vary from trigonal bipyramidal, also through a series of distorted structures, to tetragonal pyramidal. Geometries of compounds for coordination number 6 are invariably distorted octahedral structures. Perfect tetrahedral and perfect octahedral structures are rare because the former with crystal-field electronic structure $(e_g)^4(t_{2g})^5$ and the latter with electronic structure $(t_{2g})^6(e_g)^3$ are orbitally degenerate and are subject to the Jahn-Teller effect. Theoretically, either an axial elongation or an axial compression would remove the degeneracy of the octahedral structure. However, nearly all six-coordinate copper(II) compounds exhibit axial elongation. A chain which apparently exhibits axial compression will be discussed in a subsequent section of this review. Six-coordinate copper(II) environments may depart significantly from a quasi-octahedral structure. An important example will be discussed below. Chains with copper(II) in distorted environments lead to significant problems when attempts are made to ascertain mechanisms of super-exchange and super-exchange pathways. Such problems are fascinating and stimulate additional research activity with chains of copper(II) compounds.

In the following sections, examples of copper(II) chain compounds with different types of structures will be presented. Space does not allow an exhaustive coverage, so examples have been selected to illustrate the rich variety of properties which may be encountered in working with copper(II) chain compounds.

A. The Uniformly Spaced Chain

The classical work on antiferromagnetically exchange-coupled spin 1/2 chain compounds was done by Bonner and Fisher, [8] and much subsequent work has been carried out in the spirit of this pioneering work. It may be said with some confidence that the physics of uniformly spaced copper(II) compounds is reasonably well understood,[11] but much remains to be explained concerning the chemical and structural features that determine the sign and magnitude of exchange coupling constants in most uniformly spaced chains. The difficulty arises from the large array of parameters, both chemical and structural, that are encompassed in the problem. The most reasonable approach is to study chains in which the number of parameters may be minimized, that is, those with a nearly constant structure and similar ligands, with the hope that some of the derived parameters may be transported from system to system. These studies are necessarily largely phenomenological and empirical in nature.

The uniformly spaced chain may be sketched as follows:

$$\text{——M——M——M——M——M——M——M——M——M——M——}$$
$$\backslash\_J\_/$$

where one exchange coupling constant J describes the nearest-neighbor exchange interaction, and J may be positive, denoting ferromagnetic exchange, or negative, denoting antiferromagnetic exchange. The appropriate Hamiltonian for the infinite chain is given in Eq. 1.1. There are only approximate solutions to the problem, except at 0 K, but it is known for the antiferromagnetically exchange-coupled chain that the ground state is a band or continuum of degenerate states.[6,7] Hall[12] fit the numerical results of Bonner and Fisher[8] for the magnetic susceptibility of the $S = 1/2$ uniform chain to the expression

$$\chi = \frac{Ng^2\mu_\beta^2}{k_BT} \times \frac{0.25 + 0.14995x + 0.30094x^2}{1 + 1.9862x + 0.68854x^2 + 6.0626x^3}$$

where $x = |J|/k_BT$. This closed-form expression is especially useful for the analysis of magnetic data for Heisenberg uniform-chain compounds, and it has been used extensively. Several important examples of antiferromagnetically exchange coupled uniform Heisenberg chains will be given in the section on Results.

Detecting ferromagnetic interactions in linear chain compounds requires more care, although the procedure is not difficult. First, assume that there are no exchange interactions in the assembly of molecules. The system would obey the Curie Law, $\chi =$

C/T. A plot of χT versus T will yield a straight, horizontal line. If there are ferromagnetic interactions, the plot of χT versus T will follow the line described above at high temperatures, but at low temperatures, there will be a divergence from the horizontal line to higher values of χT as the temperature decreases. In a like manner, if antiferromagnetic interactions are small, then a plot of χT versus T will diverge from the horizontal line to smaller values of χT as the temperature decreases. This procedure must be used with care, especially if the spin is greater than 1/2.

The Padé approximation has been used by Baker, et al[13] to derive the following expression for the reduced magnetic susceptibility:

$$\chi(K) = [(1.0 + 5.7979916K + 16.902653K^2 + 29.376885K^3$$
$$+ 29.832959K^4 + 14.036918K^5)/(1.0 + 2.7979916K$$
$$+ 7.0086780K^2 + 8.6538644K^3 + 4.5743114K^4)]^{2/3}$$

High-temperature series expansion techniques have also been used.[14] Ferromagnetically exchange-coupled linear chains have been thoroughly treated in a review by Willet and co-workers.[15]

B. The Alternatingly Spaced or Alternatingly Bridged Chains

Alternating chain compounds continue to be of considerable interest because of the remaining theoretical problem concerning the way the singlet-state band gap varies with the alternation parameter. Also, there is an inadequate number of characterized examples, and only preliminary magneto-chemical structural correlations have been generated.

The structure of alternatingly spaced or alternatingly bridged chains may be schematically represented as

```
---M——M------M——M------M——M------M——M------M——M------M——M---
   \_J_/\_αJ_/
```

where **J** and α**J** alternate along the chain. The appropriate Hamiltonian for the Heisenberg alternating chain is

$$H = -2J \sum_{i=1}^{(N-1)/2} S_{2i-1} \cdot S_{2i} - 2\alpha J \sum_{i=1}^{(N-1)/2} S_{2i} \cdot S_{2i+1}$$

Assignment of **J** and α**J** to specific links in alternating chains is difficult and depends on magneto-structural correlations. These are not always infallible. Exchange coupling constants for important examples in which the assignment of **J** and **αJ** are well established will be discussed in the Results section.

C. Next-Nearest Neighbor Chains

The next nearest neighbor chain may be displayed schematically as follows:

```
——M——X——M——X——M——X——M——X——M——X——M——X——M——
  | J |    |     |     |     |     |     |     |     |     |     |
——X——M——X——M——X——M——X——M——X——M——X——M——X——
      \___αJ___/
```

Here there is the possibility of nearest neighbor exchange between ions in adjacent chains, as gauged by the exchange coupling constant J, as well as next-nearest neighbor exchange along a single linear chain as indicated by the exchange coupling constant **αJ**. The appropriate Hamiltonian for the description of exchange in these next-nearest neighbor chains is

$$H = -2J\sum_{i=1}^{N-1} S_i \cdot S_{i+1} - 2J\alpha\sum_{i=1}^{N-2} S_i \cdot S_{i+2}$$

Theoretical work on linear chain systems with nearest-neighbor and next-nearest neighbor exchange interactions includes finite-size ring and chain calculations,[3] Hartree-Fock level approximations,[16] 1/n expansion techniques,[17] and pair approximation methods.[18] These treatments have been aimed at determining the characteristics of the eigenvalue spectrum and calculating a variety of thermodynamic properties. Uniform linear chain systems with next-nearest neighbor interactions present a tractable model for studying the effects of increased lattice dimensionality. Ultimately, a better understanding of interchain interactions is expected to result from these studies, along with modifications of existing theories. An interesting situation arises in the ordered state as a direct result of competition between exchange coupled sites. The idea of topological spin frustration was first brought forth by Tolouse,[19] and this concept may be illustrated for this system in a very simplistic way as follows:

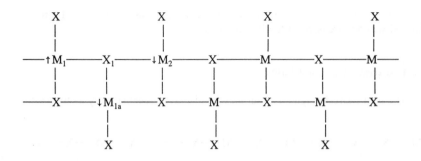

Assume the interaction between the unpaired electrons on sites M_1 and M_2 which is transmitted through the superexchange pathway M_1-X_1-M_2 is antiferromagnetic as indication by the direction of the arrows in the schematic drawing. Also, assume the interaction between the unpaired electrons on sites M_1 and M_{1a} (through the superexchange pathway M_1-X_1-M_{1a}) is antiferromagnetic, as indicated by the direction of the arrows. However, the interaction between M_{1a} and M_2 (through the superexchange pathway M_{1a}-X_1-M_2) is identical to that of M_1 and M_{1a} (through the superexchange pathway M_1-X_1-M_{1a}), yet the direction of the arrows indicate a ferromagnetic interaction. Clearly such a simple picture can not adequately describe a quantum event, but it serves the purpose of predicting spin frustration and has stimulated much research.

Real examples of transition metal complexes displaying next-nearest neighbor interactions in addition to nearest-neighbor interactions are systems of significant interest and theoretical importance. The ladder-like chain compounds $Cu(HzH)Cl_3$,[20,21] $Cu(amp)Cl_2$,[22] and $Cu(Me_3en)Br_2$[23] represent examples discovered in this research. As noted above, a variety of models have been used previously in the analysis of the magnetic data for these specific compounds. Unfortunately, there are few experimental examples, and efforts are underway to amass a body of data which will permit the construction of magneto-chemical structural correlations and a self-consistent explanation of the exchange interactions in compounds in this structural class.

D. Spin-Ladder Chain

The next most complicated chain structure is presented by the spin-ladder chain. There may be ligand bridges between the metal ions along the "runners" (or sides) of the ladder as well bridges between metal ions that form the "rungs" (or steps) of the ladder. This geometry presents very interesting problems from several points of view. First, the calculations are complex, but from a magneto-structural point of view, it is difficult to assign the exchange coupling constants to a specific interaction. A schematic representation of the spin-ladder is shown here.

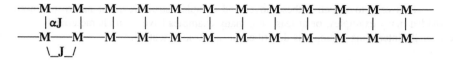

The appropriate Hamiltonian may be written as follows:

$$H = -2J\sum_{i=1}^{N/2} S_{2i-1}{\cdot}S_{2i} - 2\alpha J\sum_{i=1}^{N-2} S_i{\cdot}S_{i+2}$$

There appear to be few examples of copper(II) chain compounds that exhibit structures that would give rise to spin-ladder exchange behavior. The properties of one compound in which spin-ladder exchange has been proposed are described in the Results section.

E. Alternating Next-Nearest Neighbor Chains

Exchange in alternating next-nearest neighbor chains presents an interesting case for study. A schematic representation of a structure that would support alternating next-nearest neighbor exchange is shown here.

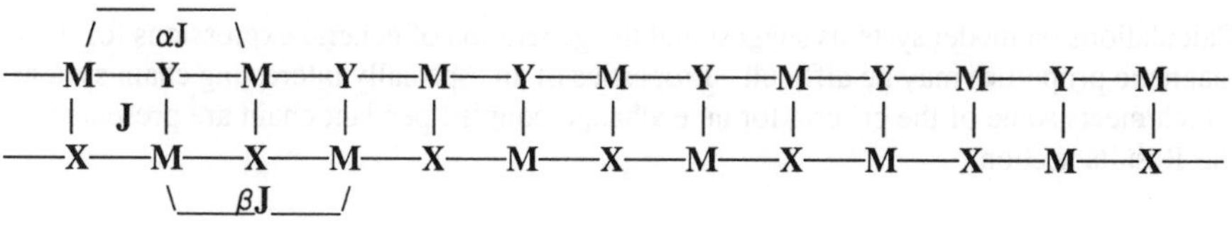

$$
\begin{array}{c}
\overline{\;\alpha J\;\overline{}} \\
\text{—M—Y—M—Y—M—Y—M—Y—M—Y—M—Y—M—} \\
\text{| J |\ \ |\ \ \ |\ \ |\ \ \ |\ \ |\ \ \ |\ \ |\ \ \ |\ \ |\ \ \ |\ \ |} \\
\text{—X—M—X—M—X—M—X—M—X—M—X—M—X—} \\
\underline{\beta J}
\end{array}
$$

Here, three exchange coupling constants are required as indicated in the sketch and in the Hamiltonian for the problem.

$$H = -2J\sum_{i=1}^{N-1} S_i{\cdot}S_{i+1} - 2\alpha J\sum_{i=1}^{(N/2)-1} S_{2i-1}{\cdot}S_{2i+1} - 2\beta J\sum_{i=1}^{(N/2)-1} S_{2i}{\cdot}S_{2i+2}$$

It is easy to visualize these novel and complex structures, but nature has not provided many examples, or at least, not many examples have been characterized. An especially interesting example is described in the Results section.

F. Pendant Chains

Pendant chains, as sketched below, have been the synthetic target in many laboratories. Unfortunately, well-characterized examples are rare, but the opportunities for magneto-structural advances and materials with unusual properties in this series of compounds are well appreciated.

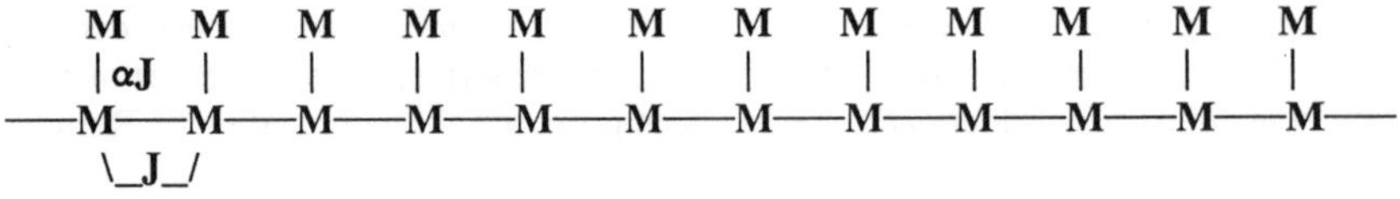

```
M    M    M    M    M    M    M    M    M    M    M    M
|αJ  |    |    |    |    |    |    |    |    |    |    |
—M—M—M—M—M—M—M—M—M—M—M—M—
\_J_/
```

The appropriate Hamiltonian for a pendant chain is

$$H = -2J\sum_{i=1}^{N/2} S_{2i-1} \cdot S_{2i} - 2J\alpha \sum_{i=1}^{(N/2)-1} S_{2i-1} \cdot S_{2i+1}$$

Calculations on model systems suggest that the generation of general expressions for their magnetic properties may be difficult. Properties of an especially interesting chain system which meets some of the criteria for an exchange-coupled pendant chain are presented in the Results section.

G. Other Chains with Novel and Complicated Structures

Nature has provided us with a vast array of chains of copper(II) with novel and complicated structures. For example, there are exchange-coupled chains with no ligand bridges, but these are invariably uniformly spaced and adequate theory exists for the description of their properties. There are other complicated chains such as helical chains, but their unusual properties are normally manifested in the ordered state and coverage of them is not in the scope of this review.

Chains of oligomers present challenging problems. Some discussion of the simplest of these, that is, chains of trimers is given in the Results section. Chains of higher oligomers are also known. Results and discussion of a wide variety of these may be found in the work of Willett and co-workers.

RESULTS

A. Uniformly Spaced Chains

Catena-di-μ-chloro[bis-pyridinecopper(II)]. The linear chain compound $Cu(py)_2Cl_2$ has received much attention since well-defined, large single crystals may be obtained by chemists with some skill in crystal growth. The compound crystallizes in the monoclinic system, space group $P2_1/n$, and is stable in the laboratory atmosphere. The structure may be thought of as square planar trans-$Cu(py)_2Cl_2$ units which are stacked in a "slipped" manner so that the chloro-ligands occupy the apical coordination positions of adjacent $Cu(py)_2Cl_2$ units, thus resulting in an infinite chain with $\{Cu_2Cl_2\}$ exchange coupling pathways. The $\{Cu_2Cl_2\}$ unit is nearly rectangular with a short copper-chloride bond of 2.28 Å occurring in the planar $Cu(py)_2Cl_2$ unit, and a long copper-chloride out-of-plane bond of 3.05 Å connecting the $Cu(py)_2Cl_2$ into the resultant chain. There is not general agreement that copper-chloride contacts of 3.05 Å should be called a bond, but the contact is short enough to transmit superexchange interactions.

High temperature magnetic susceptibility measurements by Takeda et al.[24] were followed by more extensive measurements by Hatfield and Jeter.[25] These studies establish that catena-di-μ-chloro[bis-pyridinecopper(II)] is an antiferromagnetically coupled Heisenberg chain with $J = -9.15$ cm^{-1} and g = 2.05. Subsequent work verified the earlier observations by Hatfield and Jeter.[25] There are significant interactions between the chains as documented by the heat capacity anomaly at 1.130(5) K.[26]

A large number of related di-μ-chloro- and di-μ-bromocopper(II) chains are known.[27] An extensive list, albeit somewhat dated, is given in a review from our laboratory.[11]

Catena-μ-pyrazine[bis-nitratocopper(II)]. A very interesting and important chain compound is formed by pyrazine and copper(II) nitrate. The copper ions are linked by pyrazine bridges and are separated by 6.712 Å. Each copper ion achieves a coordination number of six by bidentate coordination of the two nitrate counterions. The coordination environment is very distorted as a result of the limited angle permitted by the bidentate nitrate ion. If this coordination is considered to be of the "4+2" class, then of the four short bonds, two occur between copper and the pyrazine bridge nitrogen atoms, and the other two occur between copper and one oxygen from each of the two nitrato ligands. The two "out-of-plane" bonds are to the second oxygen of the bidentate nitrato ligand.

The compound crystallizes in the orthorhombic system and crystals grow as long slender needles with the chain axis parallel to the needle axis, that being the a axis. The first magnetic measurements were made on a bundle of needles, and while the data along the needle axis were unique, the results of the measurements perpendicular to the bundle were an average of the magnetic susceptibilities along the b and c crystallographic directions. The results clearly identified the system as being an antiferromagnetically exchange-coupled Heisenberg chain with $J = -3.7$ cm^{-1} and $g_a = 2.275$. It was not possible to obtain unique values for the perpendicular g values from the measurements on the bundle of crystals, $<g_{\perp}>$ was found to be 2.05. Boyd and Mitra[28] used the Krishan critical torque technique on a single micro-needle and confirmed the J and g_a values. In addition they were able to determine that the g_b and g_c values were equal to 2.05, values in agreement with the average determined earlier and with EPR measurements on a single crystal. Subsequent measurements[29] made to 0.18 K revealed no anomalies in the magnetic susceptibility or heat capacity. Thus, catena-μ-pyrazine[bis-nitratocopper(II)] appears to be one of the best examples of a linear chain compound, and does not exhibit long-range order above 0.18 K.

Chain Compounds with Large Metal-Metal Separations. The unusual compound (3-chloroanilinium)$_8$[CuCl$_6$]Cl$_4$ has tetragonally compressed [CuCl$_6$]$^{4-}$ ions aligned along the a axis.[30] The superexchange pathway of Cu-Cl$_{ax}$-Cl$_{ax}$-Cu is 8.5 Å, yet the copper ions are exchange coupled with $J = -5.6$ cm^{-1}. This may be understood since the ground state of copper is d$_{z2}$, and the magnetic orbitals are oriented along the superexchange pathway. The 3-bromoanilinium analog has also been prepared. Since the van der Waals radius of the bromo-substituent is 0.15 Å greater than that of the chloro-substituent, it was expected that the nature of the axial compression would be affected and the exchange coupling modulated, with the expectation that the larger volume would result in longer chloride-chloride contacts and diminished exchange coupling. This has now been confirmed.

B. Alternatingly Spaced Chains

Although organic compounds with alternatingly spaced linear-chain structures have been known and studied for many years,[11] exchange-coupled transition-metal compounds with alternatingly spaced structures and comparable magnetic properties have received much less attention. The most thoroughly studied transition-metal compound which exhibits alternating Heisenberg antiferromagnetic behavior is Cu(NO$_3$)$_2$·2.5H$_2$O,[31] but it is important to note that the compound has a ladder-like structure at room temperature.[32] Other examples of compounds that are known to have alternating chain structures include the pyrazine-bridged bimetallic copper acetate chain, [Cu$_2$(OAc)$_4$(pyz)]$_\infty$,[33] and the carcinostatic agent [3-ethoxy-2-oxobutraldehyde bis(thiosemicarbazonato)]copper(II), CuKTS.[34,35] On the basis of their magnetic properties, the compounds catena-bis(μ-bromo)bis(N-methylimidazole)copper(II)[36,37] and catena-bis(μ-chloro)bis(4-methyl-pyridine)copper(II)[38] have been suggested to have alternatingly spaced linear-chain structures at low temperatures. These compounds have uniformly spaced linear-chain structures at room temperature.[39,40] However, the compound catena-dichloro(3,6-dithiaoctane)copper(II), Cu(3,6-DTO)Cl$_2$, has an alternatingly spaced structure at 140 K, and the compound exhibits magnetic properties that may be explained by alternating Heisenberg chain theory.

Catena-Dichloro(3,6-dithiaoctane)copper(II). The compound Cu(3,6-DTO)Cl$_2$ crystallizes in the triclinic system.[41] Formula units of the compound stack along the a axis to form an alternatingly spaced chain in which the copper(II) ions have coordination number six by means of polymerization. Two chloride ligands and two sulfur atoms from the bidentate thioether constitute the equatorial plane with a sulfur atom and chloro-ligand from adjacent planar units loosely bound in the two axial positions. Copper-chlorides distances in the plane are 2.264 and 2.230 Å, while the copper-chloride out-of-plane interatomic distance is 3.234 Å. The copper-sulfur bond distances in the plane are 2.311 and 2.327 Å with the out-of-plane copper-sulfur distance being 3.361 Å. The chain is kinked with the Cu-S(bridge)-Cu bridging bond angle being 109.3°. The magnetic susceptibility of this unique chain exhibits a maximum at 4.2 K. The maximum is indicative of an antiferromagnetic intrachain exchange interaction between the alternatingly spaced and alternatingly bridged copper(II) ions. The best fit of the magnetic susceptibility data by AHC theory for S = 1/2 ions yields a J value of -2.73 cm^{-1} and α = 0.69. This represents the first example of a structurally and magnetically characterized antiferromagnetic alternating Heisenberg chain system.

Aqua[N-(salicylaldiminato)glycinato]copper(II) hemihydrate, CuNSG. Although the compound CuNSG has a uniformly spaced structure at room temperature,[2] it is most appropriately discussed in this section on alternating chains. High-field magnetization and magnetic susceptibility data for CuNSG have been analyzed within the cluster approach. The basic conclusions follow: The magnetic susceptibility data follow Heisenberg chain theory to below 10 K, at which point the data deviate from the Bonner-Fisher[8] prediction for a uniform Heisenberg chain. This deviation suggests that the chain is beginning to dimerize, and the magnetic data may be fit by alternating chain theory with an increase in the alternation parameter $0 < J_1/J_2 < 1$ as the temperature decreases. Below the knee in the magnetic susceptibility versus temperature plot, the alternation parameter becomes nearly constant. This is interpreted as being the limit of dimerization as a result of the physical constraints of the atomic repulsions between constituents of the chain.

Systems Related to Copper-NSG. Criteria for the identification of new compounds which may show properties similar to those exhibited by Cu-NSG[42] have been used to guide systematic studies on other carboxylato- or carbonato-bridged copper compounds. Observations for compounds of the general formula [Cu(L⌒L)CO$_3$] (where L⌒L is a bidentate amine) yield extremely interesting behavior. Data for two of these compounds will be reviewed here. The two chain compounds are catena-μ-[{carbonato-O,O':O"}-2,2'-dipyridylaminecopper(II) trihydrate], [Cu(dpa)CO$_3$·3H$_3$O], and the bipyridine analog, [Cu(bpy)CO$_3$·3H$_3$O]. The structure of the carbonato-bridged chain

14

in [Cu(dpa)CO$_3$·3H$_3$O], as determined by Sletten,[43] is shown in Figure 1. In this compound, the coordination about the copper(II) ion is slightly distorted from square pyramidal with the basal plane consisting of two nitrogens from the 2,2'-dipyridylamine ligand and two oxygens from a bidentate carbonate. An oxygen from the carbonate of an adjacent molecule occupies the apical position. Adjacent chains are interwoven as shown in the schematic in Figure 2. X-ray diffraction studies on the bipyridine analog yield the same space group and similar metrical parameters for the unit cell.[44]

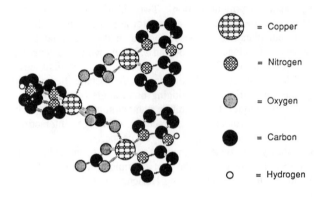

= Copper

= Nitrogen

= Oxygen

= Carbon

= Hydrogen

Figure 1. Structure of [Cu(dpa)CO$_3$]

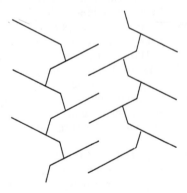

Figure 2. Schematic Diagram of [Cu(dpa)CO$_3$] extended structure.

The magnetic properties of these compounds are complicated, and suggest the unusual situation of antiferromagnetic intrachain behavior with ferromagnetic interchain behavior setting in at lower temperatures. Variable temperature magnetic susceptibility data for [Cu(dpa)CO₃·3H₃O], which were collected under an applied magnetic field of 50 Oe, are shown in Figure 3. Similar behavior is exhibited by the [Cu(bpy)CO₃·3H₃O] analog except at the lowest temperatures there is a decrease in magnetic susceptibility.

The abrupt increase in magnetic susceptibility near 3 K is the result of the onset of ferromagnetic behavior with saturation occurring below 2.5 K.

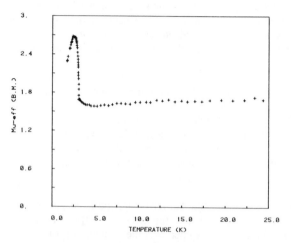

Figure 3. μ_{eff} vs. T for [Cu(dpa)CO₃]

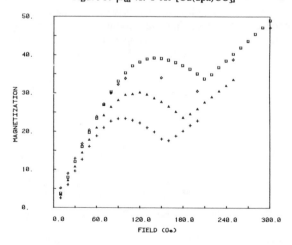

Figure 4. Magnetization of [Cu(dpa)CO₃] at □ 1.80 K, ◊ 1.90 K, ᴀ 2.11 K, + 2.48 K.

16

The observed decrease in magnetic moment prior to the rapid increase near 3 K suggests antiferromagnetic intrachain interactions. Analysis of the data from powdered samples in the temperature range several degrees above the apparent 3-D ferromagnetic ordering temperature yields $|J_{AF}| < 4$ cm^{-1} and $J_F < 1.5$ cm^{-1} with the ferromagnetic state being the ground state for [Cu(dpa)CO$_3$·3H$_3$O]. The decrease in susceptibility at the lowest temperatures reveal complicated behavior for [Cu(bpy)CO$_3$·3H$_3$O]. The observed behavior is not weak ferromagnetism as a result of canted antiferromagnetism since the parity condition[45] precludes any bulk net moment in these systems. Hidden canted antiferromagnetism would be a possibility except that the ferromagnetic state exists down to fields as low as 0.1 Oe. For hidden canted antiferromagnetism, a singlet ground state would be present at these low fields and no net moment would be evident.[46]

The small intrachain exchange coupling constants in these carbonato bridged chains may be understood in terms of the bridging geometry and the results of ab initio Gaussian 88 calculations.[47] The calculations reveal the nature of the carbonate orbitals which transmit the superexchange interactions between the magnetic d_{x2-y2} orbital on one copper(II) ion and the filled d_{z2} orbital on the adjacent copper(II) ion. If orthogonal, the interaction should be ferromagnetic, but a tilting of the dpa (or bpy) planes with respect to one another permits mixing, the advent of antiferromagnetic components, and the resultant antiferromagnetic interactions in the chains.

Alternating Ferromagnetic and Antiferromagnetic Exchange in [Cu$_2$(bpm)(OH)$_2$-(ClO$_4$)$_2$(H$_2$O)$_4$]$_n$. The compound [Cu$_2$(bpm)(OH)$_2$(ClO$_4$)$_2$(H$_2$O)$_4$]$_n$ (bpm = 2,2'-bipyrimidine) is an infinite chain polymer with the structure of a fragment of the chain being shown in Figure 5. The chains are well isolated from one another with perchlorate anions and waters of hydration filling the channels between the chains. The most striking feature of this compound is the alternating hydroxo-bipyrimidine bridging arrangement in the tetragonal plane of the copper ions. The copper-copper separation in the hydroxo-bridged fragment is 2.869 Å, while that through the bipyrimidine bridge is 5.469 Å.

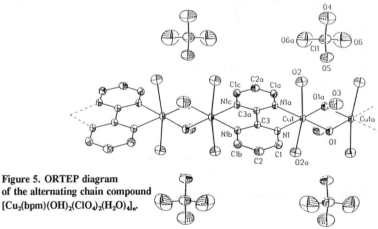

Figure 5. ORTEP diagram
of the alternating chain compound
[Cu$_2$(bpm)(OH)$_2$(ClO$_4$)$_2$(H$_2$O)$_4$]$_n$.

As shown by Petersen and co-workers,[48] the geometry about the individual copper ions is six-coordinate distorted octahedral. In the tetragonal plane about copper, the hydroxo-copper bond distances are 1.938 Å, and the copper-nitrogen bond distances are 2.048 Å. The axial sites are occupied by aqua ligands with copper-oxygen bond distances of 2.459 Å. The bipyrimidine ligands are planar, and the copper ion, the two bridging hydroxo ligands, and the bipyrimidine ligand lie nearly in the same plane. The axial aqua ligands are tilted toward the bipyrimidine ligand. A metrical parameter of particular interest is the Cu-O-Cu angle in the hydroxo-bridged fragment. This angle is 95.5°.

The magnetic susceptibility displays a rounded maximum at 113 K. The magnetic moment increases with temperature and does not reach a limiting value at the highest temperature of the measurement, 300 K, where the magnetic moment is 1.86 B. M.[49]

The magnetic susceptibility data were analyzed with a 12-spin ring approximation. This model resulted in an excellent fit to the data with the best-fit magnetic parameters being J_1 = +167.6 cm^{-1}, J_2 = -79.8 cm^{-1}, and g = 2.25. Magnetostructural relationships may be used to assign the exchange coupling constants to the appropriate bridged unit of the alternating chain.

Bipyrimidine is well known to facilitate antiferromagnetic interactions between divalent transition metals ions.[50] For the most part, the exchange coupling constants are large, ranging from -77.8 cm^{-1} to -118 cm^{-1}. In these complexes, the bipyrimidine bridge is coplanar with the copper(II) d_{x2-y2} magnetic orbital. This is an important structural feature since the delocalized π network of the bridging ligand is prohibited by symmetry to be an effective contributor in the orbital exchange mechanism. The σ system is the sole conveyor of magnetic exchange in these bimetallic units.

In order to ascertain the orbital nature of this superexchange pathway, Gaussian 88 ab $initio$ molecular orbital calculations were performed on a pyrimidine unit.[51] The resulting HOMO's show good overlap between copper d_{x2-y2} and pσ orbitals of the bridging ligand. These results, along with the exchange coupling constants that have been observed for other bipyrimidine copper(II) complexes, permit an assignment of the best-fit exchange coupling constant of -79.8 cm^{-1} to the bipyrimidine-bridged fragment of the alternating chain. The effects of bridge geometry on the sign and magnitude of exchange coupling constants in di-μ-hydroxo bridged dimers is well known.[52] There is a linear dependence of the singlet-triplet splitting in five-coordinate dimers on the angle ϕ at the hydroxo bridge given by $2J$ (cm^{-1}) = -74.53(ϕ) + 7270. The singlet-triplet splittings for six-coordinate hydroxo-bridged copper(II) dimers deviate strongly toward more positive values from that predicted from the above equation. Clearly the best-fit exchange coupling constant of +167.6 cm^{-1} corresponds to the exchange coupling in the di-μ-hydroxo fragment of the alternating chain. [Cu$_2$(bpm)(OH)$_2$(ClO$_4$)$_2$(H$_2$O)$_4$]$_n$ represents only the third example of alternating ferromagnetic and antiferromagnetic exchange coupling in a one-dimensional chain. The other two examples are the alternatingly spaced compound $CuCl_3 \cdot (4\text{-Bzpip})$[40] and the alternatingly bridged compound Cu(hfac)-TEMPOL,[53] which have $(\mathbf{J}, \alpha)$ values of (21 cm^{-1}, -0.1) and (12.9 cm^{-1}, -0.004), respectively. In both of these previous examples, the antiferromagnetic exchange coupling constants were determined by using mean-field theory or by fitting the magnetic susceptibility or high-field magnetization data to an antiferromagnetic effective spin S = 1 uniform chain.

C. Next-Nearest Exchange in Linear-Chain Compounds

<u>Exchange in the Next-Nearest Neighbor Chain Cu(HzH)CuCl$_3$</u>. The compound hydrazinium trichlorocuprate crystallizes as an infinite chain polymer with the structure shown in Figure 6. The coordination polyhedron about the copper(II) ion is best described as a 4+2 coordinate tetragonally elongated octahedron. The in-plane ligation consists of three bound chlorides and a nitrogen atom from the hydrazinium ion. The copper(II) ion, equatorial chlorides, and the coordinated and non-coordinated nitrogen atoms of the hydrazinium ligand are all coplanar. The chains are relatively isolated from one another. The copper(II) ions are related by a two-fold screw axis along <u>b</u>, which is the chain direction. As a consequence

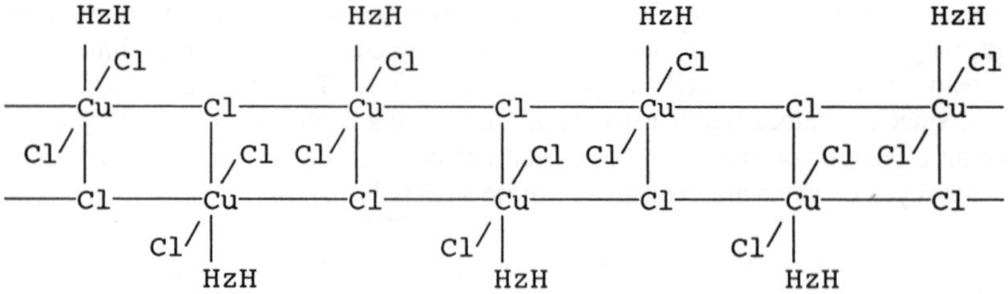

```
     HzH              HzH              HzH              HzH
      | ,Cl             | ,Cl             | ,Cl             | ,Cl
 -----Cu------Cl--------Cu------Cl--------Cu------Cl--------Cu--
  Cl/ |          | ,Cl Cl/ |        | ,Cl Cl/ |        | ,Cl Cl/ |
 -----Cl------Cu--------Cl------Cu--------Cl------Cu--------Cl--
        Cl/ |             Cl/ |             Cl/ |
           HzH              HzH              HzH
```

Figure 6. Sketch of the structure of (hydrazinium)CuCl$_3$.

of this structural feature, the axial ligation consists of two chlorides from the tetragonal plane of adjacent copper(II) ions along the chain. The bridging chloride ligands are therefore bound to three different copper(II) centers. This feature has an important bearing on magneto-chemical structural correlations. The copper-copper distance is 3.751 Å, the in-plane bridging Cu-Cl distance is 2.297 Å, and the out-of-plane Cu-Cl distance is 2.856 Å. The Cu-Cl-Cu bridge angle in the quasi-dimeric unit is 92.78°.

Ananthakrishna, et al.[54] were among the first to utilize finite sized chain and ring calculations to determine the thermodynamics of a linear chain system with next-nearest neighbor interactions. The extrapolated magnetic susceptibility results for the case $\mathbf{J} < 0$, $0 \leq \alpha \leq =0.2$ in the limit H = 0 and T = 0 indicate a gapless excitation spectrum since the susceptibility is finite. For the antiferromagnetic case, there is a singularity at $\alpha = 0.5$. This singularity has been confirmed.[3]

The magnetic properties of HzHCuCl₃ have been investigated extensively. Above 6 K, the magnetic susceptibility of HzHCuCl₃ exhibits chain-like behavior which may be fit with closed-form expressions for a next-nearest neighbor chain with J_1 (~90° interaction) = -4.84 cm⁻¹ and J_2 (~180° interaction) = -5.66 cm⁻¹, with g = 2.13. There are no magneto-structural correlations for a tricoordinate (bridging) chloride ligand, but J_1 is consistent with correlations for the ~90° interaction. Below 6 K there are complications in the magnetic behavior which have yet to be sorted out.

Cu(amp)Cl₂ and Cu(amp)Br₂. The linear-chain compound dichloro[2-(2-aminomethyl)pyridine]copper(II), Cu(amp)Cl₂, and the bromo analog have chain structures[55] which are similar to that of HzHCuCl₃, although the kinked chain of Cu(amp)X₂ is more severely distorted than that of HzHCuCl₃. The coordination environment about the copper(II) ion is a 4+2 tetragonally elongated octahedron. The in-plane ligation consists of pyridyl and amine nitrogen donor atoms from the 2-(2-aminomethyl)pyridine ligand and two halides. The individual chains are not well isolated from one another and there is interleaving of the substituted pyridines on adjacent chains.

In the initial study on Cu(amp)Cl₂, only dimer and uniform chain models were used in the analysis of the magnetic susceptibility data.[55] Next-nearest neighbor exchange coupling was realized, but recent results[3] were not available at that time for an appropriate analysis of the data. The best fit was obtained with the Bonner-Fisher uniform chain model modified with a mean field term. New data fit with the new closed-form expressions will be described later. It is important to call attention to the kink in susceptibility that was observed at 2.42 K. It was speculated that this may correspond to a long-range three-dimensional antiferromagnetic ordering or to a spin-Peierls transition. The probability of a frustration-relieving event may now be added to the list of possible explanations.

New magnetic susceptibility data for Cu(amp)Cl₂ may be fit nearly exactly by the closed form expressions for a next-nearest neighbor linear chain with J_1 = -2.28 cm⁻¹, J_2 = -4.0 cm⁻¹, and g = 2.10.[51] This set of data for a second next-nearest neighbor chain, when added to the catalog of data for members of its structural class, may be used in the generation of magneto-chemical structural correlations. The kink at 2.4 K is of especial interest, in view of the magnetic properties of HzHCuCl₃, which are severely dependent on the applied magnetic field.

Cu(Me₃en)Br₂. The chain compound dibromo(N,N,N'-trimethylethylenediamine) copper(II), Cu(Me₃en)Br₂, also has the next-nearest neighbor chain structure shown schematically above, although the chain is significantly more distorted than [Cu(HzH)-Cl₃].[22] In this case the chains are well isolated from one another, and form stacked alternating layers along the c axis. Analysis of new magnetic susceptibility data for Cu(Me₃en)Br₂ with the closed form expressions yield J_1 = -2.99 cm⁻¹, J_2 = -3.75 cm⁻¹, and g = 2.06. Of course, since there is only one member of this bromo-bridged class, it is not possible to make comparisons and look for correlations. It is important to note that such compounds exist, synthesis of additional members of the class may be envisaged, and there are opportunities for fundamental studies. It is also important to note that no discontinuities were observed in the magnetic properties of this compound, but any discontinuities may be outside the temperature range studied, or may not be detectable by the experimental techniques used.

D. Exchange in a Chain with Spin-Ladder Structure

Bis-μ-chloro-bis[chloro(4-methylthiazole)dimethylformamidecopper(II)]. The complex [Cu(N-Me-tz)(dmf)Cl$_2$]$_\infty$ crystallizes in the monoclinic system with space group P2$_1$/n, and is the first example of a mixed ligand bis-μ-chloro-copper(II) complex.[56] The copper ion is five-coordinate, with the coordination geometry being rather severely distorted from an idealized tetragonal pyramid. The Cu$_2$Cl$_2$ bridging unit is constrained to be planar by the presence of a center of symmetry, and the copper-chloride distances are 2.296 and 2.724 Å. The Cu-Cl---Cu' angle is 95.3°. Magnetic susceptibility measurements reveal antiferromagnetic exchange coupling. The dimeric formula units pack in the solid state to form a ladder-like structure, with copper(II)-sulfur contacts of 3.906 Å. Calculations on a closed ring of 10 spins, which included both nearest and next-nearest neighbor interactions, as described previously for spin ladder chains were performed. The best fit exchange coupling constants for this spin ladder system were determined to be J$_1$ = -1.17 cm^{-1} and J$_2$ = -1.0 cm^{-1}.

E. Alternating Next-Nearest Exchange in Linear-Chain Compounds

Copper(II) Phthalate Monohydrate. A compound which exhibits alternating next-nearest neighbor interactions is copper(II) phthalate monohydrate.[57] The excitation spectrum of this system is not understood since alternating next-nearest chains have received little attention. The magnetic array and independent exchange coupling constants of the alternating next-nearest neighbor chain is represented schematically in the previous section. Copper(II) phthalate monohydrate exhibits a structure for which alternating next-nearest neighbor interactions are possible. As shown in Figure 7, Cu(H$_2$O)(pht) crystallizes with two copper ions in different coordination environments alternating along the chain direction.[57] One of the copper ions is located on an inversion center, while the other copper ion is on a two-fold screw axis along b. The nearest-neighbor copper(II) sites are separated by 3.2 Å, and the next-nearest-neighbor copper(II)-copper(II) separation is 11.2 Å.

The coordination environment about both copper atoms is distorted octahedral, consisting of four oxygen atoms from four different phthalate groups and two oxygen atoms from water molecules. The point symmetry of the copper atom Cu(1) is near

21

perfect D_{4h} with the Jahn-Teller axis described by O(3)-Cu(1)-O(3'). Copper(2) is more highly distorted with its Jahn-Teller axis along H_2O-Cu(2)-H_2O. The Cu(1)-Cu(2) nearest-neighbor segments of the chain are linked via three oxygen atoms, two of these being O(2) and O(3), originating from the carboxylate groups of two independent phthalates, and the third from a water molecule. The Cu(1)-Cu(1') next-nearest-neighbor linkages are formed by two carboxylate groups from the same two phthalate ligands.

Structural evidence suggests that only one next-nearest-neighbor interaction is operative, and $g_a \approx g_b$. The Heisenberg Hamiltonian for a linear chain exhibiting nearest-neighbor and alternating next-nearest-neighbor exchange interactions is given above. This Hamiltonian, with β set to zero, since there is no exchange pathway connecting even sites, was used. The best-fit calculations of the theoretical results to the experimental magnetic susceptibility data yielded the parameters $J = -12.3$ cm^{-1}, $\alpha = 0.06$, and g = 2.19. The best-fit g value is in excellent agreement with the average g value determined from EPR measurements. Since there are two different copper(II) ions in the structure, one may have expected a divergence of the magnetic susceptibility at low temperatures as a result of incomplete compensation of the spin. However, the similar 4+2 coordination environment about the copper ions and the propensity for copper to display nearly isotropic g values precludes measurable ferrimagnetic behavior. The confirmation of nearest-neighbor as well as alternating next-nearest-neighbor exchange in Cu(H$_2$O)(pht) defines a new magnetic system.

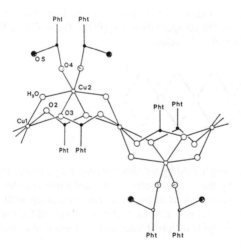

Figure 7. Structure of [Cu(pht)(H$_2$O)]$_n$.

F. Mixed Valence Chains

The discovery of high temperature superconducting materials has stimulated an interest in the properties of mixed valence chains, since it is widely believed that the superconducting mechanism involves low-dimensional structural units, either chains, layers, or combinations of the two. An interesting compound resulting in this regard is $(N_2H_5)_2[Cu_3Cl_6]$,[20] which has a deep black color and is likely an example of a Robin and Day[58] Class II system, as suggested in recent work.[59] Dark colors continue to be signals for the possibility of Class II or Class III systems, and these are of interest for this review.

Studies of the magnetic properties of Class III compounds, which frequently exhibit metallic-like electrical conductivity, usually reveal Pauli paramagnetism, and temperature dependent studies may yield information about phase transitions and charge density waves. A good example is KCu_4S_3.[60] Magnetic studies of Class II compounds usually yield a magnetic moment near or slightly larger than the spin-only moment per pair of copper ions, and the presence of metal-metal interactions are more difficult to discern.

G. Chains of Trimers

Drillon and coworkers have collected specific heat, magnetic susceptibility, and magnetization data for the compounds $A_3Cu_3(PO_4)_4$ ($A = Ca^{2+}$, Sr^{2+}) which have the structure shown schematically below:

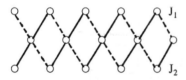

The work is complicated by the onset of long-range order, but comparisons of the analysis of the data by the Ising-transfer matrix technique and the cluster approach are encouraging. Both models yield strong intra-trimer coupling with significantly weaker inter-trimer exchange coupling. However, the magnitude of the exchange coupling constants are, as expected, quite different since different electronic structures form the basis for the two approaches.

Other compounds in this class include $Cu_3(PO_4)_2$ which has a structure with the superexchange pathway sketched below:

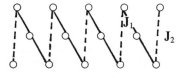

and $Cu_2O(SO_4)$ which has the following superexchange pathway:

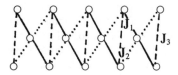

Comparison of Heisenberg and Ising Treatments. Drillon[61] uses an Ising model and the transfer matrix technique which permits the treatment of infinite systems. The cluster approach that has been used to describe the properties of the chains in this review recognizes that many systems typically exhibit Heisenberg or nearly isotropic exchange, but there is a limit to the number of spins that can be treated, and extrapolations to the infinite limit are necessary. An analysis of specific heat, magnetic susceptibility, and magnetization data on selected compounds is of interest in that it is anticipated that the advantages and limitations of each will be identified.

CONCLUSIONS

Low-dimensional materials are of interest because of their theoretical importance. These chemical systems provide real examples with properties which may be used to verify theoretical models and their properties are often used to refine the abstract models. In addition, unusual and unexpected properties are often encountered, and the explanation of these properties require expanded theoretical expeditions.

Low-dimensional materials have also been selected for study because of their promise for technological applications and advances. Such systems may have techno-logical applications because they frequently exhibit bistability. That is, under certain applied conditions they exhibit one set of properties, which may change when the conditions are altered. This permits them to be used in such applications as switches, actua-tors, sensors, display devices, and in logic circuits. Low-dimensional materials selected for discussion in this review have been considered for applications such as these.

Low-dimensional materials often exhibit properties that are of potential techno-logical value. A dramatic example is provided by the discovery of high temperature superconductivity. Other properties that may be exploited in technological applications

24

include their magnetic and dielectric properties.

Experimental procedures to be employed include synthesis of new novel compounds, crystal and molecular structure determinations of selected members of the series, characterization of powdered and single crystal samples by electronic spectroscopy, magnetic susceptibility measurements at a range of magnetic fields, electron paramagnetic and nuclear magnetic resonance, and heat capacities.

There are many more examples of chains with complex structures. However, the complexity of the structures leads to a large number of exchange parameters, and unique solutions of the magnetic susceptibility to a Hamiltonian is not possible at this time. The major goal of this research is the generation of magneto-structural correlations and the determination of the transferability of the correlations from one chemical chain to other chemical systems. Realization of these goals will be a triumph for magnetochemical and magnetostructural research.

REFERENCES

1. Hatfield, W. E.; Helms, J. H. "From the Magnetic Susceptibility of Clusters to the Magnetic Susceptibility of Chains with Novel Structures", *Materials Science*, in press.
2. Ueki, T.; Ashida, T.; Sasada, Y.; Kakudo, M. *Acta Crystallogr.* **22** (1967) 870.
3. Hatfield, W. E.; Helms, J. H.; Kirk, M. L. unpublished observations.
4. Heisenberg, W. *Z. Physik.* **38** (1926) 411.
5. Dirac, P. A. M. *Proc. Roy. Soc.* **112A** (1926) 661.
6. Bonner, J. C.; Sutherland, B.; Richards, P. M. Magnetism and Magnetic Materials-1974, Proceedings of the 20th Annual Conference on Magnetism and Magnetic Materials; Graham, C. D.; Lander, G. H.; Rhyne, J. J., Eds., AIP: New York; 1965.
7. Müller, G,; Beck, H.; Bonner, J. C. *Phys. Rev. Lett.* **43** (1979) 75.
8. Bonner, J. C.; Fisher, M. E. *Phys. Rev. A* **133** (1964) 768.
9. Levine, L. N. Quantum Chemistry, 3rd Ed.; Allyn and Bacon: New York; 1983.
10. Kirk, M. L.; Hatfield, W. E. unpublished observations.
11. Hatfield, W. E.; Estes, W. E.; Marsh, W. E.; Pickens, M. W.; ter Haar, L. W.; Weller, R. R. Extended Linear Chain Compounds; Miller, J. S., Eds., Plenum Press: New York; 1983.
12. Hall, J. W. *Ph. D. Dissertation*, University of North Carolina, Chapel Hill, NC 1977.
13. Baker, G. H. Jr.; Rushbrooke, G. S.; Gilbert, H. E. *Phys. Rev. A* **135** (1964) 1272.
14. Rushbrooke, G. S.; Wood, P. J. *Mol. Phys.* **1** (1958) 257.

15. Willett, R. D.; Gaura, R. M.; Landee, C. P. Extended Linear Chain Compounds; Miller, J. S., Eds., Plenum Press: New York; 1983.
16. Neimeijer T. *J. Math. Phys.* **12** (1969) 1487.
17. Selke W. *Z. Physik.* **B27** (1977) 81.
18. Murasaki, T.; Tanaka, Y.; Uryu, N. *J. Phys. Soc. Japan* **52** (1983) 2192.
19. Tolouse, G. *Commun. Phys.* **2** (1977) 115.
20. Brown, D. B.; Donner, J. A.; Hall, J. W.; Wilson, S. R.; Wilson, R. B.; Hodgson, D. J.; Hatfield W. E. *Inorg. Chem.* **18** (1979) 2635.
21. Helms, J. H. *Ph. D. Dissertation*, University of North Carolina, Chapel Hill, NC 1988.
22. O'Conner, C. J.; Eduok, E. E.; Fronczek, F. R.; Kahn, O. *Inorg. Chim. Acta* **105** (1985) 107.
23. Estes, W. E. *Ph. D. Dissertation*, University of North Carolina, Chapel Hill, NC 1977.
24. Takeda, K.; Matsukwa, S.; Haseda, T. *J. Phys. Soc. Jap.* **30** (1971) 1330.
25. Jeter, D. Y.; Hatfield, W. E. *J. Inorg. Nucl. Chem.* **34** (1972) 3055.
26. Duffy, W.; Venneman, J. E.; Strandberg, D. L.; Richards, P. M. *Phys. Rev. B.* **9** (1974) 2220.
27. Hatfield, W. E. unpublished observations.
28. Boyd, D. W.; Mitra, S. *Inorg. Chem.* **19** (1980) 3547.
29. Koyama, M.; Suzuki, H.; Watanabe, T. *J. Phys. Soc. Jpn.* **40** (1976) 1564.
30. Tucker, D. A.; White, P. S.; Trojan, K. L.; Kirk, M. L.; Hatfield, W. E. *Inorg. Chem.* **20** (1991) 823.
31. Bonner, J. C.; Friedberg, S. A.; Kobayashi, H.; Meier, D. L.; Blöte, H. W. *Phys. Rev. B.* **27** (1983) 248.
32. Morosin, B. *Acta Crystallogr. Sect. B* **B26** (1970) 1203.
33. Valentine, J. S.; Silverstein, A. J.; Soos, Z. G. *J. Amer. Chem. Soc.* **96** (1974) 97.
34. Taylor, M. R.; Glusker, J. P.; Gabe, E. J.; Minkin, J. A. *Bioinorg. Chem.* **3** (1974) 189.
35. Hatfield, W. E.; ter Haar, L. W. In Biological and Inorganic Copper Chemistry; Karlin, K. D.; Zubieta, J., Eds., Adenine Press: Guilderland, New York; 1986.
36. Smits, J. J.; deJongh, L. J.; van Ooijen, J. A. C.; Reedijk, J.; Bonner, J. C. *Physica B+C* **97** (1979) 229.
37. Jansen, J. C.; van Koningsveld, H.; van Ooijen, J. A. C. *Cryst. Struct. Commun.* **7** (1978) 637.
38. Crawford, V. H.; Hatfield, W. E. *Inorg. Chem.* **16** (1977) 1336.
39. Marsh, W. E.; Valente, E. J.; Hodgson, D. J. *Inorg. Chim. Acta* **51** (1981) 49.
40. De Groot, H. J. M.; de Jongh, L. J.; Willett, R. D.; Reedijk J. *J. Appl. Phys.* **53** (1982) 8038.
41. Olmstead, M. M.; Musker, W. K.; ter Haar, L. W.; Hatfield, W. E. *J. Amer. Chem. Soc.* **104** (1982) 6627.
42. Hatfield, W. E.; Helms, J. H.; Rohrs, B. R.; ter Haar, L. W. *J. Am. Chem. Soc.* **108** (1986) 542.

26

43. Sletten, J. *Acta Chem. Scand.* A **38** (1984) 491.
44. Rohrs, B. R., *Ph. D. Dissertation*, University of North Carolina, Chapel Hill, NC 1988.
45. Turov, E. A. Physical Properties of Magnetically Ordered Crystals; Academic Press: New York; 1965.
46. Morish, A. H. Physical Principles of Magnetism; Krieger: Huntington, New York; 1980.
47. GAUSSIAN 88 (TM); © Gaussian, Inc., 1988.
48. Petersen, J. D.; Morgan, L. W., personal communication.
49. Kirk, M. L.; Hatfield, W. E.; Lah, M. -S.; Kessissoglou, D.; Pecoraro, V. L.; Morgan, L. W.; Petersen J. D. *J. Appl. Phys.* **69** (1991) 6013.
50. Julve, M.; De Munno, G.; Bruno, G.; Verdaguer, M. *Inorg. Chem.* **27** (1988) 3160 and references therein.
51. Kirk, M. L. *Ph. D. Dissertation,* University of North Carolina, Chapel Hill, NC 1990.
52. (a) Hatfield, W. E. *Comments Inorg. Chem.* **1** (1981) 108. (b) Hodgson, D. J. *Prog. Inorg. Chem.* **19** (1975) 173.
53. Benelli, C.; Gatteschi, D.; Carnegie, D. W., Jr.; Carlin R. L. *J. Am. Chem. Soc.* **107** (1985) 2560.
54. Ananthakrishna, G.; Weiss, L. F.; Foyt, D. C.; Klein, D. J. *Physica B* **81B** (1976) 275.
55. Wilson, R. B. *Ph. D. Dissertation*, University of North Carolina, Chapel Hill, NC (1978).
56. Marsh, W. E.; Helms, J. H.; Hatfield, W. E.; Hodgson, D. J. *Inorg. Chim. Acta* **150** (1988) 35.
57. Prout, C. K.; Carruthers, J. R.; Rossotti, F. J. C. *J. Chem. Soc. A* (1971) 3350.
58. Robin, M. B.; Day, P. *Ad. Inorg. Chem. Radiochem.* **10** (1967) 247.
59. Scott, B.; Willett, R. W. *Inorg. Chem.* **30** (1991) 110.
60. (a) ter Haar, L. W.; DiSalvo, F. J.; Bair, H. E.; Fleming, R. M.; Waszczak, J. V.; Hatfield, W. E. *Phys. Rev. B Conds. Matter* **35** (1987) 1932. (b) Brown, D. B.; Zubieta, J.A.; Vella, P. A.; Wobleski, J. T.; Watt, T.; Hatfield, W. E.; Day, P. *Inorg. Chem.* **18** (1979) 2635.
61. Drillon, M. personal communications.

FERRIMAGNETIC CHAINS: MODELS AND MATERIALS

E. Coronado[1], M. Drillon[2] and R. Georges[3]

[1]Dept. Química Inorgánica, Univ. Valencia, Doctor Moliner 50, 46100 Burjasot , Spain
[2]Groupe des Matériaux Inorganiques, IPCMS-EHICS, 1 rue Blaise Pascal 67008 Strasbourg, France
[3]Lab. de Chimie du Solide, CNRS, 351 cours de la Libération, 33405 Talence, France.

1. Introduction

One of the major advances in recent magnetochemistry has been the preparation and study of one-dimensional (1D) ferrimagnets, or so-called ferrimagnetic chains. This class of low dimensional systems has been discovered in the 1980's in compounds showing structurally ordered bimetallic chains[1]. From a chemical point of view, bimetallic chains represent the natural extension of the bimetallic dimers and other polynuclear species, driven by the aim of the chemists to prepare and characterize more and more complex systems. From a physical point of view, the ferrimagnetic chains are made of two different magnetic moments located on alternating sites and coupled by an antiferromagnetic exchange coupling. Due to the non-compensation of these moments, bimetallic chains show a distinctive magnetic behavior generally exhibiting a minimum in the χT vs. T plot and a divergence at lower temperatures.

This kind of behavior is in fact not an exclusive property of bimetallic systems. Indeed, several examples of homometallic 1D ferrimagnets have been reported in the last few years. This behavior may occur for instance in systems with alternating g-factors or for a particular stacking of the metal ions, such as intertwining double chains. We will refer in those cases to topological 1D ferrimagnets, since the ferrimagnetic behavior results as a consequence from the topology of the metal network.

In the present review we discuss some representative results obtained in these classes of one-dimensional ferrimagnets. The paper is organized as follows: in the first part, we briefly review the different theoretical models developed for analyzing the magnetic properties of ferrimagnetic chains; the second part is devoted to the magnetochemistry of a series of bimetallic compounds. We focus on the EDTA bimetallic complexes, which furnish model systems for investigating a wide variety of ferrimagnetic chains with different spin and/or anisotropy values; finally, we present in the third part the magnetic properties of some topological 1D ferrimagnets and other complex systems.

2. General models

The most authoritative reviews of exchange coupling in 1D systems are restricted to ferro (F) and antiferromagnetic (AF) spin chains[2]. The discovery of ferrimagnetic chains has motivated a rapid development of theoretical models, which has been favored by the existing background in F and AF spin chains. In fact, most of the calculation procedures required to get the thermodynamic properties of ferrimagnetic chains are similar to those for the F or AF chains.

In the simplest case, a bimetallic chain is constituted by two distinct metal ions namely A and B, which alternate according to the sequence $\cdots$-A-B-A-B-$\cdots$. The exchange interactions are between nearest neighbors, only. On the other hand, two different magnetic moments are associated to the two sublattices with generally different spin values (S_a and S_b), Landé factors (g_a and g_b) and local anisotropies (D_a and D_b). The magnetic system may be schematized as

$$\cdots\text{-}S_a g_a(D_a)\text{-}S_b g_b(D_b)\text{-}S_a g_a(D_a)\text{-}S_b g_b(D_b)\text{-}\cdots$$

and it is described by the overall spin Hamiltonian

$$H = H_{ex} + H_a + H_z \tag{1}$$

where H_{ex}, H_a and H_z represent the exchange contribution, single-ion anisotropy and Zeeman term, respectively.

The exchange coupling within a pair A-B may be written as

$$H_{ex} = -J_x S_{xa} S_{xb} - J_y S_{ya} S_{yb} - J_z S_{za} S_{zb} \tag{2}$$

where J_x, J_y and J_z are the components of the exchange constant, which define the dimensionality of the interaction. The particular cases $J_x = J_y = J_z$, $J_x = J_y \neq 0$, $J_z = 0$, and $J_x = J_y = 0$, $J_z \neq 0$ correspond to the common models used in magnetism, namely Heisenberg, XY and Ising models, respectively. For the ferrimagnetic chains, theoretical treatments have been restricted to the fully isotropic (Heisenberg), and to the anisotropic (Ising) cases. This has been motivated by the possibility of obtaining analytical solutions for the thermodynamic properties, in connection with the symmetry of the ground spin states of the interacting metal ions[3]. Basically, it has been shown that the exchange model is imposed by the ion of lower symmetry. Thus, when the bimetallic chain is made of an isotropic ion, such as manganese(II) or copper(II), alternating with an anisotropic one, such as cobalt(II), the exchange coupling is Ising-like, since the latter typically shows a ground spin state with $g_z \gg (g_x, g_y)$.

For an infinite chain, the exchange hamiltonian involves the summation between nearest neighbors. According to the structure of the material, different kinds of chains will result by reference to the exchange network. We will use the term "linear chain" when each magnetic center interacts with two nearest neighbors only. In this case, the terms "uniform chain", "alternating chain" and "random chain" will refer respectively to the presence of a unique interaction (Fig.1-a), two alternating interactions (Fig.1-b), or a random distribution of the interactions (Fig. 1-c). The term "double chain" will deal with magnetic systems made essentially of two interacting linear chains (Fig. 1-d). These often give rise to striking magnetic properties, such as spin frustration (1-e), topological ferrimagnetism (1-f), or both effects (1-g). It is to be noticed that such complex chains are the first step towards the 2D systems.

Owing to the large diversity of theoretical situations that can be encountered, we will restrict here the discussion to the most representative models used for describing the magnetic properties of the existing 1D materials. A detailed discussion of the theories developed in connection with one-dimensional ferrimagnets is reported elsewhere[4].

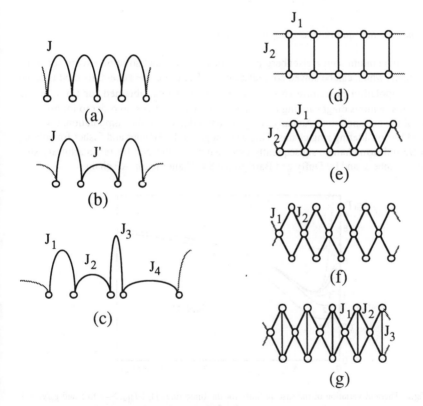

Fig.1. Different kinds of exchange networks examined in the present review: a) uniform chain, b) alternating chain, c) random chain, d) ladder-like chain, e) spin frustrated double chain, f) topological 1D ferrimagnet, g) spin frustrated topological 1D ferrimagnet.

30

2.1. Heisenberg Model

Closed expressions of the thermodynamic functions of interest can be obtained for some convenient 1D magnetic systems. However, analytical solutions for the Heisenberg chain are only available when, at least, part of the spins are treated in the classical approximation. This consists in replacing a spin operator by a classical vector, avoiding the difficulties associated with non commuting operator algebra. The classical approximation is allowed for large spin quantum numbers since then the spin component commutators, (S_x, S_y), etc.., which are of the order of magnitude of S, are negligible compared to the products $S_x S_y$, etc.., which are of the order of S^2. According to the nature of the two kinds of interacting spins, quantum-quantum, classical-classical, and quantum-classical approaches are discussed.

2.1.1. The quantum-quantum ferrimagnetic chains

For the quantum spin Heisenberg chains, there is no exact solution except for finite length chains which can be solved computationally. Therefore, the general method is based on an extrapolation procedure (N→∞) of the exact results, obtained for finite length chains. This requires diagonalizing larger and larger real matrices, in order to improve the quality of the extrapolation. Actually, a significant reduction of the computational work is provided by considering ring chains. This was suggested by Bonner and Fisher for solving the S=1/2 uniform chain[5], subsequently extended by Weng[6] and Blöte[7] to arbitrary spin quantum numbers, and by Duffy and Barr[8] to the S=1/2 alternating AF chain.

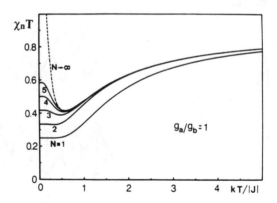

Fig.2. Thermal variation of the susceptibility for the finite rings $[1/2\text{-}1]_N$, N=1 to 5 and g_a/g_b = 1. The normalized χT product is given in units of $N_A g^2 \mu_B^2 / k_B$. The dashed curve corresponds to the infinite ring extrapolation (from ref. 10).

For the alternating spin chains, made up of two quantum spin sublattices S_a and S_b and abbreviated as $[S_a\text{-}S_b]$, the ring chain procedure has allowed to derive the thermal and magnetic properties of uniform chains with $S_a=1/2$ and $S_b= 1$ to $5/2$. For the simplest case, i. e. the [1/2-1] chain with $g_a = g_b$, the main features are conveniently given by extrapolating the results obtained up to 10-spin rings $[1/2\text{-}1]_N$ where N stands for the number of spin pairs[9] (Fig. 2). A well defined minimum of χT is observed at intermediate temperature. This results from two conflicting tendencies: on one hand, short range correlations, due to AF coupling, tend to reduce the resulting moment of a pair of neighboring spins; on the other hand, long range correlations bring an increasing number of moments into a coherent behavior., giving rise to quasi rigid and independent magnetic blocks of increasing amplitude. Such a competition between short range and long range correlations does occur, so far as AF exchange and distinct magnetic moments are involved, independently of the exact nature of the spin hamiltonian. Therefore, the χT minimum may be considered as a signature of ferrimagnetism in 1D systems. Notice that the minimum becomes less deeper as the difference between the two magnetic moments increases, and vanishes for very small, but finite, values of the moments ratio.

For the [1/2-1] system, the influence of alternating Landé factors, g_a and g_b, has also been investigated. Rational expressions fitting the limiting behavior of the susceptibility for regular $(g_a=g_b)$ and alternating (specifically $g_a = 2g_b$) Landé factors have been calculated (Table I). While for $g_a = g_b$ the extrapolated susceptibility diverges at low temperature in a way similar to the Heisenberg S=1/2 ferromagnetic chain, an antiferromagnetic-like behavior (no susceptibility divergence at low temperature), is predicted in the low temperature limit for the critical ratio $g_a/g_b=2.67$ (Fig. 3). This value is in between the Néel state value (2.0) and the spin wave theory value (3.56). This phenomenon recalls the magnetization compesation observed in some rare earth iron garnets. For this reason we will refer to it as compensation phenomenon.

Table I. Polynomials giving the behavior of {1/2-1} Heisenberg chains for regular $(g_a = g_b)$, and alternating $(g_a = 2g_b)$ Landé factors (from ref. 10)

g_a / g_b	$[\chi T] / [(N_a\mu_B^2/k_B)g_b^2]$
1	$[Ax^3 + Bx^2 +Cx +D] / [Ex^2 + Fx + G]^*$ A = -0.034146801, B = 2.8169306411 C = -7.2310013697, D = 11, E = 1.29663274, F = 0.69719013595, G = 12
2	$[Ax^2 + Bx +C] / [Dx^3 + Ex^2 + Fx + G] + Hx$ A = 2.944723391, B = -8.643216582 C = 20, D = 2.207977566, E = 2.210070570, F = 5.150935691, G = 12, H = 0.00232325

$^*x = |J|/k_BT$

32

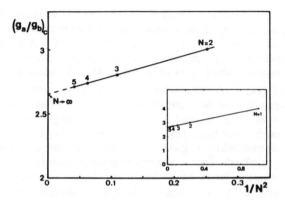

Fig.3. Critical ratio for the $[1/2\text{-}1]_N$ chain; extrapolation to the infinite length chain (from ref. 10).

Finally, it can be pointed out[10] that the reduction of the chain length imposed by the computer capacities may entail questionable extrapolations of the numerical data (for example, N=1 to 3 in the [1/2-5/2] system). Anyway, a swift convergence of energy levels is noted when S increases, so that a reasonable description of the magnetic and specific heat features of the infinite chain is expected. These features are summarized in Table II. Further, we note in this case a spin dependence of the critical value of the ratio g_a/g_b, which seems to be closely given by $4(S+1)/3$.

Table II. Thermal and magnetic properties of [1/2-S] Heisenberg ferrimagnetic chains (from ref. 10).

| System | Specific heat | | Magnetic susceptibility[a] | |
	$C_{p_{max}}/R$	$k_B T_{max} / |J|$	$(\chi T)_{min}$	$(k_B T)_{min} / |J|$
[1/2,1]	0.337	0.705	1.507	0.577
[1/2,3/2]	0.356	0.96	2.221	1.190
[1/2,2]	0.351	1.2	2.609	2.1
[1/2,5/2]	0.35	1.4	2.80	2.9
$[1/2,\infty]^b$	-	-	2.83	2.98

[a]values for $g_a = g_b$
[b]classical spin scaled to S=5/2

2.1.2. The classical-classical ferrimagnetic chains

Let us consider now ferrimagnetic chains made of classical spins, only. For this class of chains analytical expressions of the thermodynamic properties are generally available. The classical spin model was first introduced by Fisher[11] to solve the uniform F or AF Heisenberg chain. In that case the zero-field magnetic susceptibility is given by the expression

$$\chi_0 = (N_A \mu_B^2 \beta M^2/3)\ (1+P)/(1-P) \tag{3}$$

where M is the magnetic moment, β is the Boltzmann factor $(1/k_B T)$, N_A, μ_B and k_B have their usual meanings, and P is the Langevin function:

$$P(\gamma) = \coth(\gamma) - (\gamma)^{-1}, \qquad \gamma = \beta J \tag{4}$$

To fit experimental data, scaling factors must be introduced in Eq. 3:

$$J \to J(S(S+1))\ ;\ M = g(S(S+1))^{1/2}$$

This method has been extended to a large variety of more complex 1D magnetic systems, most of them being ferrimagnetic. A first study was achieved by Thorpe[12] who considered chains involving two randomly distributed magnetic species, A and B, with concentrations c and $1-c$. Distinct interactions were introduced to account for the different kinds of pairs $(J_{aa}, J_{ab} = J_{ba}, J_{bb})$. In such a model the two magnetic species are implicitly assumed to carry equal magnetic moments. In the special case in which one of the species is non magnetic, the system reduces to a chain of isolated fragments for which the susceptibility expression is

$$\chi_0 = (N_A \mu_B^2 \beta M^2/3)\ ((1+P)/(1-P) - (2P/N)(1-P^N)/(1-P^2)) \tag{5}$$

where N is now the mean length of the chain, and P is defined as previously.

Such systems cannot be considered as really 1D ferrimagnets since, for instance, they do not show any net magnetic moment in the ground state (except for ferromagnetic coupling). The first theoretical approach concerning really ferrimagnetic isotropic chains was motivated by experiments performed on Mn-Ni ordered bimetallic chains[13]. The zero-field magnetic susceptibility (per pair of sites) of a uniform chain with two alternating magnetic sublattices (moments M_a and M_b) considered from a classical point of view may be written as

$$\chi_0 = (N_A \mu_B^2 \beta/6)\ (M^2(1+P)/(1-P) + \delta M^2(1-P)/(1+P))$$

with P defined as previously, and $M = M_a + M_b$, $\delta M = M_a - M_b$. This expression reduces to Fisher's one when the two sublattices are equivalent. In the present case, the following scaling factors must be introduced to fit experimental data:

$$J \rightarrow J[S_a(S_a+1)S_b(S_b+1)]^{1/2} \quad ; \quad M_i = g_i(S_i(S_i+1))^{1/2} \ (i = a, b)$$

Eq. 6 is valid for both F and AF couplings. As expected, in the AF case and for distinct moments, Eq. 6 may give rise to the typical minimum in the $\chi_0 T$ vs. T plot.

The classical spin approximation also allows to introduce alternating or random exchange coupling[14]. The expression for the random two-sublattice chain is given by

$$\chi_0 = (N_a \mu_B^2 \beta/6) \ (\ M^2(1+\overline{P})/(1-\overline{P}) + \delta M^2(1+\overline{P})/(1-\overline{P}) \ ) \tag{7}$$

where $\overline{P}$ is the average P value over the J distribution. Assuming a uniform distribution of the exchange constant over a domain ranging from J-λ to J+λ ($\lambda \geq 0$), $\overline{P}$ is given by

$$\overline{P} = (2\beta\lambda)^{-1} \ln\{(J-\lambda) \sinh[\beta(J+\lambda)]/(J+\lambda) \sinh[\beta(J-\lambda)]\} \tag{8}$$

Further developments with more than two spin sublattices or more complex chains, as for example chains made of connected dimers[15] triangles[16] or rings[17] have been investigated within the classical spin approximation.

2.1.3. The quantum-classical ferrimagnetic chains

The first study of a quantum-classical Heisenberg chain was reported by Dembinski and Widro[18] for the treatment of the correlation functions and specific heat. Further, Blöte[19] proposed an expression of the susceptibility, but he neglected the magnetic contribution of the quantum sublattice, making his result useless for real systems. The first analysis involving alternating quantum and classical spins was reported by Seiden, Verdaguer et al.[20], who considered the uniform chain $... S_{i-1}, s_i, S_i, s_{i+1}...$, where s_i is a $S=1/2$ quantum operator, and S_i a classical spin. The calculated zero-field susceptibility is given by

$$\chi_0 = (N_a \mu_B^2 \beta/3) \left(M^2(S+1)/S + g^2 s(s+1) + 2(1-P)^{-1}(PM^2 - 2Q \ gsM + Q^2 g^2 s^2) \right) \tag{9}$$

where P and Q are given in terms of γ ($= \beta J$) by

$$P = \{(1+12\gamma^2) \ \text{sh}\gamma - (5\gamma^{-1}-12\gamma^{-3}) \ \text{ch}\gamma - \gamma^{-1} + 12\gamma^{-3})\}/(\text{sh}\gamma - \gamma^{-1}\text{ch}\gamma + \gamma^{-1}) \tag{10}$$

$$Q = ((1+2\gamma^2) \ \text{ch}\gamma - 2\gamma^{-1} \ \text{sh}\gamma - 2\gamma^{-2})/(\text{sh}\gamma - \gamma^{-1}\text{ch}\gamma + \gamma^{-1}) \tag{11}$$

This model has been extended to linear chains showing various random characters (exchange constant, spin quantum number, classical moment, quantum spin Landé factor)[21]. In the particular case of alternating exchange parameters, the zero-field susceptibility is given by[22]

$$\chi_0 = (N_A\mu_B^2\beta/3) \left(M^2 + g^2s(s+1)+2(1-P)^{-1}(M^2P + g^2 Q_+Q_- + Mg(Q_++Q_-))\right) \quad (12)$$

where Q_+ and Q_- are related to γ_+ (= βJ_+) and γ_- (= βJ_-), with $J_\pm = J(1\pm\alpha)$. Clearly, Eqs. 9 and 12 have to be conveniently scaled to fit experimental results.

Further generalizations of this model have been proposed for chains showing an alternation of a classical spin and a more or less complex quantum moiety[21].

2.2. Ising Model

In systems showing some amount of exchange or local anisotropy, restrictions to obtain exact solutions are met, similarly to the fully isotropic case. Only the Ising-like anisotropic coupling allows a rigorous treatment for quantum as well as for classical spins. In this case, advantage is taken of the commutation rules of the spin operators. This arises from the fact that only the z-component is involved for each spin. As a result, exact solutions of the parallel susceptibility ($\chi_{//}$) and magnetization (with the applied field along the z direction) can be easily derived for a large variety of 1D systems. Conversely, the perpendicular susceptibility ($\chi_\perp$) has only been calculated in few cases. It appears to remain finite at low temperature. Thus, it will be negligible compared to $\chi_{//}$, which diverges upon cooling, in most ferrimagnetic chains. Accordingly, the lake of exact expression for $\chi_\perp$ does not really hinder the fitting of experimental data.

The general mathematical procedure is based on the so-called transfer matrix method proposed by Kramers and Wannier[23] for linear 1/2 spin chains. It is straightforward to generalize this method to multi-sublattice chains with more or less complex exchange networks.

One of the most representative ferrimagnetic chains is built up from S = 1/2 spins with both alternating Landé factors (g_a and g_b) and exchange interactions (J and J')[24]. The field dependent partition function and the related thermodynamic quantities are derived by solving a 2x2 matrix. The zero field parallel susceptibility is given by

$$\chi_{//}= (N_A\mu_B^2\beta/2)(g_+^2\exp(\beta J_+/2) + g_-^2\exp(\beta J_-/2))/\cosh(\beta J_-/2) \quad (13)$$

where $g_\pm = (g_a\pm g_b)/2$, and $J_\pm = (J\pm J')/2$. This expression is valid for both F and AF couplings. As in the isotropic case, a minimum generally occurs in the thermal variation of the $\chi_{//}T$ product (for $g_a \neq g_b$ and AF couplings) with a divergence in the low temperature range.

The same procedure has been achieved for the [1/2-S] alternating chains with arbitrary S[25]. The method allows introducing single-ion anisotropy on S. Exact expressions for specific heat, parallel susceptibility and magnetization have been obtained (Table III). Single-ion anisotropies have also been introduced in the [1-1] Ising chain. A detailed discussion of the influence of alternating zero-field splitting parameters (D_a and D_b) on the magnetic behavior has been reported[26].

On the other hand, the compensation phenomenon has been discussed in the framework of the Ising model for general [S_a-S_b] chains[27]. Compensation is observed for all $S_a = S_b$, and when both spin numbers are smaller than 2. For equal magnetic moments, a divergence of $\chi_{//}T$ is observed when, at least, one of the spin numbers is larger than 2 (Fig.4). For the critical value S=2, $\chi_{//}$ approaches a Curie law ($\chi_{//}T$ has a finite limit) at low temperatures.

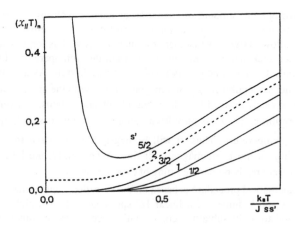

Fig.4. Magnetic behavior of [1/2-S] Ising chains for compensated magnetic moments (from ref. 21).

It is worth-mentionning that under definite conditions, the transfer matrix method may also apply to random linear chains. This aspect has been thoroughly examined for alternating quantum-classical Ising chains with a random exchange distribution[16].

Finally, specific treatments based on the transfer matrix method have been reported for studying various complex 1D systems with unusual exchange topologies as for example the so-called ladder-like or frustrated double chains[28]. Such systems are treated within the simple Ising framework because, to date, a Heisenberg type numerical treatment should require too much computational work. Anyway, the Ising model furnishes useful answers to important questions, as for example the influence of weak interchain coupling, or the ground spin configuration when competing exchange interactions are present. When considering the same spin on different sites coupled by AF interactions, it appears that the magnetic ground-state is either antiferromagnetic (in ladder-like chains) or paramagnetic (in frustrated chains). In turn, 1D ferrimagnetism may be observed in more complex homometallic systems in which spin sublattices do not compensate. This occurs in the trimeric chain displayed in Fig. 1.f. The exact expressions of the parallel susceptibility derived for these three kinds of S=1/2 chains are given in Table IV.

Table III. Magnetic susceptibility and specific heat for the alternating [1/2-S] Ising chain.

Specific heat $\qquad C_p/R = \beta^2(Z''/Z - Z'^2/Z^2)$

Magnetic susceptibility

$$\chi_{\parallel} = \left(\frac{N_A}{\beta}\right) \frac{\left(S''_o + \dfrac{S_0 S''_0 - 2P''_0}{(S_0^2 - 4P_0)^{1/2}}\right)}{S_0 + (S_0 - 4P_0)^{1/2}}$$

$$Z = \sum_{j}^{S} (2 - \delta_{jo})\left[\cosh(\beta jJ_+) + \cosh(\beta jJ_-)\right]$$

$$Z' = \sum_{j}^{S} j(2 - \delta_{jo})\left[J_+\sinh(\beta jJ_+) + J_-\sinh(\beta jJ_-)\right]$$

$$Z'' = \sum_{j}^{S} j^2(2 - \delta_{jo})\left[J_+^2\cosh(\beta jJ_+) + J_-^2\cosh(\beta jJ_-)\right]$$

$$S_o = 2\sum_{j}^{S} (2 - \delta_{jo})\cosh(\beta jJ_+)$$

$$S''_o = 2\sum_{j}^{S} (2 - \delta_{jo})\left\{\left[g_a\mu_B\beta/2\right]^2 + j^2(g_b\mu_B\beta)^2\right\}\cosh(\beta jJ_+) + jg_ag_b\mu_B^2\beta^2\sinh(\beta jJ_+)$$

$$P_o = \sum_{j,k}^{S} (2 - \delta_{jo})(2 - \delta_{ko})\left[\cosh(\beta jJ_+)\cosh(\beta kJ_+) - \cosh(\beta jJ_-)\cosh(\beta kJ_-)\right]$$

$$P''_o = (g_b\mu_B\beta)^2 \sum_{j,k}^{S} (2 - \delta_{jo})(2 - \delta_{ko})(j^2 + k^2)\left[\cosh(\beta jJ_+)\cosh(\beta kJ_+) - \cosh(\beta jJ_-)\cosh(\beta kJ_-)\right]$$
$$\quad - 2jk\left[\sinh(\beta jJ_+)\sinh(\beta kJ_+) - \sinh(\beta jJ_-)\sinh(\beta kJ_-)\right]$$

with $J_{\pm} = (J \pm J')/2$. The summations extend over $j, k = 0$ (1/2) to S; j_o and k_o correspond to Kronecker symbols.

Table IV. Magnetic susceptibility and specific heat for Ising double chains with topologies d, e and f (see Fig. 1).

I. Ladder like chain:

$$\chi_{||} = \left(\frac{N_A g^2 \mu_B^2}{2 k_B T}\right)\left(\frac{ab E_+ - 2a\,\sinh(\alpha)}{E_+^2 - 2E_+(ab - \cosh(\alpha)\sinh(\alpha)) + 2(1 + b^2)\sinh(\alpha)\cosh(\alpha)}\right)$$

$$E_+ = 2\cosh(\alpha)\cosh(\beta) + 2[\cosh^2(\alpha)\sinh^2(\beta) + 1]^{1/2}$$

$$\alpha = J_1/2k_B T$$
$$\beta = J_2/4k_B T$$
$$a = \exp(\alpha)$$
$$b = \exp(\beta)$$

II. Topological 1D Ferrimagnet:

Magnetic susceptibility: $\chi_{//} = [N_A g^2 \mu_B^2/(2k_B T)](A/B)$

$$A = U\exp(K_3)[4\exp(K_+) + \cosh(K_+) + \exp(-2K_3)] - 4[\cosh(K_+) - \cosh(K_-)]$$
$$B = 2U\{U - \exp(K_3)[\cosh(K_+) + \exp(-2K_3)]\}$$

with $\quad U = \exp(K_3)\cosh(K_+) + \exp(-K_3)\cosh(K_-) + 2\cosh(K_3)$
$$K_\pm = (J_1 \pm J_2)/2k_B T; \quad K_3 = J_3/4k_B T$$

Specific Heat: $C_p/R = (P/U - Q^2/U^2)$

$$P = 16K_3^2[\exp(K_3)\cosh(K_+) + \exp(-K_3)\cosh(K_-) + 2\cosh(K_3)] +$$
$$+ 8K_3[K_+\exp(K_3)\sinh(K_+) - K_-\exp(-K_3)\sinh(K_-)] +$$
$$+ K_+^2\exp(K_3)\cosh(K_+) + K_-^2\exp(-K_3)\cosh(K_-)$$

$$Q = 4K_3[\exp(K_3)\cosh(K_+) - \exp(-K_3)\cosh(K_-) + 2\sinh(K_3)] +$$
$$+ K_+\exp(K_3)\sinh(K_+) + K_-\exp(-K_3)\sinh(K_-)$$

Table IV. Cont.

III. Frustrated Double Chain:

Magnetic susceptibility: $\chi_{//} = [N_A g^2 \mu_B^2/(8k_B T)](A/B)$

$A = U^2 \exp(K_1)\exp(K_2) + 2[1 - \exp(2K_1)]U + \exp(3K_1)\exp(-K_2) +$
$\qquad \exp(-K_1)\exp(K_2) - 2\exp(K_1)\exp(-K_2)$

$B = ER$

with $E = \exp(-2K_1) - \cosh(K_2) + \exp(-K_1)R$

$R = [\exp(2K_1)\sinh^2(K_2) + 2\cosh(K_2) + 2]^{1/2}$

$U = \exp(K_1)\cosh(K_2) + \exp(-K_1) + [\exp(2K_1)\sinh^2(K_2) +$
$\qquad + 2\cosh(K_2)-+-2]^{1/2}$

Specific heat: $C_p/R = (B/Z - A^2/Z^2)$

$Z = \exp(-K_1) + \exp(K_1)\cosh(K_2) + U^{1/2}$

$A = V + P/U^{1/2}$

$B = W + (UQ - P^2)/U^{3/2}$

$U = \exp(2K_2)\sinh^2(K_1) + 2\cosh(K_1) + 2$

$V = -K_1\exp(-K_1) + K_1\exp(K_1)\cosh(K_2) + K_2\exp(K_1)\sinh(K_2)$

$W = K_2^2[\exp(K_2)\cosh(K_1) + \exp(-K_2)] + K_1 K_2 \exp(K_1 + K_2) +$
$\qquad + K_1\exp(K_2)\sinh(K_1)$

$P = K_2\exp(2K_2)\sinh^2(K_1) + (1/2)K_1\exp(2K_2)\sinh(2K_1) + K_1\sinh(K_1)$

$Q = 2K_2^2[\exp(K_2)\sinh(K_1) + 2K_1 K_2 \exp(2K_2)\sinh(2K_1) +$
$\qquad + K_1^2\exp(2K_2)\cosh(2K_1) + K_1^2\cosh(K_1)$

with $K_i = J_i/2k_B T \qquad i = 1, 2$

---

3. 1D Magnetic Materials

3.1. Alternating site ferrimagnets

Several families of chain compounds exhibiting two alternating magnetic sites are known to date. In the particular case of coordination chemistry, we should mention the bimetallic family of dithiooxalato and oxamato derivatives reported by Kahn's group[29], the metal-radical family reported by Gatteschi's and Rey's groups[30], and the EDTA family reported by our groups[31]. The former two families are currently under study as a means to stabilize bulk molecular-based ferro- and ferrimagnets. Indeed, the presence of non negligible interchain interactions, together with relatively strong intrachain interactions, lead, in the first family, to the occurrence of a 3D magnetic ordering in the temperature range 2-14 K. In some cases this ordering gives rise to a spontaneous magnetization, and the critical temperature can be raised up to 30 K by bringing nearer the chains[32]. For the metal-radical family, which shows stronger intrachain interactions but better isolation of the chains, the critical temperatures are of the same order of magnitude.

Here, we focus on the EDTA family which deals with a large variety of low-dimensional magnetic model systems. EDTA refers to the hexadentate ligand ethylenediamine-NNN'N'-tetra-acetate (Fig. 5). This ligand furnishes several isostructural series of bimetallic compounds of variable dimensionality. The most thoroughly studied series is the hexahydrate one MM'(EDTA).$6H_2O$ containing ferrimagnetic chains (with [MM'] = [MnCo], [MnNi], [MnCu], [CoCo], [CoNi], [CoCu], [NiNi]). By conducting the synthesis under high temperature and pressure conditions[33], some water molecules are eliminated. Then, the chains may intersect, and the dimensionality increases from 1D to 3D. The corresponding series, of general formula $M^tM(M'EDTA)_2.4H_2O$ (with [M^tMM'] = [CoCoCo], [CoCoNi], [CoNiNi], [ZnNiNi]), is formed by bimetallic layers.[MM'] of octahedral sites linked by intermediate tetrahedral sites M^t. Except for the [ZnNiNi], these compounds exhibit a three-dimensional ferrimagnetic ordering with bulk magnetization[34]. A detailed discussion of this series is reported in ref. 31. Here we consider only the EDTA compounds showing chain structures. Their magnetic features are listed in Table V.

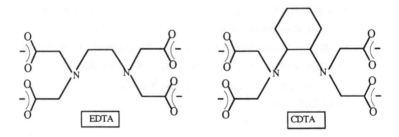

Fig. 5. Organic ligands EDTA and CDTA

3.1.1. Structural features

The structure of the hexahydrate series[35] consists of zigzag chains built up from two alternating octahedral sites referred to as "chelated" or "hydrated" according to their local environment (Fig. 6.top). In the chelated site, the metal ion is hexacoordinated by EDTA, forming thus the anionic moiety $[M'(EDTA)]^{-2}$. In the hydrated one, M^{+2} is in a less distorted octahedron formed by four water molecules and two oxygen atoms belonging to two carboxylate bridges of two adjacent $[M'(EDTA)]^{-2}$ complexes. In bimetallic compounds both sites are selectively occupied by distinct metal ions giving rise to cationic ordering in the chains. On the other hand, alternating anti-anti and anti-syn configurations of the carboxylate bridges occur along the chain (Fig. 6.bottom), giving rise to slightly alternating M-M' distances. The chain may then be schematized as

$$—M(H_2O)_4O_2---M'(EDTA)—M(H_2O)_4O_2---M'(EDTA)—$$

where dashed and full lines refer to alternating carboxylate bridges.

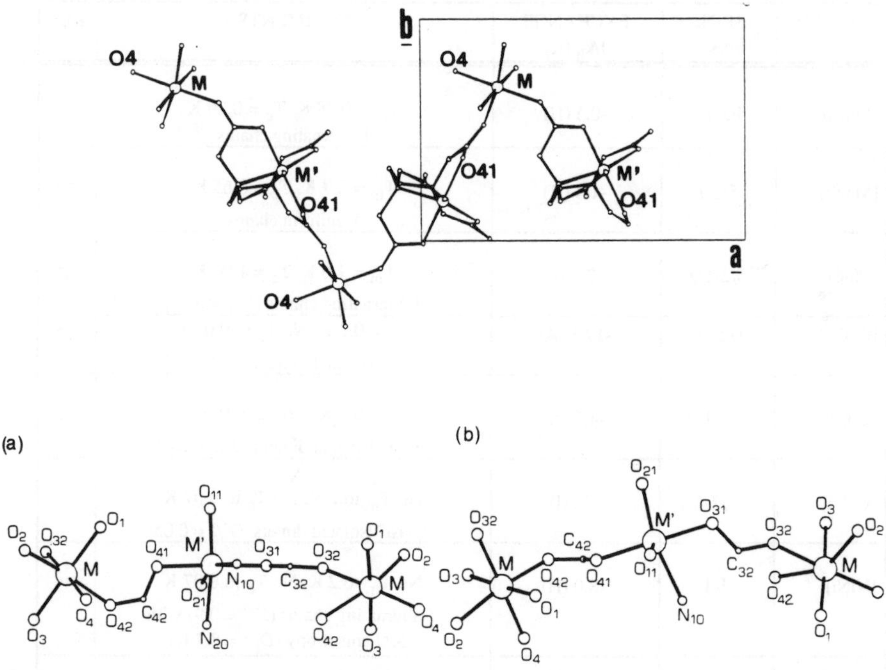

Fig. 6. Top: Structure of the series MM'(EDTA).6H$_2$O (M, M'= Mn, Co, Ni, Cu) showing the two different metal sites. Bottom: Views showing the anti-syn (a) and anti-anti (b) configurations of the carboxylate bridges alternating along the chain (from ref. 36).

From the magnetic point of view, this results in both an alternation of the magnetic moments and of the exchange coupling. Furthermore, according to the magnetic anisotropy of the sites, the exchange coupling between nearest-neighbors ions may be either isotropic or anisotropic. An estimate of this anisotropy may be deduced from the measured g-values through the expression

$$J_\perp/J_{//} \approx (g_{\perp a} \cdot g_{\perp b})/(g_{//a} \cdot g_{//b}) \tag{14}$$

In view of the EPR spectra of cobalt(II) in both sites (Fig. 7)[36], for the bimetallic chains containing cobalt at the chelated site, this ratio is about 0.2, which is very close to the Ising limit and justifies so the use of the Ising model in such cases. In turn, this ratio is about 0.7 when the cobalt is at the hydrated site, and the exchange anisotropy is expected to be in between the Ising and Heisenberg limits.

Table V. Magnetic characterization of the ferrimagnetic chains of the EDTA family (crystalline phase). The letters inside the parentheses refer to the symmetry of the exchange hamiltonian: (H) = Heisenberg, (A) = anisotropic, (I) = Ising. T_m and T_c are defined in the text.

COMPD	SPINS (S-S')	EXCHANGE J/k_B (K)	COMMENTS	REF
[MnCu]	5/2-1/2	-0.5 (H)	$T_m = 0.28$ K. $T_c = 0.20$ K J-alternating chains	37
[MnNi]	5/2-1	-1.5 (H)	$T_m = 2.7$ K. $T_c = 0.65$ K J-uniform chains	37
[MnCo]	5/2-1/2	-2.7 (I)	$T_m = 1.6$ K. $T_c = 1.06$ K J-alternating chains (J'/J = 0.22)	37
[CoCo]	1/2-1/2	-17.5 (A)	$T_m = 0.2$ K. No T_c to 0.07 K "quasi" isolated dimers (J'/J < 0.01)	24
[CoCu]	1/2-1/2	-4.5 (A)	$T_m = 0.2$ K. No T_c to 0.07 K "quasi" isolated dimers (J'/J < 0.01)	24
[CoNi]	1/2-1	-26 (I)	No T_m to.2 K. No T_c to 0.07 K "quasi" isolated dimers (J'/J ≈ 0.05)	38
[NiNi]	1-1	-8.0 (H)	No T_m to.2 K. No T_c to 0.07 K J-alternating chains (J'/J = 0.9) with local anisotropy ($D_{Ni} = 8.7$ K)	35
[MnMn]	5/2-5/2	-0.72 (H)	$T_c = 0.65$ K J-uniform chains. Weak ferromagnetism	40

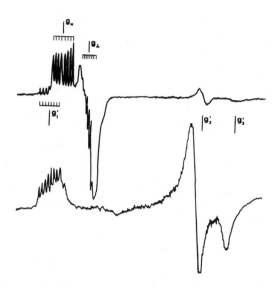

Fig. 7. Powder EPR spectra at 4 K of Co(II) comparing the spin-anisotropy of the lowest Kramers doublet in the hydrated ($g_{//}$ = 5.79, $g_{\perp}$= 3.8) and chelated (g_1'= 8.0, g_2'= 1.45, g_3'= 1.23) sites (from ref. 36).

3.1.2. The bimetallic [MnM'] series (M' = Cu, Ni, Co)

The a.c. magnetic susceptibility data for these compounds[37] show the typical features of 1D ferrimagnets, namely a minimum of χT at T_m and a rapid divergence at lower temperature (Figs. 8-10. Table V). This divergence is limited by 3D antiferromagnetic ordering due to weak interchain interactions. The critical temperature, T_c, is clearly indicated by a sharp peak in both susceptibility and specific heat data (Fig. 11). Another interesting feature of the specific heat is the slight bump observed above T_c, which is attributed, to the remaining 1D short-range magnetic ordering.

In the low dimensional regime, the magnetic data of [MnCu] have been discussed on the basis of the quantum-quantum Heisenberg approach (section 2.1.1). In Fig. 8, the experimental data are compared with the numerical results obtained for a ring of three [5/2-1/2] pairs and a ratio g_{Cu}/g_{Mn} = 1.15. The fitting above 1K gives values of the exchange interaction ranging from -0.35 K and -0.5 K, which reproduce the experiment above 1 K, have been obtained However, important discrepancies are observed at lower temperature. They have been attributed to the effect of a small J alternation, which is not taken into account in the model. On the other hand, from the position of the specific heat maximum leads to J/k_B= 0.45 K, compatible with the magnetic data.

44

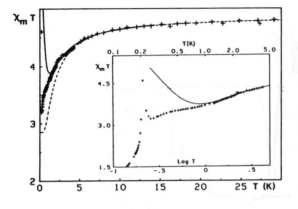

Fig. 8. Magnetic behavior of [MnCu]. The variation of a [5/2-1/2] Heisenberg closed chain (N = 3) is reported as solid line. The behavior of a dimer (J/k_B = -0.45 K) is given in dashed line (from ref. 37).

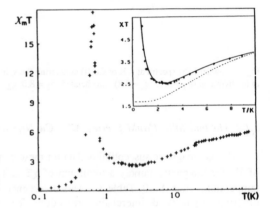

Fig. 9. Magnetic behavior of [MnNi]. The solid line corresponds to the fit from the uniform chain model. The behavior of an isolated dimer (J/k_B = -1.5 K) is given in dashed line (from ref. 37).

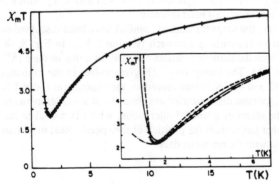

Fig. 10. Magnetic behavior of [MnCo]. Solid line corresponds to the best fit from the J-alternating [5/2-1/2] Ising model. Comparison to uniform chain (- -, J/k_B = -1.6 K) and dimer (-.-, J/k_B = -3.4 K) limits is given in the inset (from ref. 37).

In order to fit the magnetic data for [MnNi] over the whole 1D regime the classical-classical spin approach previously reported (section 2.1.2) has been used. The fact of attributing a classical spin to the nickel(II) may be questionable. Actually, comparing the results of classical and quantum treatments evidences that both the height and position of the minimum are in close coincidence whatever the ratio between Landé factors (Fig. 12). This justifies the classical spin approach in the present case. A very satisfying description of the ferrimagnetic behavior has been obtained with the following set of parameters: J/k_B = -1.5 K, g_{Mn} =1.96, and g_{Ni}/g_{Mn} = 1.23 (Fig. 9). On the other hand, the magnetic specific heat strongly depends on the spin nature, in such a way that the classical approximation must be avoided for the analysis of the thermal data. Using thus the numerical quantum approach, the analysis of the Schottky anomaly observed at ca. 5 K, gives $J/k_B \approx$ -2.2 K, in reasonable agreement with the susceptibility results.

For [MnCo], the magnetic data in the 1D temperature range have been analyzed within the Ising framework (Table III). A very satisfying description of the magnetic susceptibility data, in the region of the minimum has been obtained with the parameters J/k_B = -2.7 K, J'/k_B = -0.6 K, g_{Mn}= 1.86, g_{Co}= 5.5 (Fig. 10). This indicates a significant J alternation along the chain.

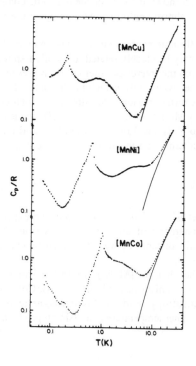

Fig. 11. Specific heat of the [MnM'] series (M' = Cu, Ni, Co). Full line corresponds to the specific heat of the isomorphous nonmagnetic compound [ZnZn] (from ref. 37).

46

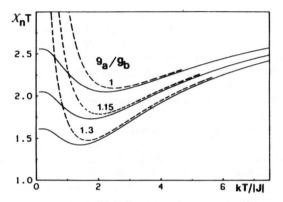

Fig. 12. Comparison between quantum (N=3; solid line) and classical (dashed line) [5/2-1] Heisenberg chains showing the influence of a g-alternation (from ref. 37).

The very weak exchange interactions found in these compounds come from the anti-anti and anti-syn configurations of the carboxylate bridges, which are little favourable to transmit exchange interactions. This has been shown to facilitate the magnetic contribution to the specific heat, since the lattice part is almost negligible in the temperature range of interest (below 10 K).

Another feature is the occurrence of a 3D antiferromagnetic ordering, only observed in Mn(II) containing bimetallic chains. This may be closely related to the high-spin ground state configuration of these chains, for which the interchain interactions represent a first-order effect, while they act to second order when only g factors alternate (the ground configuration of the chain is then S = 0). A further reason is the stronger dipolar coupling due to the large magnetic moment carried by Mn(II). An estimate of the 1D character of the manganese chains is provided by the ratio j/J (j and J refer to interchain and intrachain interactions). For a Heisenberg chain with alternating spins this ratio may be deduced from T_c by

$$T_c = (2J/k_B)[S_a(S_a+1)S_b(S_b+1)(j/J)]^{1/2} \qquad (15)$$

For both [MnCu] and [MnNi] we get $j/J \approx 0.2\text{-}0.3.10^{-2}$, as could be expected for isostructural compounds. The good 1D character of this series results from the large interchain distances (the shortest intermetallic distance between two adjacent chains is ca. 7.3 Å).

In the case of [MnCo] T_c is 60% higher than for [MnNi] (Table V). This relatively large value probably comes from the exchange-anisotropy exhibited in [MnCo]. Indeed, it is well established that any kind of anisotropy tends to favor magnetic ordering; consequently, 3D ordering should occur at higher temperature in an Ising type chain than in the corresponding Heisenberg one.

3.1.3. The [CoM'] series (M' = Co, Cu, Ni)

1D ferrimagnetic behavior is possible even in regular spin chains, provided the Landé tensors on the two alternating sites are different. The first experimental examples showing this type of behavior are provided by [CoCo] and [CoCu], which may be viewed below 30 K as regular S=1/2 spin-chains with alternating g-tensors[24].

These compounds exhibit the usual χT minimum, near 0.2 K (Figs. 13 and 14). In both cases, the analysis of the data, using the two-sublattice Ising model, suggests the presence of a very strong J-alternation, with a ratio J'/J < 0.01 (Table V). Owing to the large number of adjustable parameters used in the magnetic analysis, the values of J and J' are unreliable. In such cases the specific heat measurements have shown to be very useful for providing a more reliable value for J. In Fig. 15 are shown the specific heat data of [CoCo]. The asymmetric Schottky-type anomaly around 3 K, related to the low dimensional character of the system, can be nicely fit, using an anisotropic dimer model, and gives $J_{//}/k_B$ = -17.5 K and $J_{\perp}/J_{//}$ = 0.22. The amount of exchange anisotropy is in full agreement with that predicted from Eq. 14. Although the J-alternating Ising model also reproduces the C_p results, the resulting values of the exchange parameters do not agree with the magnetic results. The large dimerization of these chains has been recently confirmed by a single-crystal EPR study on [CoCu][36]. It has shown the presence of exchange-coupled S=1/2 dimers with a large zero-field splitting ($|D|/k_B$ = 0.43 K) in the spin-triplet state. With this value it has been possible to calculate the amount of exchange anisotropy, $J_{\perp}/J_{//} \approx 0.7$, in good agreement with the value deduced from Eq. 14.

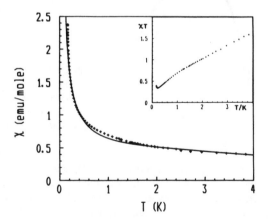

Fig. 13. Magnetic susceptibility of [CoCo]. Solid line corresponds to the best fit from the J-alternating [1/2-1/2] Ising model. The temperature dependence of χT is given in the inset (from ref. 24).

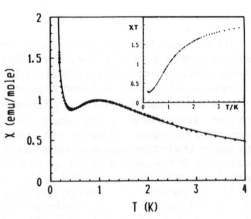

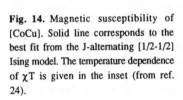

Fig. 14. Magnetic susceptibility of [CoCu]. Solid line corresponds to the best fit from the J-alternating [1/2-1/2] Ising model. The temperature dependence of χT is given in the inset (from ref. 24).

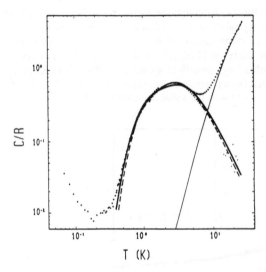

Fig. 15. Specific heat of [CoCo]. Full and dashed lines refer to the best fit from the J-alternating Ising chain ad the anisotropic dimer models, respectively (from ref. 24).

We have just seen situations in which alternating Landé factors in a regular spin chain gives rise to 1D ferrimagnetism. Conversely, theory predicts that, for convenient Landé factor ratio, 1D antiferromagnetism is expected even for alternating spin chains This results from some accidental compensation of the magnetic moments of the two sublattices, and depends on the exchange anisotropy. For instance, for the [1/2-1] chain, the critical ratio g_a/g_b increases from 2 to 2.67 when going from the Ising to the Heisenberg limits.

The first bimetallic chain approaching this kind of compensation is provided by [CoNi][38], which shows a continuous decrease of χT down to 1.5 K (Fig. 16). Owing to the good 1D character of the system, the lack of minimum of χT cannot be attributed to interchain interactions, but better to a magnetic moment compensation. Indeed, the g_{Co}/g_{Ni} ratio (5.8 / 2.3). falls into the critical value range. In the discussion of the magnetic susceptibility and specific heat data some discrepancies arise concerning the dimensionality of the exchange interactions. Thus, the magnetic results have been explained on the basis of the [1/2-1] Ising model assuming a local anisotropy on the Ni(II) site, $D_{Ni}/k_B = -9$ K, and a uniform exchange constant ($J/k_B = -20$ K). On the other hand, the specific heat behavior exhibits two Schottky anomalies at ca. 0.5 K and 6 K (Fig. 16). Using the parameters derived from the magnetic data only the high-temperature anomaly is reproduced, whereas both peaks.are fairly well reproduced if a strong dimerization is assumed. However, the corresponding exchange parameters ($J/k_B = -26$ K, $J'/k_B=-1$ K) lead to a poor agreement with the magnetic susceptibility data. This may be of minor relevance since (i) some cationic disorder is chemically possible, able to introduce a paramagnetic contribution, (ii) an Ising model has been assumed, whereas, in fact, the interactions are likely intermediate between the Ising and Heisenberg limits.

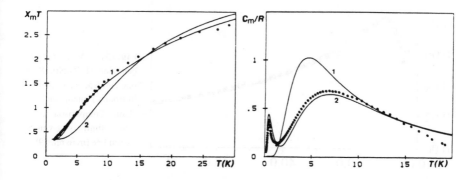

Fig. 16. Magnetic susceptibility and specific heat data for [CoNi]. Curves 1 and 2 correspond to the theoretical variations of chain and dimer limits, respectively (from ref. 38).

The strong dimerization observed in these chains has been attributed to the large anisotropy of Co(II). Thus, due to the zigzag structure of the chain, a given metal site has the principal axes of its g-tensor differently orientated with respect to those of its right and left neighbors. Such orbital effect may be at the origin of the observed very different exchange values[36].

3.1.4. The chain [NiNi]

In this compound the magnetic susceptibility exhibits a maximum around 10 K and a rapid decrease at lower temperature (Fig. 17). No divergence is seen down to 2 K, despite this compound may be viewed as a 1D ferrimagnet, provided the local g-factors and D-anisotropies of the alternating sites are different. The data were firstly analyzed in terms of a regular spin chain ($S_a=S_b=1$) with alternating g-factors and uniform Heisenberg or Ising exchange coupling[35]. Surprisingly, the Ising model better fits the overall data; this may be related to the large local anisotropy constant of the chelated Ni(II) ion, which in the present case is very close to the exchange coupling. Recently, this system has been reinvestigated[39] assuming alternating Heisenberg interactions and local anisotropy. This analysis has shown that the chain is only weakly dimerized ($J'/J = 0.9$), with $J/k_B = -8.0$ K and $D_{Ni}/|J| = 1$. An additional proof of this weak dimerization is provided by the specific heat data (Fig. 18). Indeed, the Schottky anomaly at ca. 6 K is satisfactorily accounted for by the J-alternating Heisenberg model using the previous exchange parameters.

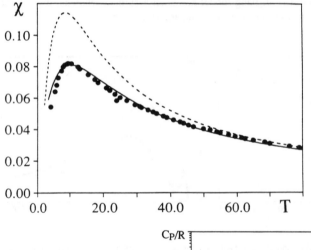

Fig.17. Magnetic behavior of [NiNi]. Solid line corresponds to the best fit from a J-alternating [1-1] chain. Comparison to the dimer limit is given as dashed line (from ref. 39).

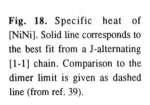

Fig. 18. Specific heat of [NiNi]. Solid line corresponds to the best fit from a J-alternating [1-1] chain. Comparison to the dimer limit is given as dashed line (from ref. 39).

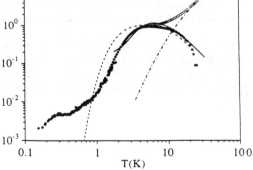

3.1.5. The chain [MnMn]

A further example of two-sublattice homometallic chain is provided by $MnMn(EDTA).9H_2O$. Although this nonahydrate salt is not isostructural to the hexahydrate series, the structural features of the chains are very similar[40]. Thus, in [MnMn] the zigzag chains show alternating Mn(II) sites bridged by carboxylate groups (Fig. 19). One manganese ion (Mn1) is hexacoordinated to four water molecules and two oxygen atoms of two carboxylate bridges, while the other one (Mn2) is heptacoordinated by EDTA and to one water molecule. These different manganese environments result in different local anisotropies as supported by EPR measurements ($D_1/k_B \approx 0.008$ K and $D_2/k_B \approx 0.16$-0.19 K, for Mn1 and Mn2, respectively).

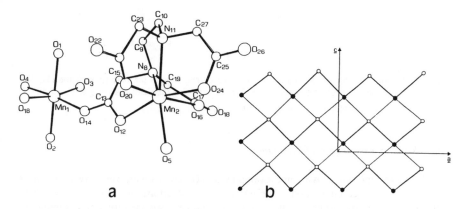

Fig. 19. Structure of [MnMn]: (a) view of the two Mn sites; (b) view of the magnetic lattice showing the connections between the chains. Filled and open circles correspond to Mn1 and Mn2 ions, respectively.

The magnetic features of this chain are reported in Table V and Fig. 20. The magnetic susceptibility shows a rounded maximum at about 3 K, in agreement with AF intrachain interactions, and a sharp peak at $T_c = 1.489$ K, attributed to 3D magnetic ordering. The data above T_c have been closely fit by the classical-spin Heisenberg model, giving an intrachain exchange interaction of $J/k_B = -0.72$ K. It is worth noticing that it has not been necessary to introduce any anisotropy for this fitting, despite the fact that zero-field splitting of chelated manganese is of the order of magnitude of the exchange interaction. The effect of local anisotropies has been clearly observed in the ordered state. Thus, the weak ferromagnetism exhibited by this compound below T_c has been attributed to a spin canting arising from the different local anisotropies of the two magnetic sites. Using Eq. 15.the ratio j/J has been estimated to be 10^{-2}, which is one order of magnitude larger than for [MnNi] and [MnCu]. This agrees with the better 1D character of the hexahydrate series, for which the shortest interchain Mn-Mn distance is 7.34 Å, compared to 6.12 Å in [MnMn].

52

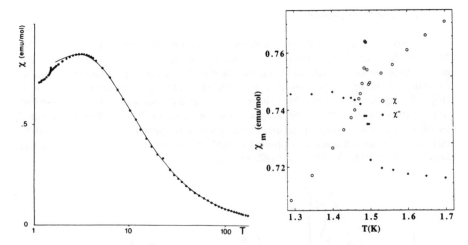

Fig.20. Magnetic behavior of [MnMn]. The solid line corresponds to the best fit with the classical Heisenberg chain model. Right: Low-temperature behavior showing the sharp peak of χ at $T_c = 1.489$ K, and an out-of-phase susceptibility (χ'') at the same temperature (from ref. 40).

3.1.6. Random exchange 1D ferrimagnets

We have reported above some examples of 1D ferrimagnets showing uniform or alternating exchange interactions. The possibility of obtaining the MM'(EDTA).$6H_2O$ compounds as amorphous phases, has allowed us to investigate the problem of alternating magnetic moments in presence of random-exchange interactions. We present here the magneto-structural results obtained in these phases.

Local structure investigations of the amorphous complexes have been carried out in [NiNi] and [CoNi] by Large Angle X-ray Scattering (LAXS)[14]. The following results have been brought out: (i) zigzag chains are present as in the crystalline state; (ii) the EDTA coordination scheme is identical, with this ligand wraping around one metal atom in a hexacoordinated fashion. The only noteworthy difference between the amorphous and crystalline phases is a random distribution of the intermetallic distances M-M', which may be related to a bond angle distribution between connected ions. On the other hand, it has been shown by optical measurements (UV-visible) that the selective occupation of the two types of metal sites is preserved in the amorphous compounds. Hence, from the magnetic point of view the amorphous phase may be viewed as made from alternating magnetic moments coupled by a random exchange interaction.

The low temperature magnetic behavior of these compounds, investigated down to 4 K, show clear differences with that of the crystalline parent compounds[41]. Thus, below 20 K, they show the typical behavior of random exchange 1D antiferromagnets: $\chi \sim T^{-\alpha}$, with α varying in the range 0.6-0.9 (Fig. 21, Table VI). A quantitative analysis has been performed by means of the classical random exchange model (section 2.1.2. Eqs. 7 and 8). After introducing the corresponding scaling factors, very satisfying descriptions of the experimental data have been obtained in all cases (Fig. 22). The corresponding parameters, reported in Table VI, call for the following comments:

(i) The use of the simple classical spin Heisenberg exchange model in all cases may be questionable, as has been pointed out in the study of the crystalline phases. As a result, the obtained exchange parameters only have a qualitative significance.

(ii) The width of the exchange distribution, λ, which implies the coexistence of F and AF couplings, may seem overestimated. This may result from the square-type of distribution assumed by the model, which is not realistic. Furthermore, a paramagnetic contribution arising from possible defects in these phases could also explain the overestimation of λ. Anyway, this parameter may be a useful tool for comparing the amounts of disorder in these systems.

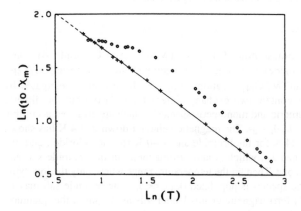

Fig.21. Low temperature magnetic behaviors (T < 20 K) of crystallized (o) and amorphous (+) [CoCo] complexes. The observed variation for the latter is typical of a random exchange chain system (from ref. 41.a).

Fig. 22. Magnetic behavior of [CoNi] complex in the amorphous state. Solid line corresponds to the best fit with a random exchange Heisenberg chain model. The values of the parameters are given in Table VI (from ref. 14).

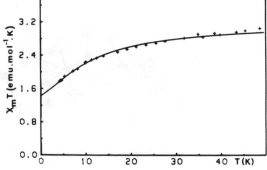

Table VI. Magnetic parameters of the random chains of the EDTA family (amorphous phase)

COMPD	J/k_B (K)	λ/k_B (K)	g_a	g_b
[NiNi]	-2.4	7.1	2.03	2.26
[CoNi]	-3.6	8.0	4.76	2.10
[CoCo]	-3.8	6.4	4.76	4.28
[CoCu]	-6.6	8.8	4.50	2.39
[MnNi]	-3.4	4.2	2.03	2.53

3.2. Topological 1D ferrimagnets and other complex chains.

3.2.1. The TMSO series.

The isostructural compounds $CuX_2.4/3TMSO$ (X = Cl and Br, and TMSO = tetramethylene sulfoxide) consisting of alternating chains of dimeric $[Cu_2X_6]^{2-}$ anions and monomeric $[Cu(TMSO)_4]^{2+}$ cations (Fig. 23), have been investigated by Landee et al.[42] The chains contain two exchange interactions, J_d (within the dimeric species), and J_m (between dimeric and monomeric species), which are arranged according to the sequence $-J_d-J_m-J_m-J_d-J_m-J_m-J_d-$. The magnetic behavior down to 1.4 K has shown that J_d is ferromagnetic ($\approx$ 24 K for the bromide and $\approx$ 40 K for the chloride), and one order of magnitude larger than J_m, which is antiferromagnetic in the bromide salt and ferromagnetic in the chloride one. Then, in the temperature range where only the low-lying spin-state S=1 of the dimeric species is populated ($k_B T \ll J_d$), the bromide salt may be viewed as a made of [1/2-1] ferrimagnetic chains Under this assumption, the quantum-quantum Heisenberg model (section 2.1.1, Eq. 1 of Table I) gives an antiferromagnetic exchange interaction J_m/k_B = -0.94 K and predicts a χT minimum near 0.7 K.

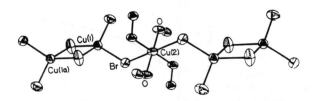

Fig.23. Structure of the chains in $CuBr_2.4/3TMSO$ (from ref. 42).

3.2.2. MnMn(CDTA).7H$_2$O

The first example of spin frustrated chain has been reported[16] in the manganese compound MnMn(CDTA).7H$_2$O, where CDTA is the tetra-anion of the trans-cyclohexane-1,2 diamine NNN'N'-tetra-acetic acid (Fig. 5). Its structure comprises chains of dimeric manganese molecules connected through carboxylate bridges. These dimers are formed by hexacoordinated (hydrated) and heptacoordinated (chelated) metal sites sharing two oxygens of two carboxylate groups (Fig. 24). From a magnetic point of view, the exchange network may be viewed as a chain of triangles sharing a corner

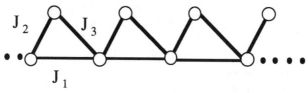

J$_2$ refers to the intradimer exchange interaction, and J$_1$ and J$_3$ to the interdimer ones. When these interactions are antiferromagnetic, we are dealing with a spin frustrated 1D system since three spins cannot be two by two oppositely oriented.

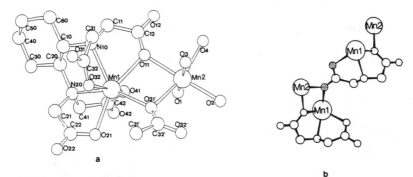

Fig.24. Structure of MnMn(CDTA).7H$_2$O. (a) view of the Mn1-Mn2 dimer; (b) connections between the dimers (from ref. 15).

The magnetic properties of this compound show a continuous decrease of χT upon cooling, indicating the presence of antiferromagnetic interactions between the manganese ions (Fig. 25). However, χ(T) does not show the characteristic antiferromagnetic maximum down to 2 K. This result is in agreement with the fact that frustration tends to reduce short-range order. The effect of the spin-frustration on the zero-field magnetic susceptibility can be estimated from the classical spin approach. As can be seen (Fig. 26), the antiferromagnetic maximum of the linear chain observed at $k_BT/|J| \approx 0.5$, tends to disappear when spin-frustration is introduced (antiferromagnetic J$_1$). Thus, for J$_1$/J$_2$ larger than 0.4 one observes a continuous increase of χ upon cooling.This model, gives a satisfying fit of the experimental data with the following parameters: $J_1/k_B = -0.94$ K, $J_2/k_B = -1.10$ K and $J_3/k_B = -0.23$ K.

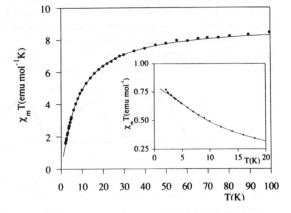

Fig.25. Magnetic behavior of $MnMn(CDTA).7H_2O$. Solid line represents the best fit to the triangular chain model (from ref. 16).

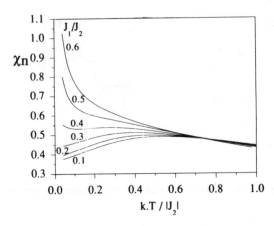

Fig.26. Calculated magnetic behavior of the triangular Heisenberg chain for $J_2 = J_3 < 0$ and different J_1/J_2 ratios (from ref. 16).

3.2.3. $Co(OH)(NO_3).H_2O$

A recent example of 1D frustration has been reported by Angelov et al.[43] in the hydroxide nitrate $Co(OH)(NO_3).H_2O$, whose structure exhibits infinite double chains of edge-sharing octahedra CoO_6 (Fig. 27).

The magnetic behavior shows a χT minimum around 30 K, and a strong increase on cooling (Fig. 28), the origin of which is rather different from that of previously described 1D ferrimagnets. At high temperature, the decrease of χT reflects the behavior of isolated cobalt(II); indeed, the $^4T_2$ ground-term is splitted under the combined action of spin-orbit coupling and non-cubic crystal field terms, giving six Kramers doublets, with an effective low-lying spin state $S = 1/2$. The divergence of χT below 30 K results from ferromagnetic exchange couplings between the effective spins of adjacent Co(II) ions. Low field susceptibility measurements exhibit a maximum at 2.5 K, which suggests the onset of a long range AF ordering. Furthermore, the magnetization field dependence shows the typical features of a metamagnetic behavior; it increases linearly up to 720 Oe., then a clearcut transition to a ferromagnetic state sets on.

The low dimensional magnetic behavior has been discussed by considering infinite double chains of $S = 1/2$ Ising chains, made of two interacting linear chains (Fig. 1-e). Each Co(II) ion is connected to its nearest neighbors through two types of exchange interactions, namely interchain (J_1) and intrachain (J_2) interactions. An exact expression of the zero-field parallel susceptibility can be derived (Table IV).

The theoretical variation of the magnetic susceptibility is plotted for different $J_1/|J_2|$ ratios ($J_1 > 0$) in Fig. 29. Spin frustration effects are expected to occur when J_2 is antiferromagnetic whatever the sign of J_1. Thus, the double chain behaves as a 1D ferromagnet when $J_1 > 2|J_2|$, or as a 1D antiferromagnet when $J_1 < 2|J_2|$. For $J_1 = 2|J_2|$ there is a full compensation between the two exchange pathways, and the system looks like a set of magnetically isolated 1/2 spins giving rise to a Curie like behavior. .

Assuming that this model holds in the range 3-30 K, where only the lowest Kramers doublet of cobalt(II) is significantly populated, the authors have fitted the experimental data of the cobalt(II) nitrate using the calculated expression. The best fit exchange parameters have been found to be $J_1/k_B = 45.6$ K and $J_2/k_B = -19.5$ K, which gives a ratio $J_1/|J_2|$ larger than 2. Accordingly, the antiferromagnetic intrachain interactions are not large enough to compensate interchain interactions, so that ferromagnetic-like behavior should dominate, as is observed.

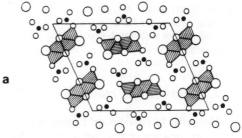

a

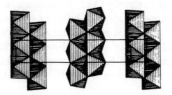

b

Fig. 27. Structure of Co(OH)(NO$_3$).H$_2$O (a) in the ac plane, and (b) in the perpendicular direction. (from ref.43)

58

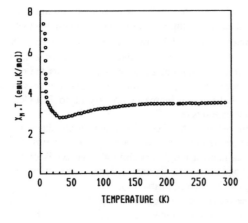

Fig. 28. Magnetic behavior of $Co(OH)(NO_3).H_2O$. (from ref.43).

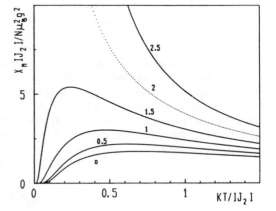

Fig. 29. Calculated magnetic behavior of the triangular double chain (Ising model) for different $J_1/|J_2|$ ratios ($J_1>0$) (from ref.43).

3.2.4. The copper(II) phosphates $A_3Cu_3(PO_4)_4$; A = Ca, Sr.

This series furnishes model systems to relate 1D ferrimagnetism to topology effects[44]. The crystal structure corresponds to infinite chains of copper(II) trimers linked by oxygen atoms. The middle copper ion, Cu(1), exhibits a square planar surrounding, while the other two, namely Cu(2), lie in irregular polyhedra of five oxygen atoms. These polyhedra are connected through bridging oxygen atoms to form infinite ribbons (Fig. 30). As a result, the metal network may be described as intertwining double chains of copper(II) ions. In view of the metal-oxygen distances and bond angles, two different exchange constants must be considered: J_1 within the trimer units, J_2 between adjacent trimers. Then, the exchange network corresponds to that reported in Fig. 1-f.

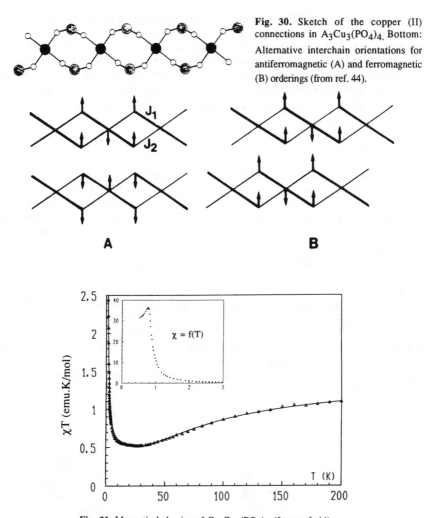

Fig. 30. Sketch of the copper (II) connections in $A_3Cu_3(PO_4)_4$. Bottom: Alternative interchain orientations for antiferromagnetic (A) and ferromagnetic (B) orderings (from ref. 44).

Fig. 31. Magnetic behavior of $Ca_3Cu_3(PO_4)_4$ (from ref. 44).

The magnetic data of these compounds show the χT minimum around 25 K (Fig. 31). Further, the onset of a 3D magnetic order is observed at 0.9 K (Sr compound), and at 0.8 K (Ca compound). The nature of the ordered state has shown to be different in both cases. Thus, while the former shows moment cancellation at the scale of the lattice (i.e., a 3D antiferromagnetic ordering), the second exhibits an out-of-phase signal in zero-field susceptibility measurements, indicating the presence of a net magnetic moment (i.e., a 3D

60

ferromagnetic ordering). This difference has been attributed to the relative positions of adjacent chains. Thus, assuming AF Cu-Cu exchange interactions, a 3D antiferromagnetic ordering is stabilized when the largest interchain interactions occur between Cu(2) ions, whereas a 3D ferromagnetic one is stabilized when these occur between Cu(1) and Cu(2) (Fig. 30 bottom). In the low-dimensional regime, the data have been fitted using the Ising model previously described (section 2.2, Table IV), with J_1/k_B =-150 K and J_2/k_B = -2.5 K for the Sr compound, and J_1/k_B = -138 K and J_2/k_B = -2.6 K for the Ca one.

3.2.5. $Cu_2O(SO_4)$

This compound shows a polymeric structure[45] formed by a sequence of two different copper(II) sites: a trigonal bipyramid site, Cu1, and a hexacoordinated one, Cu2, in which the copper shows a 4+2 coordination (Fig. 32). These sites are linked via oxygen atoms so as to form lozenge chains of dimeric Cu1-Cu1 entities alternating with Cu2 sites. The exchange network is similar to that reported for the copper phosphates but with a further interaction (J_3) between outer copper(II) ions (of the Cu1 type) belonging to adjacent trimers (Fig. 1-g)). If this interaction is also antiferromagnetic, a frustration situation occurs, and hence the system may be viewed as a spin frustrated topological 1D ferrimagnet.

For this kind of 1D system exact expressions of the parallel magnetic susceptibility and specific heat have been deduced using the Ising model (Table IV). When the three interactions are AF and $J_1 = J_2$, this model predicts a ferrimagnetic behavior for a ratio $J_1/J_3 > 0.5$. In such cases J_1 and J_2 impose a parallel alignment of the spins in the dimer, even when J_3 is antiferromagnetic. Conversely, an antiparallel alignment of the spins in the dimer is predicted for $J_1/J_3 < 0.5$; the system then nearly behaves as a set of isolated 1/2 spins (one per trimer).

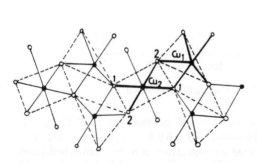

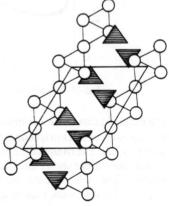

Fig. 32. Top: Structure of $Cu_2O(SO_4)$ showing the two different Cu(II) sites and the oxygen connections. Right: Sketch of the lozenge chains (from ref. 45).

The title compound does exhibit the ferrimagnetic behavior with a minimum in χT at 25 K (Fig. 33). Further, a sharp peak is observed at 16 K, which has been attributed to magnetic ordering. A good fit in the paramagnetic regime from the Ising model[46] gives similar magnitudes for the three exchange parameters (J_1/k_B = -66 K; J_2/k_B = -72 K; J_3/k_B = -68 K). A Heisenberg-type analysis[47] has roughly confirmed these values (J_1/k_B = -54.6 K; J_2/k_B = -62.7 K; J_3/k_B = -41.5 K).

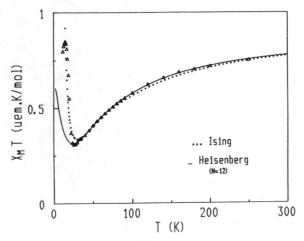

Fig. 33. Magnetic behavior of $Cu_2O(SO_4)$ (from ref. 46).

3.2.6. The series $Ba_2CaMFe_2F_{14}$ with M = Mn, Co and Cu(II)

Darriet et al. have shown[48] that usovite related compounds, such as $Ba_2CaMFe_2F_{14}$ with M = Mn, Co and Cu(II), may be good candidates to stabilize 1D ferrimagnetism. The magnetic species are stacked above each other in order to form chains of lozenges; these chains are well separated by Ca^{+2} ions (Fig. 34). In the chain each Fe(III) can interact with the two nearest neighbors M(II) ions, while each M(II) can interact with four Fe(III).

From a magnetic point of view, the cobalt and manganese compounds exhibit a rounded minimum in χT vs. T curves around 60 K (Fig. 35). At lower temperature, χT presents a maximum around 30 K, which indicates the onset of long range magnetic ordering. A Heisenberg model with classical spins, fully justified for Mn(II) or Fe(III) ions, was used to discuss the magnetic data of the Mn-Fe compound[49]. A general expression of the zero-field susceptibility has been derived. Using this model, the experimental data have been fit in the paramagnetic region, giving J_1/k_B = -6.7 K, J_2/k_B = -2.7 K and g_{Mn} = g_{Fe} = 2.0 (Fig. 35).

In the copper(II) compound, due to the location of the single electron in the $d_{x^2-y^2}$ orbital, J_2 may be neglected with respect to J_1. Accordingly, the magnetic behavior has

62

been described[50] by assuming isolated Fe-Cu-Fe linear trimers, with an antiferromagnetic coupling ($J_1/k_B = -18$ K).

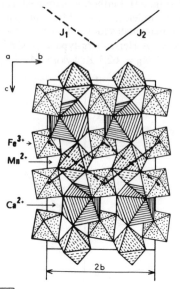

Fig. 34. Schematic representation of the usovite-type structure (from ref. 48).

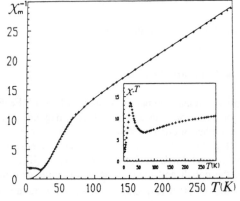

Fig. 35. Magnetic behavior of $Ba_2CaMnFe_2F_{14}$ (from ref. 49).

3.2.7. The compound $Ba_7CuFe_6F_{34}$

The structure of this compound, reported by Ferey et al.[51], consists of a stacking of bimetallic rings made of copper(II) and iron(III) ions (Fig 36). The building unit can be described as a chain of rings sharing the copper(II) ions. The Fe(III)-Fe(III) and Cu(II)-Fe(III) intrachain distances are 3.682 Å and 3.941 Å, respectively, while the shortest interchain distance is 5.134 Å, giving rise to well separated magnetic chains.

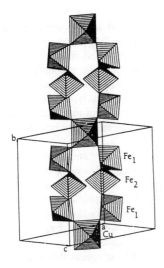

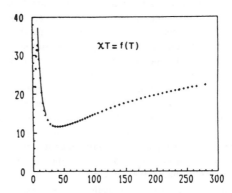

Fig. 36. Left: Perspective view of the bimetallic chain in $Ba_7CuFe_6F_{34}$. Bottom: Magnetic behavior of $Ba_7CuFe_6F_{34}$ (from ref. 51).

The thermal variation of the χT product shows the typical features of 1D ferrimagnets (Fig. 36). Furthermore, a sharp maximum of χT, observed at 8 K, suggests a phase transition to a magnetically ordered state. Such a system thus appears as a new example of bimetallic ferrimagnetic chain characterized by a very exotic magnetic network.

Owing to the topology of the chain, spin multiplicity and number of interacting ions in the unit cell, an analysis of the data using finite chain computations, as performed for linear quantum chains, is hopeless. In fact, a simplified model provides some insight into the Cu(II)-Fe(III) exchange coupling, if we assume that each [Fe₃] unit behaves as a S = 5/2 spin at low enough temperature. When only the low-lying spin states are thermally populated, it has been assumed that in each ring the 5/2-spins of the two [Fe₃] units that interact with copper (II) are roughly parallel, so that a model of ferrimagnetic chain ($S_a = 1/2$, $S_b = 5$) should be a convenient approximation. Thus, the low temperature fit of the experimental data gives the effective value $J/k_B = -6.1$ K for the coupling between Cu(II) and one adjacent [Fe₆] entity. Obviously, the quality of the fitting is restricted to the low temperature range.

4. Conclusion

We have shown in this review that more and more complexes chain systems have been isolated in the last few years which exhibit 1D ferrimagnetic properties. Basically, the observed properties result either from a non compensation of two magnetic moments (typically two different metal ions) located on a linear chain, or an exotic array of one or two metal ions connected by several competing interactions.

The common feature to these systems is the occurrence of a minimum of the χT product at a finite temperature, which is related with the strength of the exchange coupling, and a divergence at lower temperature due to the uncompensated ground-state.

Different models have been reported for describing the magnetic properties based, according to the nature of the spins and interactions, either on computational procedures or analytical treatments when available. Among the most striking results, we can note the presence of a compensation temperature, similar to that observed for rare earth garnets, predicted in some specific situations depending on the spin values and the ratio between local moments.

Further, it is to be noted that the interconnected chains are the first step towards the 2D and 3D systems. The calculations achieved for these low dimensional systems may be used for describing 3D networks made of z nearest neighbor chains interacting with a given one.

Another feature of special interest is the possibility of stabilizing a giant magnetic moment at quite high temperature, since it is driven by the magnitude of the intrachain coupling, and accordingly of the correlation length. Finally, we notice that unlike the ferromagnetic chains, the antiferromagnetic coupling in 1D ferrimagnets may range up to several wave numbers. As a result, an appropriate stacking of these chains should promote a high ordering temperature. This may be a useful way for the preparation of new molecular ferromagnets which constitutes a challenge worldwide.

5. References

1. For a recent revision see for example C. P. Landee in *Organic and Inorganic Low-Dimensional Crystalline Materials*, eds. P. Delhaes and M. Drillon. NATO ASI Series, vol. B168 (Plenum Press, New York, 1987) p. 75.
2. (a) L. J. de Jongh and A. R. Miedema, *Adv. Phys.* 23 (1974); (b) W. E. Hatfield, W. E. Estes, W. E. Marsh, M. W. Pickens, L. W. Ter Haar and R. R. Weller in *Extended Linear Chain Compounds*, ed. J. S. Miller, vol. 3 (Plenum Press, New York, 1983) p. 43; (c) J. C. Bonner in *Magneto-Structural Correlations in Exchange-Coupled Systems*, eds. R. D. Willett, D. Gatteschi and O. Kahn. NATO ASI Series (Reidel Dordrecht, 1985) p. 157; (d) R. L. Carlin, *Magnetochemistry* (Springer-Verlag, Berlin, 1986).
3. L. J. de Jongh in *Magneto-Structural Correlations in Exchange-Coupled Systems*, eds. R. D. Willett, D. Gatteschi and O. Kahn. NATO ASI Series (Reidel Dordrecht, 1985) p. 1.
4. R. Georges, M. Drillon and E. Coronado in *Magnetic Properties of Quasi-One Dimensional Magnetic Compounds*, eds. L. J. de Jongh and J. Darriet, *Physics and Chemistry of Materials with Low Dimensional Structures Series* (Ed. in chief S. Levy), in preparation.
5. J. C. Bonner and M. E. Fisher, *Phys. Rev.* **135** (1964) 640.
6. C. Y. Weng, Thesis, Carnegie-Mellon University (1968).

7. H. W. Blöte, *Physica B* **79** (1975) 427.

8. W. Duffy and K. P. Barr, *Phys. Rev.* **165** (1968) 647.

9. M. Drillon, J. C. Gianduzzo and R. Georges, *Phys. Lett.* **96-A** (1983) 413.

10. M. Drillon, E. Coronado, R. Georges, J. Curély and J. C. Gianduzzo, *Phys. Rev.B* **40** (1989) 10992.

11. M. E. Fisher, *Am. J. Phys.* **32** (1964) 343.

12. M. F. Thorpe, *J. Phys. Paris* <u>36</u> (1975) 1177.

13. M. Drillon, E. Coronado, D. Beltran and R. Georges, *Chem. Phys.* **79** (1983) 449.

14. A. Mosset, J. Galy, E. Coronado, M. Drillon and D. Beltran, *J. Am. Chem. Soc.* **106** (1984) 2864.

15. J. J. Borrás-Almenar, E. Coronado, R. Georges, C. J. Gómez-García and C. Muñoz-Roca, *Chem. Phys. Lett.* **186** (1991) 410.

16. J. J. Borrás-Almenar, E. Coronado, J. C. Gallart , R. Georges, C. J. Gómez-García *J. Magn. Magn. Mater.* **104-107** (1992) 835.

17. X. Qiang, J. Darriet and R. Georges, *J. Magn. Magn. Mater.* **74** (1988) 219.

18. T. Dembinski and T Wydro, *Phys. Status Solidi* **67** (1975) K123.

19. H. W. Blöte, *J. Appl. Phys.* **50** (1979) 7401.

20. (a) J. Seiden, J. Phys. **44** (1983) L947; (b) M. Verdaguer, A. Gleizes, J. P. Renard, J. Seiden, *Phys. Rev. B* **29** (1984) 5144 .

21. J. Curely, Doctoral Thesis, Bordeaux (1990).

22. (a)Yu Pei, O. Kahn, J. Sletten, J. P. Renard, R. Georges, J. C. Gianduzzo, J. Curély and Qiang Xu, *Inorg. Chem.* **27** (1988) 47; (b) R. Georges and O. Kahn, *Mol. Cryst. Liq. Cryst* **176** (1989) 473.

23. H. A. Kramers and G. H. Wannier, *Phys. Rev.* **60** (1941) 252.

24. E. Coronado, M. Drillon, P. R. Nugteren, L. J. de Jongh and D. Beltrán, *J. Am. Chem. Soc.* **10** (1988) 3907.

25. F. Sapiña, E. Coronado, M. Drillon, R. Georges and D. Beltrán, *J. Phys. (Paris)* **C8** (1988) 1423.

26. R. Georges, J. Curély and M. Drillon, *J. Appl. Phys.* **58** (1985) 914.

27. (a) J. Curély, R. Georges and M. Drillon, *Phys. Rev.B* **33** (1986) 6243; (b)J. Curély and R. Georges, *Phys. Rev. B*, in the press.

28. M. Belaiche, Doctoral Thesis, Strasbourg (1988).

29. (a) O. Kahn, *Struct. Bonding (Berlin)* **68** (1987) 89; (b) O. Kahn in *Magnetic Molecular Materials*, eds. D. Gatteschi, O. Kahn, J. S. Miller and F. Palacio. NATO ASI Series vol E168 (Kluwer, 1991) p. 35.

30. (a) A. Caneschi, D. Gatteschi, J. Laugier and P. Rey, *Acc. Chem. Res.* **22** (1989) 392; (b) A. Caneschi and D. Gatteschi, *Progress in Inorg. Chem.***39** (1991) 331.

31. E. Coronado in *Magnetic Molecular Materials*, eds.D. Gatteschi, O. Kahn, J. S. Miller and F. Palacio. NATO ASI Series vol E168 (Kluwer, 1991) p. 265.

32. K. Nakatani, P. Bergerat, E. Codjovi, C. Mathoniere, Y. Pei and O. Kahn, *Inorg. Chem.* **30** (1991) 3977.

33. P. Gómez-Romero, G.B.Jameson, N.Casañ-Pastor, E.Coronado and D.Beltrán, *Inorg. Chem.* **25** (1986) 3171.

34. F. Sapiña, E. Coronado, D. Beltrán and R. Burriel, *J. Am. Chem. Soc.* **113** (1991) 7940.

35. E. Coronado, M. Drillon, A. Fuertes, D. Beltrán, A. Mosset and J. Galy, *J. Am. Chem. Soc.* **108** (1986) 900.

36. J. J. Borrás-Almenar, E. Coronado, D. Gatteschi and C. Zanchini, *Inorg. Chem.* **31** (1992) 294.

37. E. Coronado, M. Drillon, P. R. Nugteren, L. J. de Jongh, D. Beltrán and R. Georges, *J. Am. Chem. Soc.* **111** (1989) 3874.

38. E. Coronado, F. Sapina, M. Drillon and L. J. de Jongh, *J. Appl. Phys.* **67** (1990) 6001.

39. J. J. Borrás-Almenar, Doctoral Thesis, Valencia (1992).

40. J. J. Borrás-Almenar, R. Burriel, E. Coronado, D. Gatteschi , C. J. Gómez-García and C. Zanchini, *Inorg. Chem.* **30** (1991) 947.

41. (a) E. Coronado, M. Drillon, D. Beltrán and J.C. Bernier, *Inorg. Chem.* **23** (1984) 4000; (b) E. Coronado, M. Drillon, D. Beltrán, A. Mosset and J. Galy, *J. Phys. (Paris)* **46** (1985) 639.

42. C. P. Landee, A. Djili, D. F. Mudgett, M. Newhall, H. Place, B. Scott and R. D. Willett, *Inorg. Chem.* **27** (1988) 620.

43. S. Angelov, M. Drillon, E. Zhecheva, R. Stoyanova, M. Belaiche, A. Derory and A. Herr, *Inorg. Chem.* **31** (1992) 1514.

44. M. Drillon, E. Coronado, M. Belaiche and R. L. Carlin, *J. Appl. Phys.* **63** (1988) 3551.

45. V. E. Flugen-Kahler, *Acta Cryst.* **16** (1963) 1009.

46. M. Drillon, M. Belaiche, J. M. Heintz, G. Villeneuve, A. Boukhari and J. Aride in *Organic and Inorganic Low Dimensional Crystalline Materials*, eds.P. Delhaes and M. Drillon. NATO ASI Series vol. B168 (Plenum Press, New York, 1987) p. 421.

47. M. Belaiche, M. Drillon, J. Aride, A. Boukhari, T. Biaz, and P. Legoll, *J. Chim. Phys.* **88** (1991) 2157.

48. J. Darriet, X. Quiang, A. Tressaud, R. Georges and J. L. Soubeyroux, *Phase Trans.* **13** (1988) 49.

49. X. Quiang, J. Darriet, J. L. Soubeyroux and R. Georges, *J. Magn. Magn. Mater.* **74** (1988) 219.

50. J. Darriet, X. Quiang, A. Tressaud and P. Hagenmuller, *Mat. Res. Bull.* **21** (1986) 1351.

51. J. Renaudin, G. Ferey, M. Zemirli, F. Varret, A. de Kozak and M. Drillon in *Organic and Inorganic Low Dimensional Crystalline Materials*, eds.P. Delhaes and M. Drillon. NATO ASI Series vol. B168 (Plenum Press, New York, 1987) p. 389.

Spin Levels of High Nuclearity Spin Clusters

Dante Gatteschi and Luca Pardi
Department of Chemistry, University of Florence
Florence, Italy

1. Introduction

Magnetochemistry is more and more interested at the investigation of systems comprising large numbers of coupled spin [1,2]. When the number tends to infinity, either low (one-, two-) dimensional or three-dimensional magnetic systems are obtained. These have been well investigated for many years now, and their properties are reasonably well understood. From the theoretical point of view the simplifications allowed by the translational symmetry of the systems has been widely used.

In the last years however there have been many reports on the synthesis, and the investigation of the magnetic properties of relatively large spin aggregates [3-16], which however comprise finite numbers of coupled spins within a single molecule. An esthetically rewarding example is shown in Figure 1.

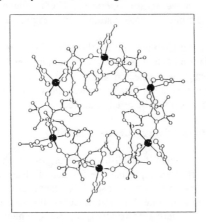

Figure 1. Sketch of the structure of $[Mn(hfac)_2(NITPh)]_6$

The molecule [17], $[Mn(hfac)_2(NITPh)]_6$ where hfac= hexafluoroacetylacetonate, and NITPh= 2-phenyl-4,4,5,5-tetramethyl-4,5-dihydro-1H-imidazol-1-oxy-3-oxide, has six manganese(II), S= 5/2, and six NITPh radicals, S= 1/2, whose structure is shown below,

which form a ring. The manganese and radical spins are strongly antiferromagnetically coupled, in such a way that the ground state is S= 12.

By analogy to the high nuclearity metal clusters which were earlier investigated as models of bulk metals [18], we like to call the large spin clusters high nuclearity spin clusters, HNSC. The understanding of their magnetic properties is a challenge, because the simplifications associated to translational symmetry, which are possible for infinite assemblies, are not possible here. However there is more than just the intellectual amusement to solve difficult problems. In fact the improvement of the synthetic methodologies affords larger and larger molecular clusters, which eventually reach the mesoscopic scale where quantum effects coexist with bulk, classic, behavior. It is the same mesoscopic scale where the process of miniaturization of electronic devices is currently heading, and where novel properties are expected.

In this Chapter we want to focus on the difficulties associated with the interpretation of the magnetic properties of HNSC, and to suggest which are the possible ways of finding efficient ways to calculate the spin levels of the clusters. In particular we advocate the use of irreducible tensor operators, ITO's, which, although somewhat elaborated allow drastic cuts in the memory storage and diagonalization time required for the calculations.

The organization of the Chapter will be the following: we will first highlight the current methods of calculation of the energy levels of spin clusters, then we will work out the ITO approach, showing with selected examples how it is possible to tackle the calculation of relatively large clusters. Finally we will show some experimental results. In an Appendix the ITO formalism will be briefly outlined.

2. Current Approaches to the Energy Levels of HNSC

Within the spin hamiltonian formalism the energies of the levels of exchange coupled systems can be expressed replacing the full set of electron coordinates by spin angular momentum operators. In the trivial case of two identical interacting systems, each characterized by the presence of one unpaired electron, the spin hamiltonian approach, when applicable, predicts that there are two low lying levels, namely a singlet and a triplet, which are separated by an energy gap J if the spin hamiltonian is expressed as:

$$H = J S_1 . S_2 \qquad (1)$$

where the singlet is below the triplet for $J > 0$. This is a well known result, which derives from the fact that the functions appropriate for the set of two spins can be written as:

$$| S_1 \, m_1 \, S_2 \, m_2 > \qquad (2)$$

where $| m_1 | \le S_1$; $| m_2 | \le S_2$.

As far as two, or three, spins are concerned it does not really matter to elaborate complex formulations to calculate the spin levels. Even the simplest approach which writes the functions appropriate to the cluster as in (2), calculates the hamiltonian, and then diagonalizes it, is adequate. However matters become rapidly untractable when the number of spins increases and when theirs values also increase. In fact the dimensions of the bases grow as $(2S+1)^n$, and for instance four coupled $S = 5/2$ spins already require a basis of 1,296 functions.

Characterizing the functions on one center as $| S_1 \, m_1 >$ corresponds to classify them according to the irreducible representations of the rotation group. The products of two functions $| S_1 \, m_1 > |S_2 \, m_2 >$ give rise to a reducible representation, which can be decomposed according to the well known rules of addition of angular momenta. Singlet and triplet are just the irreducible representations deriving from this decomposition. In general it is useful to express the functions as eigenfunctions of the total spin operator of the system, i.e. of $S = \Sigma \, S_i$. Therefore it must be expected that big advantages originate from the full exploitation of the symmetry properties of the full rotation group. These unfortunately are not as well known to the average chemist as are the corresponding properties of the finite point symmetry groups. In fact the full rotation group is a group containing an infinite number of elements, and learning to write the irreducible representations requires an elaborate algebra. However, given the big simplifications which are possible, we feel that time has come to make an additional effort in order to exploit symmetry as deep as possible.

In order to write the correct eigenfunctions of the total spin operator it is possible to improve the brute force diagonalization approach, by using appropriate coefficients, which can be easily calculated [19]. These are called Clebsch - Gordon coefficients and are formally described as

$$| S_1 \, S_2 \, S \, M > = \Sigma_{m1,m2} | S_1 \, m_1 \, S_2 \, m_2 > < S_1 \, S_2 \, m_1 \, m_2 | S \, M > \qquad (3)$$

These are related to Wigner's 3-j coefficients by

$$< S_1 \, S_2 \, m_1 \, m_2 | S \, M > = (-1)^{S_1 - S_2 + M} (2S+1)^{1/2} (S_1 S_2 S; m_1 m_2 - M) \qquad (4)$$

The 3-j symbols can be easily calculated as shown in Appendix. The procedure of construction of the functions indicated by Eq. 3 is analogous to the construction of the symmetry adapted wavefunctions of point group symmetry. In fact the total spin eigenfunctions are linear combinations of the $|S_1 m_1 S_2 m_2>$ functions, which are weighted according to the Clebsch-Gordon coefficients. In (3) $m_1+m_2 = M$, and $|S_1-S_2| \le S \le S_1+S_2$. Using Eq. 3 one can easily factorize the $(2S+1)^n$ hamiltonian matrix of n equivalent spins S into a few smaller blocks. For instance in the case of four $S = 5/2$ the $1,296 \times 1,296$ matrix is split into eleven blocks, corresponding to $S = 10, 9, 8, 7, 6, 5, 4, 3, 2, 1, 0$. The dimensions of the corresponding matrices are $n_S = 1, 3, 6, 10, 15, 21, 24, 24, 21, 15, 6$, respectively. The advantage in term of ease of diagonalization is obvious. However, under the aspect of computer memory storage things are not as brilliant. In fact the matrix elements must be calculated using the explicit form of the functions. In general a given function $|S_1 S_2 S_3 S_4 SM>$ is given by a linear combination of $|m_1 m_2 m_3 m_4 M>$ functions. The maximum number of such functions is given by the number, n_M, of $|m_1 m_2 m_3 m_4 M>$ states which have the same $M = m_1 + m_2 + m_3 + m_4$. It should be immediately clear that n_M is larger than n_S when $M = S$. For example for $S = 0$, $n_S = 6$, but the functions with $M = 0$ are given by:

$$n_{M=0} = \Sigma_{i=0,10}\, n_{S=i}$$

This is because all the S states have an $M = 0$ component. Therefore $n_{M=0} = 146$. Even if only a 6×6 matrix must be diagonalized for $S = 0$, the functions which must be stored are linear combinations of 146 states, each of them specified by four numbers. It is clear that in terms of memory storage this is an unfortunate event.

In order to reduce this requirement, and to cut down the time necessary to calculate the elements of the hamiltonian matrix, it is necessary to introduce the Irreducible Tensor Operator approach [20,21], which extends the use of symmetry to the hamiltonian operator, making it possible to calculate the matrix elements without writing the eigenfunctions of the total spin operator explicitly.

3. The Irreducible Tensor Operator Approach

The Irreducible Tensor Operator, ITO, formalism classifies the operators according to the irreducible representations of the full rotation group.

Any operator $\mathbf{O}$ can be classified according to its rank. For instance an angular momentum operator, like S_i, is a first rank operator, which has three components, which behave as the ± 1, 0 components of an $S = 1$ function. Therefore it can be indicated as T_{1q}, where $q = \pm 1, 0$.

A first-rank operator is related to the usual S_x, S_y, and S_z operators by:

$$T_{11}(S) = - 1/\sqrt{2}\, (S_x + i\, S_y)$$
$$T_{1-1}(S) = 1/\sqrt{2}\, (S_x + i\, S_y)$$
$$T_{10}\ (S) = S_z \tag{5}$$

The possible ranks of operators range from zero to infinity, but those which will be relevant here are only the zero and first rank operators. This means that in the following we will only consider isotropic exchange hamiltonians, neglecting the anisotropic and antisymmetric terms [22]. Since the ITO approach has general validity the latter might be easily taken into account, but the expressions would become much more complex, which is not appropriate at an introductory level. The zero rank operators are scalar operators and can be indicated as T_0.

Having standardized the operators it is possible to express the matrix elements using the Wigner-Eckart theorem which states that a matrix element is given by the product of two terms, one dependent and one independent of the spin components M, q, and M':

$$<SM|T_{kq}(S)|S'M'> = (-1)^{S-M} < S\|T_k(S)\|S'> \quad (SkS';Mq-M') \quad (6)$$

where the term in parenthesis is a 3-j symbol as defined above and $< S\|T_k(S)\|S'>$ is a reduced matrix element which does not depend on the k, q, M, and M' components.

The reduced matrix elements for zero and first rank tensor operators are given by:

$$< S\|T_0\|S'> = \delta_{SS'} \sqrt{(2S+1)} \quad (7)$$
$$< S\|T_1\|S'> = \delta_{SS'} \sqrt{S(S+1)(2S+1)} \quad (8)$$

where $\delta_{SS'}$ is a Dirac δ function, equal to 1 for S=S' and equal to zero otherwise.

Equations (6-8) are enough to calculate the matrix elements of any individual spin operator, but in order to calculate the matrix elements of (1) it is necessary to learn how to handle compound operators, i.e. those which are formed by products of individual spin operators.

Now both the irreducible tensor operators $T_1(S_1)$ and $T_1(S_2)$ span the irreducible representations $k_1 = k_2 = 1$ of the rotation group. Their scalar product $T_1(S_1).T_1(S_2)$ must span the k = 0 irreducible representation of the rotation group, because the scalar product is just a number. The appropriate compound operator can be obtained by using a relation analogous to Eq. 3.

$$T_1(S_1).T_1(S_2) = \Sigma_{q_1q_2} (110;q_1q_20) T_{1q_1'}(S_1) T_{1q_2}(S_2) \quad (9)$$

where the term in parenthesis is a 3-j symbol which, when made explicit corresponds to :

$$T_1(S_1).T_1(S_2) = 1/\sqrt{3} [T_{11}(S_1)T_{1-1}(S_2) - T_{10}(S_1)T_{10}(S_2) + T_{1-1}(S_1)T_{11}(S_2)] \quad (10)$$

By substituting in (10) the expressions (5) we find

$$S_1.S_2 = -\sqrt{3} \, T_1(S_1).T_1(S_2) \quad (11)$$

Therefore the hamiltonian (1) can be expressed as:

$$J_{12} \, S_1.S_2 = -\sqrt{3} \, J_{12} \, T_1(S_1).T_1(S_2) = -\sqrt{3} \, J_{12} \, T_0(S) \quad (12)$$

The matrix element of (12) can be expressed using the Wigner-Eckart theorem

$$< S_1 \, S_2 \, S \, M|T_0(S)|S_1 \, S_2 \, S' \, M' > = \delta_{SS'} \, \delta_{MM'} \, (-1)^{S-M} <S_1 \, S_2 \, S \parallel T_0(S) \parallel S_1 \, S_2 \, S >.$$
$$.(S0S;-M0M) \qquad (13)$$

The reduced matrix element is given by:

$$<S_1 S_2 S \parallel T_0(S) \parallel S_1 S_2 S> = (-1)^{S+S1+S2} <S_1 \parallel T_1(S_1) \parallel S_1><S_2 \parallel T_1(S_2) \parallel S_2>.$$
$$\{S_1 0 S_1'; S_2 S S_2'\} \qquad (14)$$

where the symbol in curly brackets is a 6-j symbol [19], i.e. a number, akin to a 3-j symbol, which can be easily calculated given the values to the numbers indicated in it. Tables of 6-j symbols are available, but they can be easily calculated with a simple computer program as becomes evident from the formulae given in Appendix. The number 0 which is present in the 6-j symbol in (14) is related to the fact that $k = 0$.

The results are rather trivial for the case of two spins, in the sense that the same conclusions can be easily reached even with more elementary approaches. However when the ITO approach is extended to systems comprising more spins the advantages become obvious. It is important to stress immediately however that the most obvious advantage of the ITO'approach is that the functions must not be explicitly written, but simply indicated, in order to calculate the matrix elements (13). This is the point that we will try to illustrate in the following.

It should be intuitively clear that the procedure outlined above can be extended step by step to larger clusters, of course exploiting the irreducible representations of the real orthogonal rotation group. As far as the hamiltonian operator is concerned nothing new is introduced because a general operator

$$H = J_{ik} \, S_i.S_k \qquad (15)$$

can be expressed in function of the appropriate ITO's using equations analogous to (11). However in order to have a closed expression for the matrix element it is necessary to exploit fully the symmetry as far as the functions are concerned. In order to write the functions explicitly it is necessary to choose a suitable coupling scheme, i.e. to decide in which order to couple pairs of spins to give the total spin S. It is clear that on increasing the number of spins the number of possible ways of coupling spins increases. However there is no problem as far as the use of ITO's is concerned, because any coupling scheme is acceptable. Therefore the use of one or the other is just a matter of opportunity, or of personal taste.

An example may help in clarifying how to work out the problem. Let us take into account a cluster of four spins, S_1, S_2, S_3, S_4, as shown below:

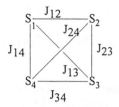

A possible scheme is that of coupling the individual spins one after the other as S_1 to S_2 to give S_{12}, S_{12} to S_3 to give S_{123} and S_{123} to S_4 to give S. The functions are then labelled as $|S_1 S_2 S_{12} S_3 S_{123} S_4 S >$. Another possibility is that of coupling first S_1 to S_2 to give S12, S_3 to S_4 to give S_{34}, and then the resulting spins together as indicated: $|S_1 S_2 S_{12} S_3 S_4 S_{34} S_{12} S_{34} S>$.

It is apparent that in order to fully indicate a function we need n individual spins, n-2 intermediate spins, and one total spin.

In order to express the matrix elements it is necessary to indicate the rank of the ITO operator associated to each individual, intermediate, and total spin operator according to the chosen coupling scheme. For instance if we have to calculate the matrix element $H = J_{23} S_2.S_3$ within the first coupling scheme we must specify the rank of the ITO's associated to S_1, k_1, S_2, k_2, S_{12}, k_{12}, S_3, k_3, S_{123}, k_{123}, S_4, k_4, S, k. The rules for assigning the values are simple. Since S_1 is not contained in the hamiltonian we are considering, $k_1 = 0$; S_2 is a rank one operator, and is present in the hamiltonian, so $k_2 = 1$, S_{12} is the result of the coupling of S_1 and S_2 and k_{12} is the result of the coupling of k_1 and k_2. It can be calculated with the simple rules:

$$0 \times 0 \rightarrow 0; 1 \times 0 \rightarrow 1; 0 \times 1 \rightarrow 1; 1 \times 1 \rightarrow 0 \qquad (16)$$

The meaning of (16) is that the product of two scalar operators is a scalar operator; the product of a scalar and a vector operator is a vector operator; the product of two vector operators is a scalar operator as long as we are interested in isotropic coupling only. From eq. 16 we conclude that $k_{12} = 1$. S_3 is rank one and is present in the hamiltonian therefore $k_3 = 1$; k_{123}, which is given by the coupling between $k_{12} = 1$ and $k_3 = 1$ is $k_{123} = 0$. k_4 is zero, because S_4 is not present in the hamiltonian, and k = 0.

In the second coupling scheme indicated above the k values would be for the same hamiltonian:

k_1	k_2	k_{12}	k_3	k_4	k_{34}	k
0	1	1	1	0	1	0

It is easy to verify that, independent of the coupling scheme, for n spins the functions and the hamiltonians are individuated by a set of 2n-1 numbers: n are related to the individual spins, n-2 to intermediate spin, and 1 to the total spin.

Now we can give a general formula for an isotropic hamiltonian $H = J_{ik} S_i.S_k$ for a set of n spins [21]

$$< S_1\, S_2....S_h....S.....SM\ |J_{ik}\ S_i \cdot S_k\ |\ S_1\, S_2.....S_n\ \ S_p\ S'.....SM > = \ -\sqrt{3}\ J/\ \sqrt{2S+1}\ (-1)^{S-M}$$

$$\Pi_{l=1,n} < S_l\ \|\ T_{kl}\ \|\ S_l >\Pi_{\alpha=1,4,3n-3}\{S_\alpha S_{\alpha'}{}^{k_\alpha};S_{\alpha+1}S_{\alpha+1}{}'{}^{k_{\alpha+1}};S_{\alpha+2}S_{\alpha+2}{}'{}^{k_{\alpha+2}}\}$$
$$[(2S_\alpha+2+1)(2S_{\alpha'}+2+1)(2k_{\alpha+2}+1)]^2 \tag{17}$$

where S_p indicates an intermediate spin, specified by the chosen coupling scheme. The term in curly brackets is a 9-j symbol [19], i.e. a number which can be calculated according to standard procedures when the nine numbers it contains are specified. It is similar to 6-j symbols, only more complicated to be calculated. In fact we will show that, provided we are interested at the calculation of *isotropic* hamiltonian $S_i \cdot S_j$, then all the 9-j symbols reduce to 6-j symbols. In fact a 9-j symbol reduces to a 6-j symbol when at least one of its numbers is zero.

In order to do this we prefer not to use abstracts arguments but simply work out an example, which should also clarify how the formidable aspect of eq. 17 reduces to actually relatively simple calculations. In particular we will work out explicitly a matrix element for a cluster of four spins in the first coupling scheme introduced above. According to (17) the matrix element of $J_{23}\ S_2 \cdot S_3$ is given by:

$$< S_1\, S_2\, S_{12}\, S_3\, S_{13}\, S_4\, SM\ |\ S_2 \cdot S_3\ |\ S_1\, S_2\, S_{12}'\, S_3\, S_{13}'\, S_4\, SM > = $$

$$< S_1\|\ T_0\|\ S_1 > < S_2\|\ T_1\|\ S_2 > < S_3\|\ T_1\|\ S_3 > < S_4\|\ T_0\|\ S_4 >.$$

$$\{S_1S_10;S_2S_21;S_{12}S_{12}'1\};\ \{S_{12}S_{12}'1;S_3S_31;S_{13}S_{13}'0\};\ \{S_{13}S_{13}'0;S_4S_40;SS0\}$$
$$\tag{18}$$

where the appropriate values of the k_α parameters can be found in Table I. Eq. 18 shows

----

TABLE I. Coupled Functions and k Parameters for a Cluster of Four Spins.

Coupling scheme: $|\ S_1\, S_2\, S_{12}\, S_3\, S_{123}\, S_4\, S >$

Ham.	k_1	k_2	k_{12}	k_3	k_{123}	k_4	k
$S_1 . S_2$	1	1	0	0	0	0	0
$S_2 . S_3$	0	1	1	1	0	0	0
$S_3 . S_4$	0	0	0	1	1	1	0
$S_4 . S_1$	1	0	1	0	1	1	0
$S_2 . S_4$	0	1	1	0	1	1	0

----

clearly that in order to calculate a given matrix element we must evaluate a product of four (n) reduced matrix elements and three (n-1) 9-j symbols. However from Eq. 18 it is obvious that all the 9-j symbols contain at least one zero. This is indeed bound to the fact that $S_2 \cdot S_3$ is an isotropic hamiltonian. The three 9-j symbols can be made simpler using the relations given in Table II. The first two 9-j symbols in Eq. 18 are then seen to reduce to 6-j symbols, and the last to a simple product of numbers. It is a trivial exercise to verify that analogous conditions apply to all the other J_j $S_i . S_j$ possible hamiltonians.

The calculation of the matrix element (17) is then easy, and also easily programmable. We have written a FORTRAN program, CLUMAG, which in principle can allow the calculation of the energy levels of any set of spins, in any coupling scheme. Of course the limitations are given by the capabilities of the computer. At the moment we can perform the calculations of systems which require matrices not exceeding ca. 4,000 × 4,000 elements. We are currently trying to introduce the point group symmetry within the scheme of calculation, thus reducing the dimensions of the matrices.

TABLE II. Relations between 9-j and 6-j Symbols in Special Cases

$\{S_aS_a'1;S_bS_b'1;S_cS_c'0\}=\delta_{S_cS_c'}(-1)^{Sa'+1+Sb+Sc}$ $[3(2S_c+1)]^{-1/2}$ $\{S_aS_a'1;S_b'S_bS_c\}$

$\{S_aS_a'1;S_bS_b'0;S_cS_c'1\}=\delta_{S_bS_b'}(-1)^{Sa+1+Sb+Sc'}$ $[3(2S_b+1)]^{-1/2}$ $\{S_a'S_a1;S_cS_c'S_b\}$

$\{S_aS_a'0;S_bS_b'1;S_cS_c'1\}=\delta_{S_aS_a'}(-1)^{Sa+1+Sb'+Sc}$ $[3(2S_a+1)]^{-1/2}$ $\{S_c'S_c1;S_bS_b'S_a\}$

$\{S_aS_a'0;S_bS_b'0;S_cS_c'0\}= \delta_{S_aS_a'}$ $\delta_{S_bS_b'}$ $\delta_{S_cS_c'}$ $[3(2S_a+1)(2S_b+1)(2S_c+1)]^{-1/2}$

----

4. A Tutorial Example

In order to learn how to use Eq. 17 to calculate the energy levels of HNSC it is useful to work out explicitly a simple example, such as the case of four S= 1/2 spins in a ring with only nearest-neighbor interactions.

The wavefunctions which are needed within the $|S_1S_2S_{12}S_3S_{123}S_4S>$ coupling scheme are given in Table III.

TABLE III. Spin Wavefunctions for a Cluster of Four S= 1/2 Spins.

$$|S_1 \; S_2 \; S_{12} \; S_3 \; S_{123} \; S_4 \; S>$$

$$|1/2 \quad 1/2 \quad 1 \quad 1/2 \quad 3/2 \quad 1/2 \; 2>$$
$$|1/2 \quad 1/2 \quad 1 \quad 1/2 \quad 3/2 \quad 1/2 \; 1>$$
$$|1/2 \quad 1/2 \quad 1 \quad 1/2 \quad 1/2 \quad 1/2 \; 1>$$
$$|1/2 \quad 1/2 \quad 0 \quad 1/2 \quad 1/2 \quad 1/2 \; 1>$$
$$|1/2 \quad 1/2 \quad 1 \quad 1/2 \quad 1/2 \quad 1/2 \; 0>$$
$$|1/2 \quad 1/2 \quad 0 \quad 1/2 \quad 1/2 \quad 1/2 \; 0>$$

The total spin manifold of 16 states is decomposed into a set of only six functions, one corresponding to S= 2, three to S= 1, and one to S= 0.

Since we are only interested at isotropic exchange hamiltonians, (17) tells us that only matrix elements between functions corresponding to the same total spin value S can be different from zero.

Therefore we can calculate

$$< 1/2 \; 1/2 \; 1 \; 1/2 \; 3/2 \; 1/2 \; 2 \;|\, J_{12}\, \mathbf{S_1.S_2}\,|\; 1/2 \; 1/2 \; 1 \; 1/2 \; 3/2 \; 1/2 \; 2 > =$$

$$= -\sqrt{3}\;(J_{12}/\sqrt{5})\;<S_1\|T_1\|S_1><S_2\|T_1\|S_2><S_3\|T_0\|S_3><S_4\|T_0\|S_4>.$$

$$[(2S_{12}+1)(2S_{123}+1)]^{1/2}\{S_1 \; S_1 \; 1; \; S_2 \; S_2 \; 1; \; S_{12} \; S_{12} \; 0\}.$$
$$.\{S_{12} \; S_{12} \; 0; \; S_3 \; S_3 \; 0; \; S_{123} \; S_{123} \; 0\}\{S_{123} \; S_{123} \; 0; \; S_4 \; S_4 0; \; S \; S \; 0\}=$$

$$= -\sqrt{3}\;(J_{12}/\sqrt{5})\;\sqrt{3}/2.\sqrt{3}/2.\sqrt{2}.\sqrt{2}.1/6.2/\sqrt{6}.5/2\sqrt{10}. \qquad (19)$$

The reduced matrix elements have been calculated with eq. (7, 8), and the 9-j symbols have been reduced to 6-j symbols according to Table II. The 6-j symbols have been calculated as outlined in the Appendix. For instance we have calculated

$$\{1/2 \; 1/2 \; 1; \; 1/2 \; 1/2 \; 1; \; 1 \; 1 \; 0\} = (-1)^3 \; [3.3]^{-1/2}\{1/2 \; 1/2 \; 1; \; 1/2 \; 1/2 \; 1\} = -1/3. \; 1/6$$

If one wants to avoid the direct calculation of the symbols one can use Table A1.

The procedure can be repeated for all the other coupling constants J_{23}, J_{34}, and J_{14} and for all the other functions. The resulting matrices are given in Table IV.

TABLE IV. Hamiltonian Matrix for Four S = 1/2 Spins in a Ring with Nearest-Neighbor Interactions[a]

S = 2 $H_{11} = 1/4\ J_{12} + 1/4\ J_{23} + 1/4\ J_{34} + 1/4\ J_{14}$

S = 1 $H_{11} = 1/4\ J_{12} + 1/4\ J_{23} - 5/12\ J_{34} - 5/12\ J_{14}$

         $H_{22} = 1/4\ J_{12} - 1/2\ J_{23} - 1/12\ J_{34} + 1/6\ J_{14}$

         $H_{33} = -3/4\ J_{12} + 1/4\ J_{34}$

         $H_{12} = \sqrt{2}/\sqrt{3}\ J_{34} - 1/3\sqrt{2}\ J_{14}$

         $H_{23} = -\sqrt{3}/4\ J_{23} - 1/4\sqrt{3}\ J_{14}$

         $H_{13} = -1/\sqrt{6}\ J_{14}$

S = 0 $H_{11} = 1/4\ J_{12} - 1/2\ J_{23} + 1/4\ J_{34} - 1/2\ J_{14}$

         $H_{22} = -3/4\ J_{12} - 3/4\ J_{34}$

         $H_{12} = -\sqrt{3}/4\ J_{23} + \sqrt{3}/4\ J_{14}$

---

[a] The labels of the matrix elements refer to the total spin functions taken in the same order as given in Table III.

---

5. Sample Calculations

    In this section we will report the results of some sample calculations, in order to make clear what is required in order to express the energies of the levels of a HNSC. As a reference point we choose clusters comprising six spins, which are already too large in order to be tackled with the elementary approaches, and still are not too demanding in terms of computer time and memory storage.

    The possible exchange topologies for six coupled spins are rather numerous. We will take into consideration a few of them, outlined in Figure 2. The first geometry corresponds to six spins on a line, which are only connected by nearest neighbor interactions (Figure 2a). The second to a regular hexagon, with only nearest neighbor interactions (Figure 2b). 2c corresponds to a trigonal prism, 2d to a triangular lattice, 2e to a bicapped tetrahedron, 2f to two tetrahedra sharing an edge, 2g to a monocapped

78

square pyramid, 2h to dewar benzene, 2i to a regular octahedron, 2j to two coupled triangles and 2k to a pentagonal pyramid.

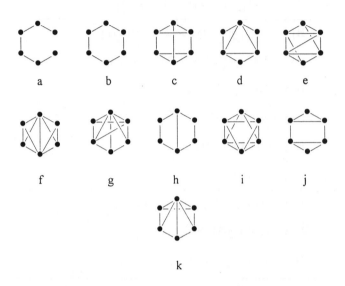

Figure 2. Possible exchange topologies for six spins.

Although these geometries do not exhaust all the possibilities they already provide a large sample. For the sake of simplicity we will assume that all the spins are identical, and that the interactions are antiferromagnetic, and equal to each other. The spins to which we will refer are $S = 5/2$.

The decomposition of the hamiltonian matrix of six $S = 5/2$ spins into blocks using total spin symmetry are shown in Table V.

We performed the calculation on an IBM RISC 6000 work station.

The ground state is $S = 0$ for all the cases of Figure 2, except e and g, which yield a ground S= 1 state. In fact, Figure 3 shows that χT for both e and g stabilizes at 1 emu mol^{-1} K at low temperature, while for all the others it goes to zero. Since we are using one coupling constant in all the calculations it is easy to check that the antiferromagnetic coupling effects are maximized for cases c and f and are minimal for d. The effect is not merely associated with the number of interactions which are present in the considered system, but depends also on spin frustration, i.e. on the number of conflicting interactions which are present in the system. A clear example of spin frustration is given by a regular triangle, where a given spin cannot obey at the same time to all the antiferromagnetic coupling interactions by which it is affected. As a result the ground state is not given simply by up-down spin orientations, but more complex arrangments are found. For instance in a regular triangle of S= 5/2 spins the ground state is a degenerate pair of spin doublets.

TABLE V. Dimensions of the Total Spin Matrices for

S	n_S	S	n_S
15	1	7	405
14	5	6	505
13	15	5	581
12	35	4	609
11	70	3	575
10	126	2	475
9	204	1	315
8	300	0	111

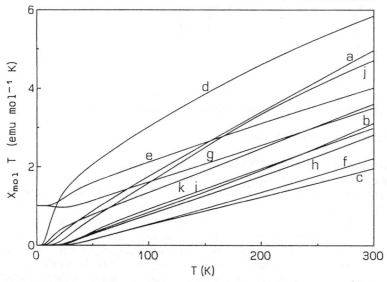

Figure 3. Calculated temperature dependence of χT for the spin topologies depicted in Figure 2. All the coupling constants have been taken equal to 100 cm^{-1}.

It is very interesting to look also at the temperature dependence of the magnetic susceptibilities of these materials, as shown in Figure 4. Except for e and g all the systems show a χ vs T plot passing for a maximum. However the position of the maxima, and the overall shapes of the curves are rather different from each other. Cases a, b, c, f, h, and i are rather similar showing a very flat χ vs T plot , with maxima at relatively high temperatures. Case k shows a maximum at lower temperature, but not a very marked one, while case d shows a sharp maximum at low temperature.

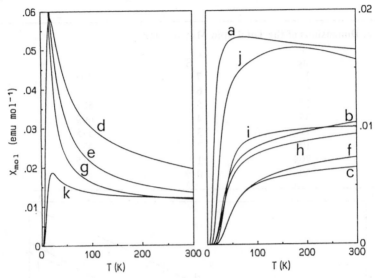

Figure 4. Calculated temperature dependence of χ for the spin topologies depicted in Figure 2. All the coupling constants have been taken equal to 100 cm^{-1}.

6. Some Reference to Real Systems

The number of reported HNSC is already fairly large, and by no means we will attempt to report all of them. Rather we will briefly discuss some results on Fe$_6$ clusters, to compare them with the calculations of the previous section. In particular we will focus on the spin topology of Figure 2j which correspond to several real complexes [4,23,24]. In these the three iron(III) ions in the two triangles are bridged by a μ_3-oxo group, and the two triangles are connected eiher by two μ_2-hydroxo bridges or by one peroxo bridge.

In a simple approach we may consider isosceles triangles, i.e. use two coupling constants, J and J', for the interactions within the triangles, and one, J$_1$, for the interaction between the triangles as shown below. We will use antiferromagnetic coupling constants

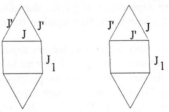

In order to understand in detail the properties of the cluster of six spins it is useful to investigate first the properties of the two isolated triangular fragments. In Figure 5 we plot the low lying energy levels for one isolated triangle as a function of the J/J' ratio. For J/J'= 1 two spin doublets are lowest in energy. As soon as the J/J' ratio becomes different from 1 the degeneracy is removed, and one spin doublet becomes lowest in energy. Relaxing further the J/J' ratio leads to S= 3/2 and eventually to S= 5/2 ground states, as shown in Figure 5.

When two such triangles are coupled according to the topology of Figure 2j, several different possibilities arise. In fact using two different coupling constants within the triangles is equivalent to assume that one center is different from the other two. If we assume a symmetry center for the clusters of six spins, two different types of clusters are possible, one with the different center on the vertex which is not connected to the second triangle, and the other with the unique vertex connected to the second triangle. The former case is relatively simple, in the sense that the antiferromagnetic coupling between the triangles leads invariably to an S= 0 ground state.

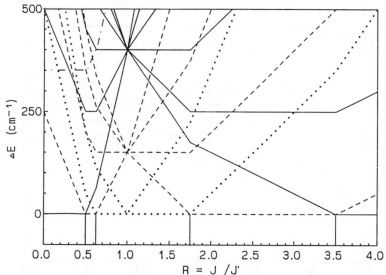

Figure 5. Energy levels for isosceles triangles of spin S= 5/2 coupled with three constants, J, J', J' as a function of the J/J' ratio. (———) S = 5/2, (- - - -) S = 3/2, (• • •) S = 1/2.

The latter case is definitely more interesting. In fact many different ground states may be stabilized depending on the relative values of the coupling constants. We performed sample calculations by keeping J' fixed at 80 cm^{-1}, J_1 at 10 cm^{-1}, and varying J between 16 and 80 cm^{-1}. For J= J', the ground state is S= 0, as shown by the plot of μ_{eff}

vs. T in Figure 6. The effective magnetic moment smoothly decreases on decreasing temperature, with a sharp increase of the slope below 20 K, and rapidly goes to zero below 10 K.

When the J/J' ratio is 0.7 the ground state becomes S= 1, and the effective magnetic moment monotonically decreases with decreasing temperature, as shown in Figure 6c. The change in slope is more gradual at low temperature and the limit value of 2.82 μ_B expected for a triplet ground state is reached at very low temperatures, close to 0 K.

When J/J'= 0.55 the ground state is S= 3. The temperature dependence of the effective magnetic moment is no longer monotonical, as shown in Figure 6b. It goes through a broad minimum around 140 K, then it increase to a maximum at ca. 20 K, and finally decreases rapidly to the value expected for S= 3.

Finally for J/J'= 0.2 the ground state becomes S= 5. The corresponding temperature variation of the effective magnetic moment goes through a broad minimum at ca. 250 K and increases monotonically to reach a plateau corresponding to μ_{eff}= 15.49 μ_B.

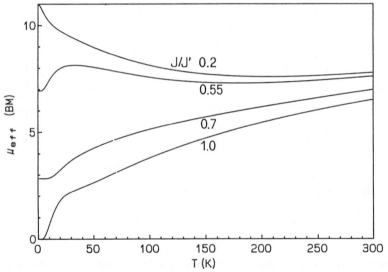

Figure 6. Calculated effective magnetic moment vs T for two coupled triangles with J'= 80 cm^{-1}, J$_1$ = 10 cm^{-1} and J varying from 16 to 80 cm^{-1}.

The experimental data reported so far give some confirmation to these calculations, since both systems with a ground S= 0 [23] and an S= 5 [4] ground states have been reported. The different ground states are the results of spin frustration effects, and show how difficult it can be to predict qualitatively the nature of the ground states in high nuclearity spin clusters in which many triangles of antiferromagnetically coupled spins are present.

7. Conclusions

The detailed investigation of the magnetic properties of high nuclearity spin clusters has just started, and already many different magnetic behaviors have been reported. The nature of the ground state is difficult to predict, but the use of more sophisticated computational techniques is now opening new possibilities for quantitative investigations. The full exploitation of the total spin symmetry through irreducible tensor operator approaches afforded large simplifications in terms of computer memory storage and dimension of matrices to be diagonalized. The next step will be that of exploiting in an equally efficient way also the symmetry associated with the invariance of the spin hamiltonians to the interchange of identical particles. We are currently working along these lines, and we have already been able to successfully calculate the energy levels of clusters comprising eight iron(III) ions.

8. References

1. Carlin R.L. *Magnetochemistry,* (Spriger Verlag, New York, 1986).
2. Willet R.D.; Gatteschi, D.; Kahn, O. *Structural Magnetic Correlation in Exchange Coupled Systems ,* (D. Reidel Publishing Company, Dordrecht, 1985).
3. Gorun S.M., Papaethymiou G.C., Frankel R.B., Lippard S.J., *J. Am. Chem. Soc.* **109** (1987) 3337.
4. J.K. Mc Cusker, C.A. Christmas, P.M. Hagen, R.K. Chadha, D.F. Harvey, D.N. Hendrickson, *J. Am. Chem. Soc.* **113** (1991) 6114.
5. K. Wieghardt, K. Pohe, I. Jibril, G. Huttner, *Angew. Chem. Int. Ed. Engl.* **23** (1984) 77.
6. W. Micklitz, S.J. Lippard, *J. Am. Chem. Soc.* **111** (1989) 6856.
7. E. Libby, K. Folting, J.C. Huffmann, G. Christou, *J. Am. Chem. Soc.* **112** (1990) 5354.
8. K. Wieghardt, *Angew. Chem. Int. Ed. Engl.* **28** (1989) 1153.
9. K.S. Hagen, W.H. Armstrong, M.M. Olmstead, *J. Am. Chem. Soc.* **111** (1989) 774.
10. C.A. Christmas, J.B. Vincent, J.C. Huffmann, G. Christou, H.R. Chang, D.N. Hendrickson, *Angew. Chem. Int. Ed. Engl.* **26** (1987) 915.
11. D. Luneau, J.M. Savariault, J.P. Tuchagues, *Inorg. Chem.* **27** (1988) 3912.
12. M.T. Pope, A. Müller, *Angew. Chem. Int. Ed. Engl.* **30** (1991) 34.
13. A. Müller, J. Döring, H. Bögye, *J. Chem. Soc. Chem. Comm.* (1981) 273.
14. D. Gatteschi, L.Pardi, A.L. Barra, A. Müller, J. Döring, *Nature* **354** (1991) 463.
15. A. Caneschi, D. Gatteschi, R. Sessoli, A.L. Barra, L.C. Brunel, M. Guillot,*J. Am. Chem. Soc.* **113** (1991) 5873.
16. P.D.W. Boyd, R. Li, J.B. Vincent, K. Folting, H.R. Chang, W.E. Straib, J.C. Huffmann, G. Christou, D.N. Hendrickson, *J. Am. Chem. Soc.* **110** (1988) 8535.
17. A. Caneschi, D. Gatteschi, J. Laugier, P.Rey, R. Sessoli, C. Zanchini, *J. Am. Chem. Soc.* **110** (1988) 2795.

84

18. G. Schmid, *Struct. Bonding (Berlin)* **62** (1985) 51.
19. A. Messiah, *Quantum Mechanics*, (North Holland Publishing Company, Amsterdam, 1969).
20. B.L. Silver *Irreducible Tensor Methods*, (Academic Press, New York, 1976).
21. D. Gatteschi, L. Pardi *Gazz. Chim. It.* in press.
22. A. Bencini, D. Gatteschi, *Electron Paramagnetic Resonance of Exchange Coupled Systems*, (Springer Verlag, Berlin, 1990).
23. W. Micklitz, S.J. Lippard, *Inorg. Chem.* **27** (1988) 3067.
24. W. Micklitz, S.G. Bott, J.G. Bontsen, S.J. Lippard, *J. Am. Chem. Soc.* **111** (1989) 372.
25. D.A. Varshalovich, A.N. Moskalev, V.K. Khersonskii *Quantum Theory of Angular Momentum*, (World Scientific, Singapore, 1988).

Appendix

A 3-j symbol [19, 25] $(S_1\ S_2\ S;\ m_1\ m_2\ m)$ must obey the triangular rule $m_1 + m_2 + m = 0$ and $|\ S_1 - S_2\ | \leq S \leq S_1 + S_2$.

It has several symmetries associated with the interchange of rows and columns which will not be mentioned here, because we are only interested at their calculation with a computer. An appropriate formula is:

$$(S_1\ S_2\ S;\ m_1\ m_2\ m) = (-1)^{S_1 - S_2 - m}\ \Gamma(S_1\ S_2\ S)^{1/2}[(S_1 + m_1)\ !(S_1 - m_1)\ !\ (S_2 + m_2)\ !$$
$$(S_2 - m_2)\ !\ (S + m)\ !\ (S - m)]^{1/2} \times \sum (-1)t\ [t\ !(S - S_2 + t + m_1)\ !\ (S - S_1 + t - m_2)\ !\ (S_1 +$$
$$S_2 - S - t)\ !\ (S_1 - t - m_1)\ !\ (S_2 - t + m_2)!]^{-1} \qquad (A1)$$

where

$$\Gamma\ (S_1\ S_2\ S) = \big[(S_1 + S_2 - S)\ !\ (S_2 + S - S_1)\ !\ (S + S_1 - S_2)\ !\big]/\ (S_1 + S_2 + 1)!$$
$$(A2)$$

and the sum extends over all the integral values of t for which the factorials are ≥ 0. The number of terms in the summation is $\mu + 1$, where μ is the smallest of the nine numbers

$$S_1 \pm m_1 \;;\; S_2 \pm m_2 \;;\; S \pm m \;;\; S_1 + S_2 - S \;;\; S_2 + S - S_1 \;;\; S + S_1 - S_2 \qquad (A3)$$

The 6-j symbols [25], which are real numbers, have also many symmetries. The formula to calculate them is:

$$(S_1 \; S_2 \; S_3 \;;\; S_4 \; S_5 \; S_6) = [\Gamma (S_1 \; S_2 \; S_3) \; \Gamma (S_1 \; S_5 \; S_6) \; \Gamma (S_4 \; S_2 \; S_6) \; \Gamma (S_4 \; S_5 \; S_6)]^{1/2}$$

$$\Sigma_t \; (-1)^t \; (t + 1)! \; [(t - S_1 - S_2 - S_3)! \; (t - S_1 - S_5 - S_6)! \; (t - S_4 - S_2 - S_6)! \; (t - S_4 - S_5 -$$

$$S_3)! \; (S_1 + S_2 + S_4 + S_5 - t)! \; (S_2 + S_3 + S_5 + S_6 - t)! \; (S_3 + S_1 + S_6 + S_4 - t)!] \; (A4)$$

where the Γ functions are defined as in eq. A2. The sum over t also runs on the $\mu + 1$ integer positive values for which the factorials are ≥ 0. μ is the smallest of the 12 numbers:

$$S_1 + S_2 + S_3 \;;\; S_1 + S_5 - S_6 \;;\; S_4 + S_2 - S_6 \;;\; S_4 + S_5 - S_3 \;;$$

$$S_2 + S_3 - S_1 \;;\; S_5 + S_6 - S_1 \;;\; S_2 + S_6 - S_4 \;;\; S_5 + S_3 - S_4 \;;$$

$$S_3 + S_1 - S_2 \;;\; S_6 + S_1 - S_5 \;;\; S_6 + S_4 - S_2 \;;\; S_3 + S_4 - S_5$$

Some 6-j symbols which involve $S = 1/2$ spin are given in Table A1.

TABLE A1. Numerical Values of some 6-j Symbols $\{abc; def\}$ involving $S = 1/2$ Spins.

---

a	b	c	d	e	f	$\{abc; def\}$
1/2	1/2	1	1/2	1/2	1	0.16667
1/2	1/2	1	1	1	1/2	-0.33333

3/2	1/2	1	1/2	1/2	1	-0.33333
3/2	1/2	1	3/2	1/2	1	-0.08333
3/2	3/2	1	1/2	1/2	1	0.263523
3/2	3/2	1	3/2	1/2	1	0.16667
3/2	1/2	2	1/2	3/2	1	0.15811
3/2	1/2	2	3/2	1/2	1	0.25000
3/2	3/2	2	3/2	1/2	1	-0.223607
3/2	1/2	2	3/2	1/2	2	0.05000
3/2	3/2	2	3/2	1/2	2	-0.10000
3/2	3/2	1	1/2	1/2	2	0.158114
1	1	1	1/2	1/2	1/2	-0.33333
1	1	1	3/2	3/2	1/2	-0.16667
1	1	1	3/2	3/2	1/2	0.263523
1	1	1	1/2	1/2	3/2	-0.16667
1	1	1	1/2	3/2	1/2	0.28868
2	1	1	1/2	3/2	3/2	-0.20412
2	1	1	3/2	3/2	1/2	0.091287

SPIN FRUSTRATION IN POLYNUCLEAR COMPLEXES

David N. Hendrickson
Department of Chemistry-0506
University of California at San Diego
La Jolla, California 92093-0506
U.S.A.

1. Introduction

In polynuclear complexes of paramagnetic transition metal ions there are magnetic exchange interactions between neighboring metal ions. If the unpaired-electron-containing orbitals of one metal ion interact with those of another metal ion, propagated perhaps by the orbitals of a bridging group, then a magnetic exchange interaction is present. The most common tendency is to pair the spins of the unpaired electrons on these two metal ions in an antiferromagnetic exchange interaction. Pairwise ferromagnetic interactions, where the unpaired spins on the two metal ions align parallel to give a ground state with the maximum spin, are much rarer than antiferromagnetic interactions.[1] When antiferromagnetic interactions are present, the ground state of the complex tends to be the state with the smallest spin value. However, even though there are only antiferromagnetic interactions present there are certain topological arrangements of metal ions in polynuclear complexes which result in a ground-state spin that is intermediate in value between the smallest and largest values possible for the complex. The topology of these complexes leads to spin frustration.[2]

In this chapter the phenomenon of spin frustration in polynuclear transition metal complexes is reviewed. Spin frustration is not a new concept, for it has been shown to be the origin of several interesting phenomena detected for extended magnetic exchange interacting compounds.[3] For example, even though the solid state structure of $CsCoCl_3$ consists of a one-dimensional chain structure formed by face-sharing $CoCl_6^{4-}$ octahedra, neutron diffraction experiments[4] show that due to interchain magnetic exchange interactions the chains of $CsCoCl_3$ form a triangular lattice and interact antiferromagnetically. The spin frustration among the magnetic chains results in phase transitions at $T_{N1} = 21.5$ K and $T_{N2} = 9.2$ K. Thermally excited domain walls, the first excited state in an Ising spin chain, are believed to travel in the frustrated chains which feel no molecular field from the neighboring chains even in the three dimensionally ordered phase.[5] Spin frustrated extended exchange interacting complexes have attracted considerable attention in recent years in part due to the possibility of new types of phase transitions.[6]

2. Trinuclear Complexes

2.1. Simple Example of Spin Frustration

The phenomenon of spin frustration in polynuclear transition metal complexes can be qualitatively understood by reference to the simple example of a hypothetical trinuclear Cu^{II} complex:

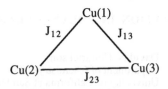

Each Cu^{II} ion has a single unpaired electron (S = 1/2). The pairwise magnetic exchange interactions can be evaluated with the Heisenberg-Dirac-van Vleck spin Hamiltonian $\hat{H}$ = $-2J_{ij}\hat{S}_i\cdot\hat{S}_j$. Antiferromagnetic pairwise interactions where the exchange parameter $J_{ij} < 0$ are the most common. Spin frustration results naturally from this triangulated topological arrangement of metal ions. If all of the pairwise interactions in the above Cu^{II}_3 triangle are antiferromagnetic, the origin of this frustration is easily understood. The spins on the Cu(1) and Cu(2) ions are paired, for example, and this frustrates the spin on Cu(3). The unpaired electron on Cu(3) wants to pair with the unpaired electrons on both Cu(1) and Cu(2), but it is "frustrated" because it cannot simultaneously spin pair with both electrons for they have an antiparallel spin alignment.

The consequences of spin frustration in a triangulated Cu^{II}_3 complexes are not profound since the ground state has S = 1/2 when all pairwise interactions are antiferromagnetic. M(SQ)$_3$ complexes, where the metal M is diamagnetic (*e.g.*, Ga^{3+}) and SQ$^-$ is the S = 1/2 o-semiquinone ligand, may exhibit this same type of spin frustration. If the metal M has unpaired electrons, a more involved case of spin frustration may develop. M(SQ)$_3$ complexes with paramagnetic metal ions are known,[7] but there has been no evidence of spin frustration, perhaps due to the apparently weak nature of SQ$^-$ $\cdots$SQ$^-$ ligand-ligand exchange interactions.

2.2. μ_3-Oxide Bridged Complexes

It is well established that the μ_3-oxide ion in $[M_3O(O_2CR)_6L_3]^{n+}$ complexes (L is some monodentate ligand) is dominant in propagating the exchange interactions.[8] These complexes have been the object of very many studies.[9] In fact, considerable effort has been expended to understand the temperature dependence of the magnetic susceptibility data for various μ_3-oxide bridged trinuclear Fe^{III} acetate complexes.[10] Regardless of the details of the data fitting, the J values for these Fe$^{III}_3$O complexes fall in the range of -20 to -31 cm^{-1}. It is relevant to note that the data for many of these Fe$^{III}_3$O complexes are best fit by employing a two J-value magnetic exchange model.

The nature of spin frustration in $[Fe_3O(O_2CR)_6L_3]^+$ complexes can be illustrated. Each Fe^{III} ion has S = 5/2. If the Fe$^{III}_3$O complex in the solid state assumes for whatever reason an isosceles triangular form, then two exchange parameters are required as schematically shown below:

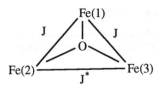

The exchange interactions in this complex are evaluated by using the Hamiltonian in Eq. 1

$$\hat{H} = -2J(\hat{S}_1 \cdot \hat{S}_2 + \hat{S}_1 \cdot \hat{S}_3) - J^*(\hat{S}_2 \cdot \hat{S}_3) \tag{1}$$

By defining the total spin operator as $\hat{S}_T = \hat{S}_A + \hat{S}_1$, where $\hat{S}_A = \hat{S}_2 + \hat{S}_3$, the operator-equivalent form of Eq. 1 is given in Eq. 2:

$$\hat{H} = -J(\hat{S}_T^2 - \hat{S}_A^2 - \hat{S}_1^2) - J^*(\hat{S}_A^2 - \hat{S}_2^2 - \hat{S}_3^2) \tag{2}$$

or more simply

$$\hat{H} = -J(\hat{S}_T^2 - \hat{S}_A^2) - J^*(\hat{S}_A^2) \tag{3}$$

where the constant value single-ion $\hat{S}_i^2$ terms have been dropped.

Plotted in Figure 1 are the energies in units of $|J^*|$ of all 27 eigenstates of Eq. 3 as a function of the ratio J/J^*. The total spin for the 27 eigenstates range from $S_T = 1/2$ to $S_T = 15/2$. This diagram was constructed for J and $J^* < 0$, i.e., both exchange interactions are antiferromagnetic. Two facts are evident from Figure 1. The most obvious fact is that the overall magnetic structure of this complex is reasonably complicated and, more importantly, is **extremely** sensitive to the ratio J/J^*. Second, there are a total of six different ground states possible depending on the J/J^* ratio. For these six different possible ground states the S_T values of 1/2, 3/2 and 5/2 are found. Thus, such a Fe^{III}_3O complex can have a ground state with any of these three S_T values.

The most useful way to interpret the diagram in Figure 1 is to consider movement along the x axis as a means of varying the degree of spin frustration present in a Fe_3O complex with respect to the two different exchange interactions J and J^*. If the J^* interaction is much strong than J ($J/J^* \leq 0.3$), the ground state is described by $S_T = 5/2$ and $S_A = 0$. This value of S_A indicates that the spin vectors S_2 and S_3 are completely paired up. The result is complete frustration of the J-coupling pathway and the five unpaired electrons on Fe(1) do **not** pair up with any of the electrons on Fe(2) or Fe(3). As the strength of the J exchange interaction increases relative to J^* (i.e., for $0.3 \leq J/J^* \leq 0.55$), a $S_T = 3/2$ ground state results. The value of $S_A = 1$ which characterizes this state indicates that the S_2 and S_3 spins have **not** completely paired up. The value of S_A can range from 0 to 5. As the relative strength of the J interaction further increases, the ground state becomes a state with $S_T = 1/2$ and $S_A = 2$ for values of J/J^* between 0.55 and 1.0. For values of J/J^* between 1.0 and 1.6 the ground state is characterized by $S_T = 1/2$ and $S_A = 3$; $S_T = 3/2$ and $S_A = 4$ are found for $1.6 \leq J/J^* \leq 2.0$. Finally at J/J^* values greater than 2.0 a $S_T = 5/2$, $S_A = 5$ ground state is found. In this last case the J-type interaction dominates and the spin on Fe(1) is antiparallel to the spins on both Fe(2) and Fe(3). In other words, the J^*-type interaction is frustrated and as a result the spins on the Fe(2) and Fe(3) ions are forced to align parallel. This parallel alignment is present even though the J^* interaction may intrinsically be antiferromagnetic.

From the above discussion it is clear that only small changes in the J/J^* ratio are needed to change the spin of the ground state in a Fe^{III}_3O complex. The S_T value of the ground state can assume intermediate values of 3/2 and 5/2. As J/J^* is increased from 0.0 to > 2.0 in Figure 1, the value of S_A increases in the series 0, 1, 2, 3, 4, and 5. The spin frustration changes from the J interaction at $S_A = 0$ to the J^* interaction at $S_A = 5$.

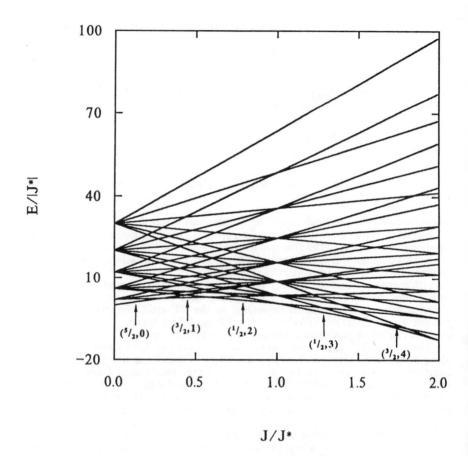

Figure 1.   Plot of the eigenvalues of Eq. 3 for a triangular array of Fe^{III}_3 ions in units of J^*. The states are labeled as (S_T, S_A).

Almost all of the triangular Fe^{III}_3O complexes which have been studied have the same bridging groups for all three sides of the triangle *i.e.*, they have the composition of $[Fe_3O(O_2CR)_6L_3]^+$. The crystal site symmetries, however, are lower than C_3 for the complexes for which X-ray structures have been reported.[9]   Thus, the variable-temperature magnetic susceptibility data for several of these complexes have been fitted using two exchange parameters (J and J^*). The most detailed study was made by Long *et al.*[11] on the complex $[Fe_3O(O_2CCH_3)_6(OH_2)_3]Cl\cdot5H_2O$. One of the two fits of the data for this complex was found with $J = -21.0$ cm^{-1} and $J^* = -33.3$ cm^{-1}. The other fit essentially interchanged the roles of J and J^*. The first fit corresponds to a ratio of $J/J^* =$

0.63, which indicates a $S_T = 1/2$, $S_A = 2$ ground state. For the 14 Fe^{III}_3O complexes where two J values were employed to fit the susceptibility data,[9] J/J^* varies from 0.58 to 1.47. Thus, two different ground states are indicated; both have $S_T = 1/2$, whereas one has $S_A = 2$ and the other $S_A = 3$. Both of these ground states have an intermediate level of frustration. It must be noted that there is still controversy about how to fit the susceptibility data for these Fe^{III}_3O complexes.

There is only one well studied asymmetric Fe^{III}_3O complex, $[Fe_3O(TIEO)_2(O_2CPh)_2Cl_3]\cdot2C_6H_6$, where the ligand TIEOH has the following structure:

The Fe^{III}_3O triangle is clearly isosceles. Gorun et al.[12] fit the susceptibility data for this complex to find $J = -8.0$ cm^{-1} and $J^* = -55$ cm^{-1}, which gives $J/J^* = 0.145$. This corresponds to a $S_T = 5/2$, $S_A = 0$ ground state, which was verified by recording the ^{57}Fe Mössbauer spectrum of the complex at 4.2 K in a 60 kG applied longitudinal magnetic field. Two signals were seen in this spectrum, one for the Fe(1) ion which has a large total magnetic field (480 kG) at the nucleus and the other signal for the Fe(2) and Fe(3) sites which have little unpaired spin density.

In Figure 2 is given a plot of the energy of each of the 19 states of a triangular Mn^{III}_3 complex in units of J^* as a function of J/J^*. Figure 2 was constructed assuming both J and J^* are negative, i.e., for antiferromagnetic interactions. When either J is zero or J^* is extremely large compared to J, then the ground state of the Mn^{III}_3 complex is characterized by $(S_T, S_A) = (2, 0)$. The unpaired electrons on the Mn(1) and Mn(3) ions are paired up, leaving the Mn(2) ion with four unpaired electrons ($S_T = 2$). From Figure 2 it can be seen that as the J/J^* ratio is increased other (S_T, S_A) states become the ground state. When J/J^* exceeds ~0.35 the $S_T = 1$, $S_A = 1$ state becomes the ground state. The $S_T = 0$, $S_A = 2$ state is the ground state when J/J^* is in the range of ~0.7 to ~1.5. For J/J^* values of ~1.5 to ~1.8 the ground state is the $S_T = 1$, $S_A = 3$ state and with J/J^* ratios exceeding ~1.8 the $S_T = 2$, $S_A = 4$ state becomes the ground state.

There are different levels of spin frustration as J/J^* is increased. With $J/J^* = 0$, the Mn(1) and Mn(3) ions are paired up and $S_A = 0$. As J/J^* is increased the spins on the Mn(1) and Mn(3) ions cannot completely couple up. The presence of the Mn(1)-Mn(2) and Mn(2)-Mn(3) antiferromagnetic interactions (J parameter) frustrates the Mn(1)-Mn(3) antiferromagnetic exchange interaction. Only intermediate values of $S_A = 1$, 2, or 3 are possible as J/J^* is increased. Finally, when J/J^* exceeds ~1.8 the value of S_A becomes the maximum value of 4. In this case $|J|$ is so large compared to $|J^*|$ that the very strong Mn(1)-Mn(2) and Mn(2)-Mn(3) antiferromagnetic interactions frustrate the Mn(1)-Mn(3) interactions so that $S_A = 4$.

In short, when either of the two antiferromagnetic interactions dominates the other, the spins in the weaker pairwise interaction cannot pair up even though the interaction is intrinsically antiferromagnetic.

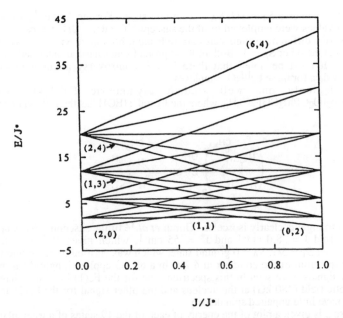

J/J*

Figure 2.   Energy levels of a triangular Mn^{III}_3 complex derived from Eq. 1. The energy of each state in units of the exchange parameter J^* is plotted as a function of J/J^*. States are labeled as (S_T, S_A).

Unfortunately magnetic susceptibility data have been presented for only a few triangular Mn^{III}_3O complexes.[9] The data for $[Mn_3O(O_2CCH_3)_6(py)_3](ClO_4)$ could be fitted by assuming an equilateral triangle to give $J = J^* = -10.2$ cm^{-1}. Obviously the ground state has $S_T = 0$ and $S_A = 2$.

McCusker et al.[13] presented detailed susceptibility and EPR data to characterize the ground state of two $Mn^{II}Mn_2^{III}O$ complexes. The data for one complex, $[Mn^{II}Mn_2^{III}O(O_2CCH_3)_6(py)_3](py)$, were shown to give $J = -5.2$ cm^{-1} and $J^* = -2.7$ cm^{-1} which corresponds to a $S_T = 3/2$, $S_A = 3$ ground state. On the other hand the data for the less symmetric complex $[Mn_3O(O_2CPh)_6(py)_2(H_2O)]\cdot^1/_2CH_3CN$ clearly indicate a different ground state where $S_T = 1/2$ and $S_A = 3$. This change in ground state resulted from small changes in the J, J^* parameters ($J = -6.5$ cm^{-1} and $J^* = -4.5$ cm^{-1} for the latter complex). The absolute values of these parameters are not so important as is the change in the J/J^* ratio.

3.   Tetranuclear Butterfly Complexes

Several tetranuclear Fe^{III} and Mn^{III} complexes which have the same $[M_4O_2]^{8+}$ core have been prepared recently. The bis-μ_3-oxo core of these "butterfly" complexes is pictured below:

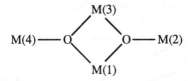

The core in some of these complexes is planar, whereas in others there are appreciable deviations from planarity. For a centrosymmetric complex the exchange interactions can be described by Eq. 3. The interaction between the M(2) and M(4) ions is assumed to be negligible.

$$\hat{H} = -2J(\hat{S}_1 \cdot \hat{S}_2 + \hat{S}_2 \cdot \hat{S}_3 + \hat{S}_3 \cdot \hat{S}_4 + \hat{S}_4 \cdot \hat{S}_1) - 2J_{13}\hat{S}_1 \cdot \hat{S}_3 \qquad (4)$$

An equivalent operator expression to replace Eq. 4 can be readily obtained by means of the Kambe method. Several of these $Mn_4^{III}O_2$ and $Fe_4^{III}O_2$ complexes have ground states and low-energy spin states which result from spin frustration.

Recently we reported[14] the preparation, X-ray structure and variable-temperature magnetic susceptibility data for $[Mn_4O_2(O_2CCH_3)_7(bpy)_2](ClO_4)$. Least-squares fitting of the susceptibility data to the theoretical susceptibility equation which can be obtained from Eq. 4 gave exchange parameters of $J_{13} = -23.5$ cm^{-1} and $J = -7.8$ cm^{-1}. The ground state for this $Mn_4^{III}O_2$ complex has $S = 3$ with the two lowest-lying excited states being two energetically degenerate $S = 2$ states at ~15 cm^{-1} above the $S = 3$ ground state. The nature of the ground and low-lying spin states were confirmed by magnetization measurements at fields up to 48kG and temperatures down to 1.8 K. The structural and magnetic characteristics of $[Mn_4O_2(O_2CCH_3)_6(bpy)_2] \cdot 2CHCl_3$ were also determined.[14] In this $[Mn_2^{III}Mn_2^{II}O_2]^{6+}$ complex the two Mn^{III} ions are located in the body of the butterfly, i.e., the M(1) and M(3) sites. Fitting the susceptibility data gave $J_{13} = -3.1$ cm^{-1} and $J(Mn^{II}-Mn^{III}) = -1.97$ cm^{-1}. This complex has a $S = 2$ ground state with six other spin states within 15 cm^{-1}, which was confirmed by variable-field magnetization data. Spin frustration is therefor also found for $[Mn^{III}_2Mn^{II}_2O_2]^{6+}$ mixed-valence complexes.

Very recently the preparation and X-ray structure of another $[Mn_4^{III}O_2]^{6+}$ complex became available.[15] A butterfly structure was found for the anion in $(NBu_4^n)[Mn_4O_2(O_2CCH_3)_7(pic)_2]$, where pic$^-$ is the monoanion of picolinic acid. Least-squares fitting of the data gave the parameters $J = -5.3$ cm^{-1}, $J_{13} = -24.6$ cm^{-1}, and $g = 1.96$. Again a $S = 3$ ground state was found.

A more global view of the ground states which are possible for a $[Mn_4^{III}O_2]^{8+}$ butterfly complex is shown in Figure 3. This figure shows which state is the ground state in the parameter space encompassing $J = +50$ to -50 cm^{-1} and $J_{13} = 0$ to -50 cm^{-1}. Depending on the values of J and J_{13} there are nine possible ground states. Each state is labeled as (S_T, S_{13}, S_{24}), where $\hat{S}_T = \hat{S}_{13} + \hat{S}_{24}$, $\hat{S}_{13} = \hat{S}_1 + \hat{S}_3$ and $\hat{S}_{24} = \hat{S}_2 + \hat{S}_4$. Thus, the two $[Mn_4O_2]^{8+}$ complexes described above both have $S_T = 3$ ground states arising from $S_{13} = 1$ and $S_{24} = 4$. Even though both types of pairwise exchange interactions are antiferromagnetic, the ground state of these complexes has 6 unpaired electrons. This results from spin frustration. The body-body antiferromagnetic interaction (J_{13}) is much larger than the body-wing tip interaction (J). The spins on the Mn(1) and Mn(3) ions have the greater tendency to pair up. However, S_{13} equals 1, not 0

94

which would be found for total coupling. Each wing-tip Mn^{III} ion [Mn(2) and Mn(4)] interacts with **both the Mn(1) and Mn(3) ions**. The net result is that the spin

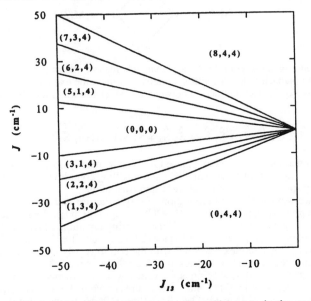

Figure 3.   Plot in J, J_{13} parameter space showing which state is the ground state for a $[Mn^{III}_4O_2]^{8+}$ complex.

alignments of the wing-tip ions are frustrated. From Figure 3 it is clear that by relatively small changes in the J and J_{13} parameters other ground states such as (2, 2, 4) or (0, 0, 0) could be realized. In summary, these $[Mn^{III}_4O_2]^{8+}$ butterfly complexes can be viewed as triangular Mn^{III}_3O units to which has been added a $Mn^{III}-O$ moiety. In principal building up molecules from such triangular Mn^{III}_3 building blocks could yield molecules with relatively large numbers of unpaired electrons.

X-ray structures and magnetic susceptibility data are available for two $[Fe^{III}_4O_2]^{8+}$ butterfly complexes: $[Fe_4O_2(O_2CCH_3)_7(bpy)_2](ClO_4)\cdot^1/_4CH_2Cl_2\cdot H_2O^{16}$ and $(Et_4N)[Fe_4O_2(O_2CPh)_7(H_2B(pz)_2)_2].^{17}$ A least-squares fit of the data for the first complex gave J = -45.5 cm^{-1} with the g-value fixed at 2.00.   It was interesting to find that the value of J_{13} is not well determined by the above data and can only be described as being more positive than -15 cm^{-1}.   The ground state for this complex has $S_T = 0$, which results from the coupling of $S_{13} = 5$ and $S_{24} = 5$.   The relative energies of this ground state and all of the thermally populated excited states depend only on J and not on J_{13}. Therein lies the origin of the indeterminancy of J_{13}.   Fitting the published susceptibility data for the $H_2B(pz)_2^-$ complex gave similar results.   Both $[Fe^{III}_4O_2]^{8+}$ complexes have the same $S_T = 0$ ground state.   In contrast to the analogous Mn^{III} complexes, the wing-body (J) antiferromagnetic interaction is dominating and the spin frustration is different for the Fe^{III} complexes.   The body-body interaction is frustrated with parallel spin alignment between the two body iron ions:

$$Fe(4)—O\underset{Fe(1)}{\overset{Fe(3)}{\diamondsuit}}O—Fe(2)$$

4. Tetranuclear Cubane Complexes

The $Mn^{IV}Mn^{III}_3$ state (the so-called S_2 state) of the photosynthetic water oxidation center has been suggested[18] to be in different spin ($S = 1/2$ or $S = 3/2$) states dependent on chloroplast treatment[19] or calcium depletion.[20] Only a few $Mn^{IV}Mn^{III}_3$ complexes have been characterized: the imidazolium salt $(H_2Im)_2[Mn_4O_3Cl_6(HIm)(O_2CCH_3)_3]\cdot^3/_2CH_3CN$[21] (cubane 1) and $[Mn_4O_3Cl_4(O_2CR)_3(py)_3]$ where $R = CH_3$[22] (cubane 2) and $R = CH_2CH_3$[23] (cubane 3). The $[Mn_4(\mu_3\text{-}O)_2(\mu_3\text{-}Cl)]^{6+}$ core in all three complexes is very similar. It is best considered as a Mn_4 pyramid with the Mn^{IV} ion at the apex, the μ_3-Cl ion bridging the basal plane and one μ_3-O^{2-} ion bridging each remaining face. Alternatively, the core can be described as a severely distorted Mn_4O_3Cl cubane. Cubanes 2 and 3 crystallize in the

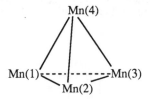

space group $R\overline{3}$, where the complex has crystallographically-imposed C_3 symmetry. The magnetic exchange interactions in these complexes can be described by the Hamiltonian in Eq. 5:

$$\hat{H} = -2J_1(\hat{S}_1\cdot\hat{S}_2 + \hat{S}_1\cdot\hat{S}_3 + \hat{S}_2\cdot\hat{S}_3) \quad -2J_2(\hat{S}_1\cdot\hat{S}_4 + \hat{S}_2\cdot\hat{S}_4 + \hat{S}_3\cdot\hat{S}_4) \tag{5}$$

The Kambe coupling scheme where $\hat{S}_T = \hat{S}_A + \hat{S}_4$ and $\hat{S}_A = \hat{S}_1 + \hat{S}_2 + \hat{S}_3$ leads to an expression for the energies of the spin states in this complex. $\hat{S}_4$ is the spin operator for the Mn^{IV} ion. In all three cases the value of μ_{eff}/molecule increases with decreasing temperature to a maximum, below which μ_{eff}/molecule decreases. These maxima occur at 60, 60 and 15 K with μ_{eff}/molecule values of 9.54, 9.28 and 9.97 μ_β for cubane complexes 1, 2, and 3, respectively. Least-squares fitting the data for each of the complexes gave the parameters tabulated below. In each case the 10 kG data set was only fit down to the temperature below which the μ_{eff}/molecule decreases rapidly. All three

Parameter	Cubane 1	Cubane 2	Cubane 3
g	1.92	1.86	2.00
$J_1(cm^{-1})$	11.2	11.6	8.58
$J_2(cm^{-1})$	-31.0	-24.0	-20.8

complexes have a S = 9/2 ground state. The lowest lying excited state has S = 7/2 and is at an energy of 227, 211, and 165 cm^{-1} above the ground state for complexes 1, 2, and 3, respectively. Thus, the rapid drop in μ_{eff}/molecule at low temperatures is attributable to a combination of zero-field and Zeeman interactions affecting the S = 9/2 ground state, for only this state is populated below ~100 K. The S = 9/2 ground state has been further established for cubanes 1 and 2 by variable-field (10-45 kG) magnetization measurements.[21]

Obviously, a $Mn^{IV}Mn_3^{III}$ complex could have a ground state with a spin from S = 1/2 to S = 15/2. Why do these complexes have a S = 9/2 ground state? The three Mn^{III} ions form the basal plane of each complex. Since cubanes 2 and 3 have C_3 site symmetry, all three Mn^{III}-Mn^{III} ferromagnetic exchange interactions are equivalent and there would be no intrinsic spin frustration in the Mn^{III}_3 triangle. However, the largest interaction present in these complexes is the Mn^{III}-Mn^{IV} interaction. This Mn^{III}-Mn^{IV} antiferromagnetic interaction dominates and frustrates the Mn^{III}-Mn^{III} interactions so that the spins of the three Mn^{III} ions are made to have a parallel orientation.

5. Hexanuclear Ferric Complexes

It is perhaps more of a challenge to prepare a polynuclear high-spin Fe^{III} complex which has an intermediate spin ground state than for manganese complexes. Nevertheless, very recently a hexanuclear Fe^{III} complex with quite interesting properties has been reported.[24] The reaction in the air of the diimidazole ligand 1 with the readily available trinuclear Fe^{III}_3O complex 2 gives Fe^{III}_6 complex 3.

The X-ray structure of complex 3 shows that it may be viewed as two μ_3-oxo-bridged Fe_3^{III} complexes bridged together by two OH^- groups and six carboxylates. Complex 3 is similar to $[Fe_6O_2(OH)_2(O_2CCMe_3)_{12}$ (4),[25] but is somewhat different than $[Fe_6O_2(OH)_2(O_3CPh)_{12}(1,4$-dioxane$(OH_2)$ (5).[26] Michlitz and Lippard[26] reported the magnetic susceptibility data for complex 5 which showed that this complex has a S = 0 ground state. No data have been reported for complex 4.

In contrast, complex 3 has definitively been established to have novel properties; it has a S = 5 ground state. In an applied field of 10.00 kG the effective magnetic moment per molecule is 9.21 μ_B at 346.1 K. This value drops slightly to 9.06 μ_B at 278.7 K, then increases gradually to a maximum of 10.89 μ_B at ~20.0 K where it plateaus before falling to 10.44 μ_B at 6.00 K. The increase in μ_{eff}/molecule with decreasing temperature is quite novel for a Fe^{III} complex. It is suggestive of a Fe^{III}_6 complex with a S = 5 ground state, which would give a spin-only value of μ_{eff} = 10.95 μ_B. It was established with ^{57}Fe Mössbauer data in the 100-300 K range that complex 3 does only have high-spin Fe^{III} ions.

Variable-field magnetization data were determined for complex 3 in applied fields of 10.00 kG (1.51 - 9.04 K), 25.00 kG (1.81 - 9.00 K) and 40.00 kG (1.57 - 9.04 K). Several different samples of 3 were examined to verify the reproducibility of the data. The data indicate that at high field and low temperature (1.57 K and 40.0 kG) the reduced magnetization plateaus at a value of ~9.2. This value is close to the saturation value of 10 expected for an isolated S = 5 state. The fact that the three isofield lines do not superimpose indicates that the ground state is split in zero field ($D\hat{S}^2_z$). The data at all three fields were simultaneously fit by a full matrix diagonalization of the Hamiltonian matrices for both parallel and perpendicular components of the applied field. A fit was obtained for g = 2.0 and D = 0.22 cm^{-1}. It is clear that complex 3 does have a S = 5 ground state.

The question at this point is why does complex 3 have the novel S = 5 ground state, whereas complex 5 has the naively expected S = 0 ground state. This difference in ground states is attributable to spin frustration. Consider complex 5. If one ascribes an antiferromagnetic coupling to each of the pairwise interactions starting with any Fe^{III} site, and continues around the periphery of the complex, there is no difficulty in identifying a spin alignment which gives S = 0:

5 $S_T = 0$

However, upon examining the resulting spin distribution, it becomes clear that two of the interactions, specifically the ions involved in both the μ_3-O^{2-} and μ_2-OH^- bridges, are spin *aligned* with respect to the μ_3-O^- pathway. This result, and the subsequent ambiguity in spin polarization, is the essence of spin frustration and is an inescapable consequence of this nuclear arrangement.

Complexes 3 and 5 can be viewed as being comprised of two $Fe_3^{III}O$ triangles. In this way we can see a possible origin of the difference in ground states. Consider an isosceles Fe^{III}_3 triangle as pictured above. This complex has 27 spin states with the total spin S_T ranging from 1/2 to 15/2. Just which state is the ground state is extremely sensitive to the ratio J/J^*. In terms of the (S_T, S_A) symbolism, six different ground states are possible: (5/2, 0), (3/2, 1), (1/2, 2), (1/2, 3), (3/2, 4) and (5/2, 5).

It is possible to rationalize the difference in ground states between complexes 3 and 5. In view of the fairly high symmetry of complex 5, the region about $J/J^* = 1$ (see Figure 1) is of importance with regard to the magnetism observed for this complex. It is by no means obvious that in the triangulated system for J = J* one should have a ground state with S_T = 1/2; spin frustration is clearly present in states with S_A = 2 or 3 and this naturally leads to difficulties in conceptualizing the origin of the ground state. However, it is clear from the above analysis of the trinuclear case that the S_T = 1/2 lowest spin state of the system is the most stable under these conditions. The origin of the S = 0 ground state in complex 5 is a manifestation in large part of the symmetry of that molecule. Each triangle in complex 5 is relatively symmetric and has S = 1/2. The net interaction between the two triangles must be antiferromagnetic to give S_T = 0 for the ground state of complex 5.

In the case of complex 3, however, the presence of the μ_2-alkoxide bridge considerably reduces the symmetry of each of the two triangles. This asymmetry in each Fe^{III}_3 triangle of complex 3 must lead to a S_T value for each triangle which is greater than 1/2, and presumably is equal to 5/2. The only other major difference in the topology of complex 3 as compared to complex 5 is the orientation of the μ_3-OH⁻ groups relative to the central plane of iron ions. However, even with this structural difference it is not clear why the two S = 5/2 triangles in complex 5 couple to give S = 5 for the ground state.

6. Higher Nuclearity Manganese Complexes

The preparation of molecules with large numbers of unpaired electrons is being pursued in many laboratories as a means to obtain building blocks for molecular-based magnetic materials.[27] Iwamura et al.[28] and Itoh et al.[29] have prepared interesting conjugated π organic molecules which have several unpaired electrons. A hydrocarbon consisting of five carbene linkages and a S = 5 ground state has the highest spin multiplicity for an organic molecule.[30] It is interesting to note that Caneschi et al.[31] reported that a [Mn_6^{II}(nitroxide)$_6$] complex has a S = 12 ground state.

It is possible that polynuclear metal complexes, which have relatively high-spin ground states as a result of spin frustration, may be good building blocks for molecular magnets. In the case of $Mn^{III}_4O_2$ butterfly complexes (vide supra) a S = 3 ground state results (A) because the Mn^1-Mn^3 body-body antiferromagnetic exchange interaction dominates the antiferromagnetic Mn(body)-Mn(wing tip) interactions. Thus, the S = 2 spins on the Mn^1 and Mn^2 ions almost spin pair and this frustrates the spins on the two wing-tip Mn^{III} ions in drawing A. Very recently Wang et al.[32] showed how spin

frustration can lead to appreciable numbers of unpaired electrons in a complex whose core is derived by the fusion of two Mn_4O_2 butterfly units at the wing tips (see drawing B). In Figure 4 it can be seen that the anion in the complex $(NEt_4)[Mn_7O_4(O_2CCH_3)_{10}(dbm)_4]\cdot3CH_2Cl_2\cdot2C_6H_{14}$ (Hdbm is dibenzoylmethane) does have the fused butterfly structure shown in B.

99

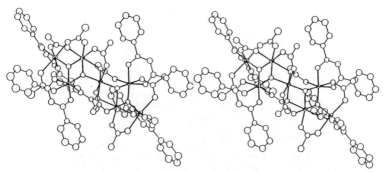

Figure 4.   Stereoview of the anion in $(NEt_4)[Mn_7O_4(O_2CCH_3)_{10}(dbm)_4]\cdot 3CH_2Cl_2\cdot 2C_6H_4$.

Wang et al.[32] studied the 5.0 - 260 K temperature dependence of the susceptibility for this $Mn^{III}_7O_4$ salt in a 10 kG field. The reduced magnetization was also determined at 50.0 kG in the 2.0 - 30.0 K range. It was concluded that this $Mn^{III}_7O_4$ complex has either a S = 3 or S = 4 ground state. This clearly establishes the presence of spin frustration in this fused butterfly complex. The antiferromagnetic couplings in the two body pairs (Mn^1-Mn^3 and Mn^5-Mn^7) probably dominate to give a resultant S = 1 (or 0) for each pair. The S = 2 spins on each of the three wing-tip ions (Mn^2, Mn^4 and Mn^6) are frustrated and cannot pair their spins totally with neighboring body manganese ions. If each body pair has S = 1, then the antiferromagnetic interactions between body pairs and wing-tip ions gives a ground state with S = 4. If the "chain" in drawing B could be extended to form a polymer, then this is a potential means of access to materials possessing a large number of unpaired electrons.

Two $Mn^{IV}_4Mn^{III}_8$ complexes have been reported. In 1980 Lis[33] reported the X-ray structure of $[Mn_{12}O_{12}(O_2CCH_3)_{16}(H_2O)_4]\cdot 2(HO_2CCH_3)\cdot 4H_2O$. A similar benzoate complex $[Mn_{12}O_{12}(O_2CPh)_{16}(H_2O)_4]$ was communicated in 1988 by Boyd et al.[34] These complexes consist of a central $Mn^{IV}_4O_4$ cubane held within a nonplanar ring of eight Mn^{III} ions by eight μ_3-oxide atoms. Overall, the structures of the acetate and benzoate complexes are similar, but not identical. A drawing of the benzoate complex is shown in Figure 5. The acetate complex has crystallographically imposed S_4 symmetry with one H_2O ligand on each of four manganese ions. In the case of the benzoate complex there are two H_2O ligands on each of two manganese ions.

In their initial communication Boyd et al.[34] reported that the benzoate Mn_{12} complex at 10.0 kG gave 12.10 μ_B for μ_{eff}/molecule at 300.8 K. As the temperature was decreased a maximum in μ_{eff}/molecule of 23.22 μ_B was observed at 10.0 K, below which μ_{eff}/molecule was seen to decrease to 18.48 μ_B at 5.0 K. With low-temperature magnetization data measured for the benzoate complex, Boyd et al.[34] concluded in the communication that this complex has a S = 14 ground state. Caneschi et al.[35] then reported that the Mn_{12} acetate complex has a S = 10 ground state in zero applied magnetic field. Since this report new susceptibility measurements have been carried out on the benzoate Mn_{12} complex.[36] It was found that because even small crystallites of the Mn_{12} benzoate complex readily torque in small external magnetic fields, the conclusions of Boyd et al.[34] are incorrect.

100

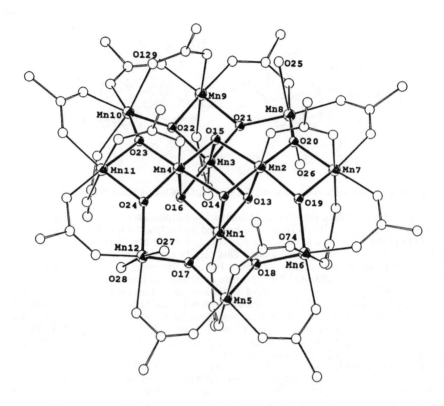

Figure 5.   ORTEP drawing of [Mn$_{12}$O$_{12}$(O$_2$CPh)$_{16}$(H$_2$O)$_4$] at 50% probability level.

Sessoli *et al.*[36] found that it was necessary to embed in parafilm a polycrystalline sample of the Mn$_{12}$ benzoate complex in order to prevent the microcrystals from torquing in an external magnetic fields. In Figure 6 are shown plots of μ_{eff}/molecule versus temperature at 10.0 kG for an unrestrained polycrystalline sample ($_O$) as well as two different (∇ and $\mathbf{\nabla}$) parafilm-embedded samples of the non-solvated Mn$_{12}$ benzoate complex. It can be seen that the maximum value of μ_{eff}/molecule for the parafilm-embedded samples is found at the smaller value of ~20.8 μ_B at ~20 K. Data are also shown ($\square$ and $\blacksquare$) in Figure 6 for a second crystalline form of the Mn$_{12}$ benzoate complex, namely the solvated form [Mn$_{12}$O$_{12}$(O$_2$CPh)$_{16}$(H$_2$O)$_4$]·PhCO$_2$H·CH$_2$Cl$_2$. This solvated form shows a maximum in μ_{eff}/molecule at ~19.8 μ_B at 15 K when embedded in parafilm. Thus, it is clear that crystallites of the Mn$_{12}$ benzoate complex exhibits considerable magnetic anisotropy. The origin of this magnetic anisotropy is not clear. Polycrystalline samples embedded in parafilm were not found to show much hysteresis.

In other words, there is no evidence of intermolecular interactions between Mn_{12} complexes in the solid state. Perhaps the magnetic anisotropy arises from the single-ion zero-field interactions of the Mn^{III} ions.

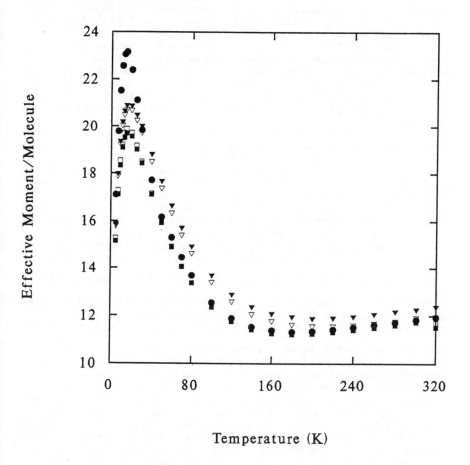

Temperature (K)

Figure 6.   Plots of effective magnetic moment per molecule versus temperature for five different samples of $[Mn_{12}O_{12}(O_2CPh)_{16}(H_2O)_4]$: (o) unrestrained polycrystalline sample; (∇ and $\blacktriangledown$) two parafilm-embedded polycrystalline samples of the non-solvated complex; ($\square$ and $\blacksquare$) two parafilm-embedded polycrystalline samples of the complex solvated with $PhCOOH \cdot CH_2Cl_2$.

The maximum seen in μ_{eff}/molecule versus temperature for both the Mn_{12} benzoate and acetate complexes likely reflects the fact that as the temperature is decreased only one (or a few) state(s) is (are) populated. Sessoli *et al.*[36] redetermined the low-temperature magnetization versus H/T characteristics of the Mn_{12} benzoate complex. Data were collected at 20.0, 30.0, 40.0 and 50.0 kG in the 2.0 - 4.0 K range. Fitting these data indicated that the non-solvated Mn_{12} benzoate complex has a S = 10 ground state, which agrees with the S = 10 ground state found by Caneschi *et al.*[35] for the Mn_{12} acetate complex.

It is probable that the S = 10 ground state observed for these Mn_{12} complexes results from spin frustration, however the detailed nature of that spin frustration has not been delineated. There are several fragments of these Mn_{12} complexes which topologically should encourage spin frustration. Obviously there are several μ_3-oxide-bridged triangular Mn_{12} fragments present. Butterfly Mn_4O_2 fragments can be identified and at the core of each complex is a $Mn^{IV}_4O_4$ cubane unit which could well exhibit spin frustration.

The AC susceptibility data measured[35,36] for the two Mn_{12} complexes are fascinating. In the AC susceptibility experiment a small external magnetic field is reversed at some set frequency. The magnetization of the sample responds to this oscillating external magnetic field. In the case of a simple paramagnet the magnetization of the sample can relax very rapidly (nanoseconds) and it can follow the oscillating external field. No out-of-phase component (imaginary part) of the AC susceptibility is expected. If the sample's magnetization relaxes (responds) slowly, then there is a non-zero imaginary component. Magnetic materials with extended magnetic exchange interactions can have domains. The kinetics associated with domain walls lead to slow relaxation of the magnetization.

The AC susceptibility of the Mn_{12} acetate complex was measured[35,36] for a polycrystalline sample in the 4 - 25 K range at frequencies of 55, 100 and 500 Hz. At all frequencies an imaginary component , χ''_M, of the AC susceptibility was detected. With the field oscillating at 500 Hz there is a maximum in χ''_M at ~6.8 K with $\chi''_M \simeq 2.7$ emu/mol. The maximum shifts to lower temperature with decreasing frequency: ~3.3 emu/mol at ~5.9 K for 100 Hz and ~3.5 emu/mol at ~5.6 K for 55 Hz. Similar observations were made for the Mn_{12} benzoate complex, where χ''_M peaks at ~5.4 K with the field oscillating at 500 Hz and at ~4.8 K with 100 Hz modulation. These AC susceptibility experiments were performed without shielding the earth's magnetic field.

Relaxation effects in zero applied field can be observed in materials with a spontaneous magnetization at the critical temperature, but in this case the temperature at which the maxima are observed are not frequency dependent. Frequency dependent maxima have been observed for superparamagnets[37] and spin glasses.[38] The former are ferromagnetic materials composed of small particles in which the magnetization competes with thermal agitation. The presence of shape anisotropy and/or magneto-crystalline anisotropy determines a barrier W to reorientation between two possible opposite easy axis directions. A relaxation time τ can be defined by Eq. 6:

$$\tau = \tau_0 \exp(-W/kT) \tag{6}$$

When $\tau \approx 1/\nu_m$, where ν_m is the operating frequency of the AC susceptometer, the sample's magnetization cannot reorient and relaxation effects become apparent. The blocking temperature is frequency dependent.

These two Mn_{12} complexes are the first two molecular complexes to exhibit a non-zero imaginary component to their AC susceptibilities. Measurements have been carried out on several other paramagnetic high nuclearity complexes, $e.g.$, $[Mn^{II}_6(nitroxide)_6]$ with a $S = 12$ ground state, and all of these complexes exhibit no imaginary component to their AC susceptibility.[39] Thus, the two Mn_{12} complexes are unique in this regard. The origin of their χ''_M responses is not known. It would be surprising to find that each $[Mn_{12}O_{12}(O_2CR)_{16}(H_2O)_4]$ complex is large enough to be a single domain. In the case of oxide magnetic materials it generally requires microcrystals with ≈ 100 metal ions in order to establish a single domain. There may be as yet uncharacterized weak intermolecular magnetic exchange interactions present in these Mn_{12} complexes. Perhaps the observed appreciable magnetic anisotropy is also important (zero-field interactions?). Heat capacity experiments would be important to see if there is a phase transition associated with the observed maximum in μ_{eff}/molecule versus temperature (Figure 6).

Very recently, Schake $et\ al.$[40] have shown that it is possible to modify the Mn_{12} complexes. In one case four of the Mn^{III} ions have been replaced by four Fe^{III} to give $[Mn_8Fe_4O_{12}(O_2CCH_3)_{16}(H_2O)_4]\cdot 2(CH_3CO_2H)\cdot 4(H_2O)$. This complex proved to be isostructural to the Mn_{12} acetate complex (tetragonal $I\bar{4}$). The four Fe^{III} ion locations are ordered and identifiable from metric parameters and the absence of Jahn-Teller distortions which are found for Mn^{III} ions. In Figure 7 is shown a plot of μ_{eff}/molecule versus temperature for this Mn_8Fe_4 complex in a 10.0 kG field. The value of μ_{eff} steadily decreases from 10.66 μ_B at 300 K to 5.29 μ_B at 9.01 K, whereupon it decreases more rapidly to 4.06 μ_B at 2.00 K. From Figure 7 it can be seen that replacing four Mn^{III} ions by four Fe^{III} ions has caused a dramatic change in the magnetic properties of this dodecanuclear complex.

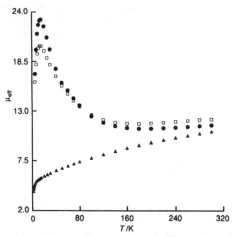

Figure 7.   Plots of effective magnetic moment per molecule versus temperature for polycrystalline samples in a 10.0 kG field for: (●) $[Mn_{12}O_{12}(O_2CPh)_{16}(H_2O)_4]$; (▲) $[Mn_8Fe_4O_{12}(O_2CCH_3)_{16}(H_2O)_4]\cdot 2(CH_3CO_2H)\cdot 4(H_2O)$; and (□) $[NPr^n_4][Mn_{12}O_{12}-(O_2CPh)_{16}(H_2O)_4]\cdot H_2O$.

Schake *et al.*[40] also reported that it was possible to prepare $(PPh_4)^+$ and $(NPr^n_4)^+$ salts of singly reduced forms of the Mn_{12} complexes. Treatment of the Mn_{12} benzoate complex in CH_2Cl_2 with one equivalent of NPr^n_4I, followed by addition of EtO_2CCH_3/Et_2O (2:1) gave $[NPr^n_4][Mn_{12}O_{12}(O_2CPh)_{16}(H_2O)_4]\cdot H_2O$. In Figure 7 it can be seen that the maximum in μ_{eff}/molecule occurs at 13.0 K for this salt and that the maximum value in μ_{eff}/molecule is appreciably less than observed for the neutral Mn_{12} complexes. The single electron added to give the Mn_{12} anionic complex likely reduces the $Mn^{IV}_4O_4$ cubane core to a $Mn^{IV}_3Mn^{III}O_4$ core. It is clear that alterations to the electron count of the Mn_{12} complexes are indeed possible, by both metal replacement and redox changes, and that this leads to major changes in the spin of the ground state.

7. Future Directions

There are, at least, three different directions that future research in this area should pursue. First, new complexes should be prepared which either alter significantly the level of spin frustration in a known topological arrangement or represent a new topological arrangement which exhibits spin frustration. A second major direction for research is to develop an understanding of the out-of-phase AC susceptibilities seen for the Mn_{12} complexes. The very recent isolation of salts of the Mn_{12} anionic complex could prove useful in this pursuit, for the bulk of the cation can be used to modulate intermolecular magnetic exchange interactions. It would also be desirable to find other high nuclearity complexes, which exhibit an out-of-phase AC susceptibility as observed for the two Mn_{12} complexes.

A third direction involves identifying approaches to connect together spin-frustrated polynuclear complexes. If the connections are made properly it would be possible to extend spin frustration in one-, two-, or three-directions. In this way the spin-frustrated polynuclear complexes can be used as molecular building blocks for new ferromagnets.

In the following proposed complex a pyridine-substituted p-dihroxybenzene dianionic binucleating ligand is suggested as the means to connect together Fe^{III}_3O triangles:

In each triangular unit the antiferromagnetic exchange interaction between the Fe_1 and Fe_2 ions should be strongest. This could frustrate the spin on the Fe_3 ion and give $S = 5/2$ per Fe^{III}_3O unit. A linear chain of $S = 5/2$ units could result. Obviously, the pyridine arms on the binucleating ligand are not unique; other chelating arms are possible.

In principal it should be possible to interconnect triangular Fe^{III}_3 or Mn^{III}_3 complexes by means of dicarboxylate bridging ligands. As shown in the following drawing this could provide a means to build a two-dimensional layer structure of triangulated, spin-frustrated complexes:

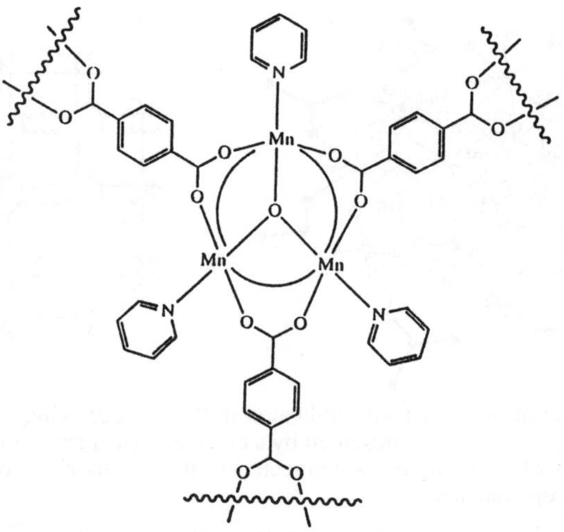

In the above proposed complex three terephthalate dianions are employed to connect one Mn^{III}_3O triangle to its neighbors. The arcs are three acetate ligands. Obviously, the above structure if it forms would not be planar. Preliminary work on a graphics workstation has indicated that the above structure could be possible. It may be necessary to use a different dicarboxylate dianion which is more flexible than the terephthalate dianion. Instead of the dicarboxylate connector the above layer structure could result by replacing the three pyridine ligands by the bidentate pyrazine ligand:

In this case all carboxylate ligands would be maintained as acetates.

The first effort to substitute carboxylate ligands by dicarboxylate ligands in solution has focused[41] on the acetate complex $[Mn^{IV}_4Mn^{III}_8O_{12}(O_2CCH_3)_{16}(H_2O)_4]$. A sample of this complex was dissolved in DMF and a DMF solution containing 0.25 equivalents of terephthalate was added. Over a period of time a solid formed. This solid is considerably less soluble in solvents which dissolve the original Mn_{12} complex. The same solid results when using different amounts of terephthalate. Chemical analysis suggests a formula where eight of the acetate ligands have been replaced by four terephthalate ligands. A composition of $[Mn_{12}O_{12}(O_2CCH_3)_8(O_2C-C_6H_4-CO_2)_4(H_2O)_4]$ does fit the chemical analysis data. Efforts are underway to grow crystals so that the X-ray structure can be determined. The powder has been screened with magnetic susceptibility and it has been found to have appreciably different properties than the monomeric complex shows. If you examine the molecular structure of the starting acetate complex shown below, a plausible structure for the terephthalate complex can be suggested. In the left drawing eight of the acetate groups are indicated with filled circles. If these are replaced by four terepthalates,

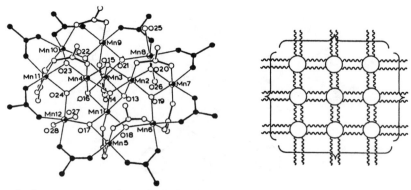

then the layer structure schematically indicated in the right drawing will result. In this drawing each Mn_{12} complex is represented by a circle and each terephthalate dianion by a wiggly line. Each Mn_{12} complex is connected to its neighboring complex in a sheet structure by two terephthalates.

8. Acknowledgments

D.N.H. is grateful for partial funding from NSF grant CHE-9115286 and National Institutes of Health grant HL13652.

9. References

1. R. D. Willett, D. Gatteschi, and O. Kahn, eds., *Magneto-Structural Correlation in Exchange-Coupled Systems* (NATO ASI Series (C140, Reidel Publishing Co., 1985).
2. J. K. McCusker, E. A. Schmitt, and D. N. Hendrickson in *Magnetic Molecular Materials*, eds. D. Gatteschi, O. Kahn, J. S. Miller and F. Palacio (NATO ASI Series E 198, Kluwer Academic Publishers, 1991) p. 297.
3. See for example: M. Tanaka, H. Iwasaki, K. Siratori, and I. Shindo, *J. Phys. Soc. Jpn.* **58** (1989) 1433 and references therein.
4. M. Mekata and K. Adachi, *J. Phys. Soc. Jpn.* **44** (1978) 806.
5. M. Mekata, Y. Ajiro, and K. Adachi, *J. Magn. Magn. Mat.* **54-57** (1986) 1267.
6. S. Fujiki, K. Shutoh, S. Inawashiro, Y. Abe, and S. Katsura, *J. Phys. Soc. Jpn.* **55** (1986) 3326.
7. M. W. Lynch, R. B. Buchanan, D. N. Hendrickson and C. G. Pierpont, *Inorg. Chem.* **20** (1981) 1038.
8. A. B. Blake, A. Yauari, W. E. Hatfield, and C. N. Sethukekshmi, *J. Chem. Soc., Dalton Trans.* (1985) 2509.
9. R. D. Cannon and R. P. White, *Prog. Inorg. Chem.* **36** (1988) 195.
10. (a) H. Güdel, *J. Chem. Phys.* **82** (1985) 2510. (b) B. S. Tsukerblat, M. I. Belinskii and B. Ya. Kuyavskaya, *Inorg. Chem.* **22** (1983) 995. (c) D. H. Jones, J. R. Sams and R. C. Thompson, *J. Chem. Phys.* **81** (1984) 440.
11. G. J. Long, W. T. Robinson, W. P. Tappmeyer and D. L. Bridges, *J. Chem. Soc., Dalton Trans.* (1973) 573.

12. S. M. Gorun, G. C. Papaefthymiou, R. B. Frankel and S. J. Lippard, *J. Am. Chem. Soc.* **109** (1987) 4244.
13. J. K. McCusker, H. G. Jang, S. Wang, G. Christou and D. N. Hendrickson, *Inorg. Chem.* **31** (1992) 1874.
14. J. B. Vincent, C. Christmas, H.-R. Chang, Q. Li, P. D. W. Boyd, J. C. Huffman, D. N. Hendrickson and G. Christou, *J. Am. Chem. Soc.* **111** (1989) 2086.
15. E. Libby, J. K. McCusker, E. A. Schmitt, K. Folting, J. C. Huffman, D. N. Hendrickson and G. Christou, *Inorg. Chem.* **30** (1991) 3486.
16. J. K. McCusker, J. B. Vincent, E. A. Schmitt, M. L. Mino, K. Shin, D. K. Coggin, P. M. Hagen, J. C. Huffman, G. Christou and D. N. Hendrickson, *J. Am. Chem. Soc.* **113** (1991) 3012.
17. W. H. Armstrong, M. E. Roth and S. J. Lippard, *J. Am. Chem. Soc.* **109** (1987) 6318.
18. D. H. Kim, R. D. Britt, M. P. Klein and K. Sauer, *J. Am. Chem. Soc.* **112** (1990) 9389.
19. G. W. Brudvig in *Metal Clusters in Proteins*, ed. L. Que, Jr. (ACS Symposium Series 372, American Chemical Society, 1988) p. 221.
20. W. F. Beck and G. W. Brudvig, *Chem. Ser.* **28A** (1988) 93.
21. J. S. Bashkin, H.-R. Chang, W. E. Streib, J. C. Huffman, D. N. Hendrickson and G. Christou, *J. Am. Chem. Soc.* **109** (1987) 6502.
22. Q. Li, J. B. Vincent, E. Libby, H.-R. Chang, J. C. Huffman, P. D. W. Boyd, G. Christou and D. N. Hendrickson, *Angew. Chem. Int. Ed. Engl.* **27** (1988) 1731.
23. This complex has been found to be isostructural with the complex in ref. 22.
24. J. K. McCusker, C. A. Christmas, P. M. Hagen, R. K. Chadha, D. F. Harvey and D. N. Hendrickson, *J. Am. Chem. Soc.* **113** (1991) 6114.
25. A. S. Batsanov, Yu. T. Struchkov, and G. A. Timko, *Koord. Khim.* **14** (1988) 266.
26. W. Micklitz and S. J. Lippard, *J. Am. Chem. Soc.* **111** (1989) 372.
27. D. Gatteschi, O. Kahn, J. S. Miller and F. Palacio, eds., *Magnetic Molecular Materials* (NATO ASI Series E198, Kluwer Academic Publishers, 1991).
28. (a) H. Iwamura, *Pure Appl. Chem.* **59** (1987) 1595. (b) H. Iwamura, *Pure Appl. Chem.* **58** (1986) 187.
29. K. Itoh, T. Takui, Y. Teki, and J. Kinoshita, *J. Mol. Elect.* **4** (1988) 181.
30. I. Fujita, Y. Teki, T. Takui, T. Kinoshita, K. Itoh, F. Miko, Y. Sawaki, H. Iwamura, A. Izuoka, and T. Sugawara. *J. Am. Chem. Soc.* **112** (1990) 4074.
31. A. Caneschi, D. Gatteschi, J. Laugier, P. Rey, R. Sessoli, and C. Zanchini, *J. Am. Chem. Soc.* **110** (1988) 2795.
32. S. Wang, H.-L. Tsai, W. E. Streib, G. Christou and D. N. Hendrickson, *J. Chem. Soc., Chem. Commun.* (1992) 677.
33. T. Lis, *Acta Cryst.* **B36** (1980) 2042.
34. P. D. W. Boyd, Q. Li, J. B. Vincent, K. Folting, H.-R. Chang, W. E. Streib, J. C. Huffman, G. Christou and D. N. Hendrickson, *J. Am. Chem. Soc.* **110** (1988) 8537.
35. A. Caneschi, D. Gatteschi, R. Sessoli, A. L. Barra, L. C. Brunel and M. Guillot, *J. Am. Chem. Soc.* **113** (1991) 5873.
36. R. Sessoli, H.-L. Tsai, A. R. Schake, S. Wang, J. B. Vincent, K. Folting, D. Gatteschi, G. Christou and D. N. Hendrickson, submitted for publication.
37. J. T. Richardson and W. D. Milligan, *Phys. Rev. B* **102** (1956) 1289.
38. S. Nagata, P. M. Keesom and H. N. Harrison, *Phys. Rev. B* **19** (1979) 1363.
39. R. Sessoli and D. Gatteschi, unpublished results.

108

40. A. R. Schake, H.-L. Tsai, N. de Vries, R. J. Webb, K. Folting, D. N. Hendrickson, and G. Christou, *J. Chem. Soc., Chem. Commun.* (1992) 181.
41. N. de Vries, H.-L. Tsai, D. N. Hendrickson and G. Christou, unpublished results.

SPIN GLASS PROPERTIES OF AMORPHOUS

INTERMETALLIC SOLIDS

Charles J. O'Connor

Department of Chemistry
University of New Orleans
New Orleans, Louisiana 70148

1. INTRODUCTION

This review will discuss several recent experiments that investigate the fascinating spin glass magnetic properties of amorphous solids. The types of materials that will be discussed in this review are amorphous intermetallic solids.

There are several methods to prepare amorphous solid materials, including techniques such as rapid thermal quenching, chemical vapor deposition, and sputtering. Recent reports from our laboratory have described the synthesis of novel molecular solids from the oxidation of Zintl anions in solution. For example, the series of solids of the formula M_2SnTe_4 is obtained from the reaction of the ternary Zintl phase material K_4SnTe_4 with transition metal halides in solution.[1-4] The series of solids $M_5(InTe_4)_2$[5,6] and $M_3(SbTe_3)_2$[7] are prepared by a similar process.

There are several reviews that describe the structure and composition of Zintl materials.[6-11] There have been relatively few reports in the literature that characterize the chemical reactivity of Zintl phases. Much of the chemistry that has been reported results from the ionic character of the Zintl phase and the reaction of Zintl anions with other salts. The reaction of solutions of these Zintl phases with transition metal cations offers a new synthetic route to novel metallic solids. The reaction consists of electron transfer from Zintl polyanions with very high chemical reactivity to transition metal cations, which results in the rapid precipitation of the neutral solid product. The metallic solids formed are amorphous and often metastable. By using this simple metathesis reaction with K_4SnTe_4, a series of ternary metal chalcogens of the formula M_2SnTe_4 (M = Cr, Mn, Fe, Co, Ni, and Cu) have been prepared and characterized. These materials exhibit some remarkable properties including resistivity that range from 10^4 to 10^{-4} ohm-cm,[12] spin glass transitions at temperatures ranging from 5 to 18 K,[1-4] a photo-magnetic effect in Fe_2SnTe_4,[13] and amorphous structure down to the 20 Å level.[14]

This review will discuss the spin glass properties of the amorphous intermetallic alloys of the general composition M:M':Te, where M is a first row transition element, M' is In, Sn, or Sb, and the stoichiometric coefficients depend on the oxidation states and electron structure of the starting materials. The physical properties of these two systems appear to be similar. Although we cannot successfully predict the spin glass freezing temperatures of a new intermetallic solid, the spin glass state has been observed in several other binary and ternary combinations of the elements [*e.g.*, MTe_2 and $M_3(AsTe_3)_2$].

In addition to the spin glass characterization, the electric properties of the semiconducting phases have been successfully characterizes as amorphous semiconductors. In the latter part of this review, the interesting photomagnetic behavior of spin glass systems is also discussed.

We will begin our presentation with a discussion of the concept of the spin glass phenomenon.

2. THE SPIN GLASS STATE

2.1 Properties of a Spin Glass

Over the past twenty years, the spin glass phenomenon has grown into a sizable area of research in solid state science.[15-26] The spin glass problem provides a testing ground for the theories that attempt to address many intriguing questions concerning amorphous structure and magnetism, the nature of the spin glass state, and the relation of the spin glass transition to critical phenomena.

There are several reviews that discuss the theory, and concept of the spin glass phenomenon,[15-26] as well as the many reports of experimental studies on spin glasses.[27-51] The spin glass phenomenon first came to prominence following reports by Mydosh and coworkers that a cusp in the ac-susceptibility of AuFe was observed at a well defined temperature.[40] These initial reports did not indicate a frequency dependence of the temperature of the cusp in the susceptibility and it was thought a phase transition to a new state had occurred.[18] Subsequent reports on other complexes showed a significant dependence of peak height and temperature on the ac frequency.

Another curiosity of spin glass systems is the marked difference between the ac and dc susceptibility. The dc susceptibility is very dependent on the manner in which the experiment is performed.[22-25] For example, a zero field cooled specimen shows much different behavior than a field cooled specimen. Even the rate of cooling of a field cooled sample has an effect on the response of the sample to a DC magnetic susceptibility measurement below the spin glass temperature. These observations are inconsistent with the usual concept of a phase transition. In addition to this curious magnetic behavior, other experiments, for example heat capacity[42-43] and conductance,[44-45] were also difficult to reconcile with a phase transition model. As a result, the term "spin glass" was coined to describe the emerging magnetic effects that had an uncanny resemblance to the behavior observed in real glass systems. Current theories are again developing that support the phase transition model, but there is still a great deal of speculation and controversy over whether an equilibrium state even exists.[17-19,29]

There are three general approaches to describing the spin glass phenomenon. The first general approach is the Neel description. This is an analogy with superparamagnetism in which domains of random moments are connected by free spins. A thermodynamic blocking equilibrium is achieved via a mechanism in which the magnetic exchange interactions have become frustrated, and a random locking of the moments occurs at T_f. The second approach to understanding the spin glass phenomenon proposes a regular lattice of spins with random degrees of coupling forces, usually a Gaussian probability of interaction strengths. The most attractive feature of this model is that it lends itself to the calculation of critical exponents. The third approach is the "window glass" analogy to the freezing of structural glasses. This model does not predict a distinct phase transition but rather a gradual change in the "viscosity" of the spins as the moments are less able to follow an external magnetic field. The major advantage of this approach is that it is easy to grasp the concept of the spin glass on an intuitive level. Although the predictions of these theories occasionally agree, the theories often contradict one another and each has serious shortcomings in its ability to explain all the facts of the spin glass state.

The spin glass state arises from a random exchange field that is experienced by the spins in the solid material. The randomness of the exchange field can be expressed as a frustration of the magnetic exchange that would be expected in a regular crystalline solid. Frustration can exist on a small scale, for example the three spins in a regular antiferromagnetic trimer experience frustration. Cluster frustration is discussed in greater detail in this volume in the review reported by Hendrickson.

There are a variety of exchange mechanism that exist. Chemists are most familiar with either direct through space exchange or superexchange mediated via a pathway composed of ligands bound to the metal. In metallic systems, an indirect exchange mechanism is also possible that is mediated by conduction electrons. This is referred to as the Rudderman-Kittel-Kasuya-Yoshida (RKKY) exchange model.[52-54] In non-crystalline transition metal solids, the exchange that is present is primarily direct exchange that will oscillate in strength and change sign as the distance between the two centers varies. With the random distribution of distances in an amorphous solid, there will be a distribution of exchanges with a variety of signs and magnitudes. Assuming a free electron model for the conduction electrons, the RKKY interaction between localized spins and the conduction electrons can be expressed as a oscillating exchange constant as illustrated in Figure 2.1.[15] For a regular crystalline solid, the net exchange parameter will be represented by one or more delta functions representing the internuclear distances. For an amorphous solid, an envelope of exchange parameters will be obtained resulting from the distribution of distances in the amorphous solid. This envelope may contain exchange constants of a variety of magnitudes and signs.

There are a variety of mechanisms that can produce the frustration of magnetic interactions that is necessary for the spin glass state to exist. Bond disorder and topological disorder, or a combination of the two phenomena are the primary causes of frustration in amorphous materials. Figure 2.2 illustrates these concepts in a two dimensional network. The random variance of the

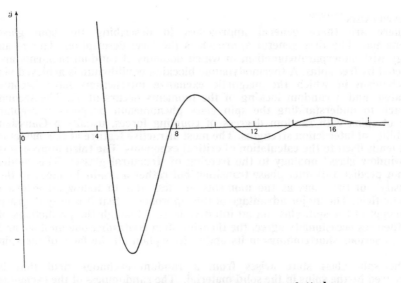

Figure 2.1 The RKKY exchange constant plotted as a function of distance.

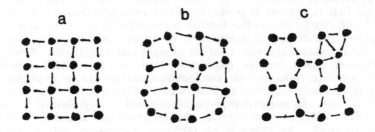

a   b   c

Figure 2.2 Frustration in a two dimensional network. a) ordered; b) bond disorder; c) topological disorder

bonding interactions in the three dimensional solid will produce the frustration that gives rise to a spin glass state. It is important to note that the appearance of the spin glass state is not restricted to amorphous materials. The frustration of exchange interaction necessary for the spin glass state may also occur as the result of impurity centers in a crystalline lattice.

2.2 Magnetic Measurements

A discussion of general magnetic susceptibility measurement and calibration techniques are reported elsewhere.[55] The magnetic properties of a spin glass are very dependent on the measurement technique employed. For many magnetic systems, ac and dc magnetic susceptibility give essentially the same result. This is because for a well behaved paramagnet, the derivative of the magnetization with respect to the magnetic field ($\chi_{ac} = \frac{\partial M}{\partial H}$) is equal to the magnetization divided by the magnetic field ($\chi_{dc} = \frac{M}{H}$). The spin glass phenomenon is characterized by some unusual behavior in the bulk magnetic properties of the materials. The electron's ability to follow a magnetic field is drastically modified by the onset of the spin glass state.

2.2.1 AC-Magnetic Susceptibility

The electron's inability to follow a magnetic field when in the spin glass state is readily apparent when ac-magnetic susceptibility measurements are made.

The inductive response of a specimen may be measured in the presence of a static magnetic field, an oscillating magnetic field, or both. If an oscillating magnetic field is present, the magnetic response depends on the frequency of oscillation. If an alternating field of sufficiently high frequency is applied, the magnetization tends to lag behind the oscillating magnetic field.

If we define the applied magnetic field as consisting of a static component H_0 and an oscillating component H_1, then we may write the magnetic field at any time t as:
$$H(t) = H_o + H_1 \cos \omega t \qquad (2.1)$$
where ω is the period of oscillation.

The resulting magnetization of a sample in the oscillating magnetic field may be written as:
$$M(t) = M_0 + M_1 \cos (\omega t - \phi) \qquad (2.2)$$
where ϕ is the phase angle by which the magnetization lags the oscillating component of the magnetic field.

We may then write
$$M(t) = \chi_o H_o + \chi' H_1 \cos (\omega t) + \chi'' H_1 \sin (\omega t) \qquad (2.3)$$
where
$$\chi_o = \frac{M_o}{H_o} \quad ; \quad \chi' = \frac{M_1 \cos(\phi)}{H_1} \quad ; \quad \chi'' = \frac{M_1 \sin(\phi)}{H_1} \qquad (2.4)$$

and χ' and χ'' depend on the frequency and magnitude of the oscillating field; χ' represents the high-frequency or in-phase component of the magnetic susceptibility, and χ'' is the paramagnetic dispersion or out-of-phase component of the magnetic susceptibility. The two components of the susceptibility may easily be sorted out by phase-sensitive detection.

2.2.2 DC-Magnetic Susceptibility

For a dc-magnetic susceptibility measurement, two different procedures are commonly used to measure the spin glass magnetic properties: zero field cooling (ZFC) and field cooling (FC). In the ZFC experiment, the sample is slowly cooled in zero field to the lowest measured temperature (well below the spin glass freezing temperature), and then the measuring magnetic field is switched on and the magnetic susceptibility is measured as temperature is raised. In the FC experiment, the measuring field is turned on at a high temperature and the sample is cooled in the magnetic field to temperatures below the spin glass freezing temperature. The magnetic susceptibility of the sample is then recorded as the temperature of the sample is raised. Figure 2.3

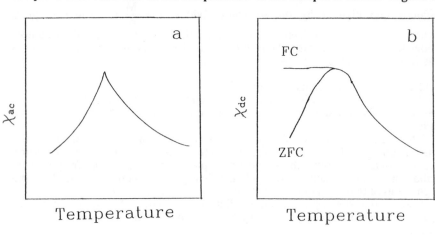

Figure 2.3 The magnetic susceptibility behavior of a spin glass measured using: a) ac techniques and b) dc techniques.

illustrates the difference in the magnetic susceptibility that is expected for ac- and dc-measurement techniques.

2.2.3 Magnetic Remanence

Perhaps the most diagnostic experiment for the characterization of the spin glass state is the analysis of the field dependence of the Isothermal Remanent Magnetization, and Thermal Remanent Magnetization. The freezing of the moments can result in a large remanent in the material.

Isothermal Remanent Magnetization (IRM): This experiment involves cooling the specimen in zero field to temperatures below the spin glass freezing point. The onset of the spin glass phase then results in a freezing of the magnetic moments in a random fashion since no external force was present for the spins to align with. The specimen is then exposed to an applied magnetic field, but the spins will tend to remain in their "frozen" random orientation. After a certain time, the magnetic field is quenched and the resultant magnetization is measured in the absence of a magnetic field. Under these conditions an ideal spin glass would show zero magnetization.

Thermal Remanent Magnetization (TRM): This experiment on the other hand exhibits vastly different magnetic behavior. In the TRM experiment, the specimen is cooled to a temperature below the spin glass freezing temperature while in an applied magnetic field. The spins, which had a tendency to align with the applied magnetic field while in the paramagnetic phase, are now frozen into a position of partial alignment with the applied field. While the specimen is in the frozen spin glass state, the magnetic field is quenched. Since the spins are frozen, their alignment is not quenched and a resultant moment is present in the material. The TRM measurement should give a large remanent relative to the IRM.

Figure 2.4 illustrates the dependence of the IRM and TRM as a function of

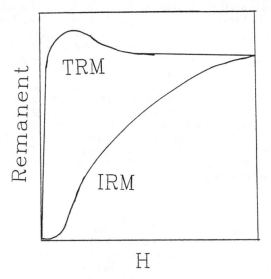

Figure 2.4 The thermal remanent magnetization (TRM) and isothermal remanent magnetization plotted as a function of magnetizing field.

magnetic field for a spin glass material. The maximum that is observed in the TRM curve is a signature of spin glass materials, although an adequate theoretical explanation for this phenomenon has not been proposed.

3. AMORPHOUS INTERMETALLIC MATERIALS

Several recent reports of the unusual magnetic behavior in amorphous intermetallic materials of relevance to this review have described the spin glass properties of M_2SnTe_4, $M_5(InTe_4)_2$, and $M_3(SbTe_3)_2$. Each of these series of compounds are prepared by the metathesis reaction between the metal chlorides or bromides and the ternary Zintl phase K_4SnTe_4,[56] K_5InTe_4,[5] or K_3SbTe_3.[57] in aqueous solution, the compounds of M_2SnTe_4,[1-4] $M_5(InTe_4)_2$,[5,6] $M_3(SbTe_3)_2$, where M = Mn, Cr, Fe, Co, Ni, are formed as a very fine and black powder.

3.1 Chemistry and Composition

The composition of the M:M':Te intermetallic materials can be illustrated by summarizing the microanalytical results of the $M_5(InTe_4)_2$ series. The elemental analysis for the solids that were prepared from the reaction of $InTe_4^{5-}$ and M^{2+} in stoichiometric ratio gave the following chemical compositions: $Cr_{2.7}InTe_{3.64}$; $Mn_{2.66}InTe_{4.03}$; $Fe_{2.86}InTe_{4.07}$; $Co_{2.31}InTe_{3.94}$ (sample A); and $Ni_{2.60}InTe_{3.99}$. Since the magnetic properties of $Co_5(InTe_4)_2$ are dependent upon preparation conditions, two samples of $Co_5(InTe_4)_2$ were prepared with Co^{2+} in two-fold excess. Sample B prepared from $InTe^{-5}$ (0.05M, 20 ml) and Co^{2+} (0.2M, 25 ml) has a composition of $Co_{2.32}InTe_{4.09}$. Sample C prepared from $InTe^{-5}$ (0.1M, 10ml) and Co^{2+} solution(0.2M, 25 ml) has a composition of $Co_{2.37}InTe_{3.97}$. Within the error of analysis, the materials can be represented by the formula corresponding to $M_5(InTe_4)_2$.

The materials resulting from rapid precipitation from solutions are amorphous. The x-ray powder diffraction patterns exhibit a lack of diffraction peaks and are consistent with a minimal amount of long range crystalline order for the freshly prepared materials. Studies on the X-ray, electron diffraction and atomic resolution transmission electron microscopy of the Ni_2SnTe_4 material did not reveal any structural order down to the 20Å level.[58] The measured magnetic and electrical properties are also consistent with those expected for an amorphous solid.

3.2 Magnetism

The intermetallic materials for each of the series M_2SnTe_4,[1-4] $M_5(InTe_4)_2$,[5,6] $M_3(SbTe_3)_2$,[7,57] $M_3(GaTe_3)_2$,[58] and MTe_2[59,60] usually show Curie-Weiss behavior at higher temperatures and non-Curie-Weiss behavior (including spin glass) at lower temperatures. The magnetic susceptibility results for $M_5(InTe_4)_2$ (M = Cr, Co, Mn, and Fe) in the temperature range of 2.5K to 300K are presented in Figure 3.1 as the plots of the reciprocal susceptibility versus temperature. The susceptibility measurements were performed at a magnetic field of 500 G on the samples that were slowly cooled at a zero magnetic field. The magnetic susceptibility of all tellurides under investigation can be interpreted on the basis of localized magnetic moments except $Ni_5(InTe_4)_2$. The high temperature magnetic data may be fit with the Curie-Weiss expression, $\chi = C/(T-\theta)$.

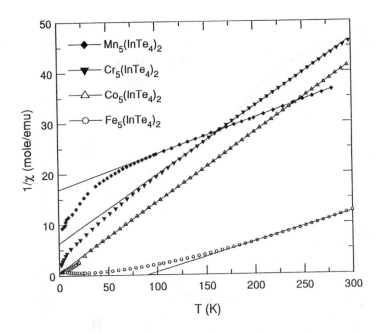

Figure 3.1 The magnetic susceptibility data of $M_5(InTe_4)_2$ (M=Cr, Mn, Co, and Fe) are plotted as $1/\chi$ vs. temperature. All data were collected at a field of 500 G on zero field coaled samples.

The parameters from least-square fits of the temperature dependence of $1/\chi$ above 100 K to Curie-Weiss equation are θ = - 235 K for $Mn_5(InTe_4)_2$, and θ = - 46.6 K for $Cr_5(InTe_4)_2$, respectively. The magnetic moment of $Mn_5(InTe_4)_2$ at room temperature is 3.60 μ_B per Mn ion, and steadily decreases to 1.0 μ_B at 2,2 K. The magnetic moment of $Cr_5(InTe_4)_2$ at room temperature is 3.2 μ_B per Cr ion and decreases rapidly below 60 K to 1.2 μ_B at 2.5 K. From the moment variation with temperature and the large negative θ values, it is apparent that the dominant magnetic exchange interaction is antiferromagnetic in these two compounds. However, the maximum in the susceptibility as a function of temperature characteristic of either long range antiferromagnetic ordering or spin glass transition has not been observed for both compounds down to temperature as low as 2.2 K.

At temperatures above 200 K, the inverse dc-magnetic susceptibility of ferrous indium telluride falls on a straight line. A fit of the high temperature magnetic data to the Curie-Weiss law $[\chi=C/(T-\theta)]$ results in a Curie constant C = 3.10emu-K/mole and the paramagnetic Curie temperature is found to be θ = +80 K. A Curie constant of 3.10 results in a g-value of g = 2.04 for iron(II) with S=2. The large positive value of θ indicates that ferromagnetic coupling between individual atomic moments occurs in this material. The magnetic

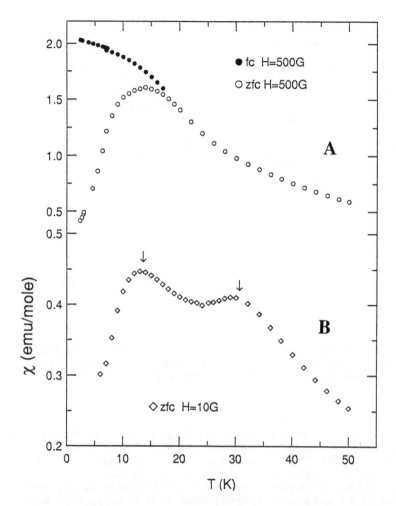

Figure 3.2 A) The dc magnetic susceptibility of $Fe_5(InTe_4)_2$ plotted as a function of temperature. Lower curve is zero field cooled and upper curve is field cooled; **B)** Low field dependence of the magnetic susceptibility of a freshly prepared sample of $Fe_5(InTe_4)_2$.

susceptibility significantly deviates from Curie-Weiss behavior below 200 K, which is attributed to the formation of clusters with large moments. In Figure 3.2, the dc-magnetic susceptibility is plotted for the temperature interval from 2 K to 50 K. The zero-field cooling dc-susceptibility rises to a maximum as the temperature is raised to 15 K, and decreases upon further increase of the temperature. The magnetic field cooled dc-susceptibility does not exhibit the same behavior in the measured temperature region. As the temperature is raised from the 2K lower limit, the field cooling dc-magnetic susceptibility decreases, albeit at a slower rate in the spin glass region. This behavior does not indicate normal ferromagnetic ordering at a temperature in accordance with the positive θ of 80 K. On the contrary, the observed dc-susceptibility of the zero-field cooled sample and of the field cooled sample suggest that this compound has a spin-glass state with a freezing temperature of about 15 K, thus antiferromagnetic interactions are present as well..

The field dependence of the dc susceptibility as a function of temperature on a zero field cooled sample has been measured to further understand the natures of the magnetic properties. The temperature dependence of dc susceptibility is shown in Figure 3.2. The temperature dependence of dc susceptibility in a low field of 10 G is very similar to the ac susceptibility result (*vide infra*). The weak maximum of susceptibility at 32 K is sensitive to a small external field and completely removed in fields > 500 G. Therefore it is likely that the hump at 32K arises from the weak ferromagnetic state which is saturated at relatively small fields. The hump at 15 K arises from the transition to a spin glass state, which was confirmed by the analysis of the field dependence of the isothermal remanent magnetization and thermal remanent magnetization.[5] As expected, the peak characteristic of spin glass is shifted to low temperatures and rounded with increasing external fields. It is apparent that $Fe_5(InTe_4)_2$ can be classified an reentrant spin glass in which exists 'a competition between ferromagnetic and antiferromagnetic interactions with 'a dominant ferromagnetism. Thus, with cooling, this compound becomes first dominated by the weak ferromagnetism at 32 K and undergoes a second transition into a spin-glass like state at 15 K.

Spin glasses are magnetic systems in which the interaction between the magnetic moments are "in conflict" with each other, due to some frozen-in disorder in the magnetic exchange fields. A spin glass state in an amorphous solid may be gradually lost due to the slow crystallization. The susceptibility on a sample sat for 6 months in a dry-box was remeasured as a function of temperature at a field of 10 G. The result is shown in Figure 3.3 along with the susceptibility measured on the fresh sample for comparison. The hump at 32 K characteristic of the para to ferromagnetism transition remains in the aged sample. However, The hump characteristic of the transition to spin glass state is shifted to 6 K as indicated by rapid decrease in the susceptibility. To confirm if this rapid decrease in the susceptibility results from antiferromagnetic ordering, the susceptibility was measured on the aged sample under zero field cooling and field cooling conditions. As shown in Figure 3.3, the susceptibility for the sample cooled in the field at low temperatures is much larger than that for the zero field cooled sample due to frozen in moments when the sample is cooled in the field. This rules out the possibility of antiferromagnetic ordering. A weaker spin glass state remains in the sample. However, the freezing temperature is shifted to a low temperature due to the growth of ferromagnetic clusters.

There are some pronounced differences between the dc- and ac-magnetic susceptibility data. The in-phase magnetic susceptibility (χ') and the out-of-phase magnetic susceptibility (χ'') are plotted together in Figure 3.4.

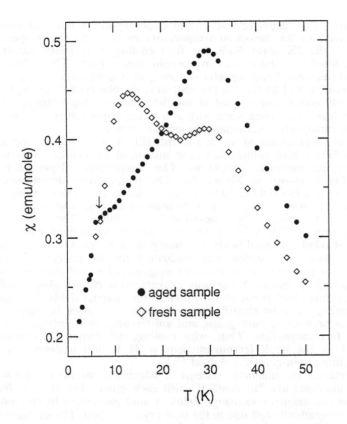

Figure 3.3 The temperature dependence of the susceptibility
collected on zero field cooled fresh and aged samples of $Fe_5(InTe_4)_2$
at a field of 10G.

The in-phase magnetic ac-susceptibility (χ') exhibit a broad maximum with two humps, one at approximately 32K and the other near 15 K. The out-of-phase component (χ'') vanishes above 40 K, but at low temperatures two maxima in accordance with the behavior of χ' are seen. The sharper peak in χ'' occurs at lower temperatures (ca. 15 K) and corresponds to the anomaly observed in the dc-magnetic measurements. The change in the frequency of the ac-experiment causes the anomalies to shift to slightly higher temperatures at higher frequencies. This effect is more pronounced at the anomaly around 15 K than at the anomaly at higher temperature. The higher temperature hump in the ac-magnetic susceptibility data likely arises from short range ferromagnetic ordering in accordance with the prediction from the high temperature Curie-Weiss fits.

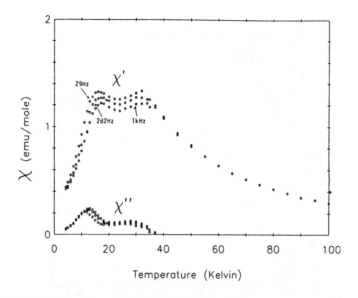

Figure 3.4 In phase (χ') and out of phase (χ'') components of the ac magnetic susceptibility plotted as a function of temperature at the frequencies indicated for $Fe_5(InTe_4)_2$.

The behavior of the χ' and χ'' data point to $Fe_5(InTe_4)_2$ being an amorphous ferromagnetic material that begins to undergo extended short range order around 35 K. This is consistent with the high temperature Curie-Weiss fit of the dc-magnetic data. At lower temperatures, a blocking of the domain wall migration sets in but is overruled by the spin-glass like anomalies around 15 K. It is important to emphasize the dc-data were measured in an applied magnetic field, while the ac-data were measured at near zero field conditions. This could account for some of the apparent discrepancy between the ac- and dc-magnetic susceptibility data.

Among the most diagnostic experiments for the characterization of the spin glass state is the analysis of the field dependence of the isothermal remanent magnetization (IRM) and thermal remanent magnetization (TRM). Both these magnetizations measured at 6.0 K for ferrous indium telluride have been illustrated on Figure 3.5. The hump at 4.0 kG in the TRM curve of this compound is characteristic of a spin-glass state and has been observed in many systems.[15-20] For a spin-glass state at weak fields the TRM measurements give large remanences relative to the IRM, while at strong fields the IRM and TRM converge to the same saturation remanence. The effect seen for the TRM is greatly enhanced by performing the experiment at lower temperatures whereas the IRM is little affected by the temperature. This is shown in Figure 3.6 where both TRM and IRM at an applied field of 3 kG are plotted as a function of temperature. The TRM and IRM curves merge near 15 K, the freezing point, which agrees with the susceptibility data at zero field

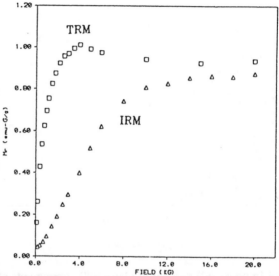

Figure 3.5 Thermal remanent magnetization and isothermal remanent magnetization plotted as a function of magnetic field at a temperature of 6 K for $Fe_5(InTe_4)_2$.

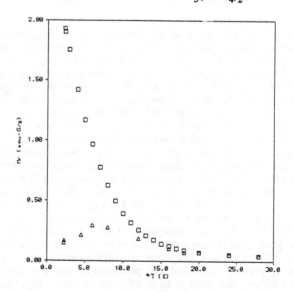

Figure 3.6 Thermal remanent magnetization and isothermal remanent magnetization plotted as a function of temperature at a magnetizing field of 3kG.

Table 3.1 Magnetic and electric parameters for amouphous intermetallic alloys.

Compound	Magnetic	$T_f(K)$	$\rho_{RT}(\Omega\text{-cm})$
Cr_2SnTe_4	SG	16	
Mn_2SnTe_4	SG	13	3×10^4
Fe_2SnTe_4	SG	12	9×10^{-4}
Co_2SnTe_4	SG*	2-5	8×10^{-4}
Ni_2SnTe_4	SG*	2-7	
$Cr_5(InTe_4)_2$			4.2×10
$Mn_5(InTe_4)_2$			3.3×10^7
$Fe_5(InTe_4)_2$	SG	15	7.0×10^{-1}
$Co_5(InTe_4)_2$	SG*	3.5	$3.7\ 3.5\times10^{-3}$
$Ni_5(InTe_4)_2$			1.3×10^{-3}
$Cr_3(SbTe_3)_2$			2.23×10
$Mn_3(SbTe_3)_2$			1.11
$Fe_3(SbTe_3)_2$	SG	4.4	2.57×10^{-3}
$Co_3(SbTe_3)_2$	SG	4.2	2.75×10^{-4}
$Ni_3(SbTe_3)_2$	SG	3.8	5.04×10^{-4}
$CrTe_2$	SG	18	7.5×10^{-1}
$MnTe_2$			7.6×10^{-1}
$FeTe_2$	SG	5	5.7×10^{-3}
$CoTe_2$	SG	6	3.22×10^{-4}
$NiTe_2$			2.00×10^{-4}

SG = spin glass behavior. Non-spin glass materials exhibit paramagnetism with deviations from Curie-Weiss behavior
* denotes transition temperature dependence on preparation conditions

124

cooling and field cooling. A summary of magnetic results for several series of intermetallic materials is illustrated in Table 1.

3.3 Conductivity

The magnitudes of specific resistivity at room temperature for $M_5(InTe_4)_2$ vary considerably with the transition metal and have a value of 3.0×10^7 ohm-cm for $Mn_5(InTe_4)_2$ and 1.3×10^{-3} for $Ni_5(InTe_4)_2$. We have observed that the trend of room temperature specific resistivity appears to parallel that of the standard reduction potential of the divalent transition metal halides that were

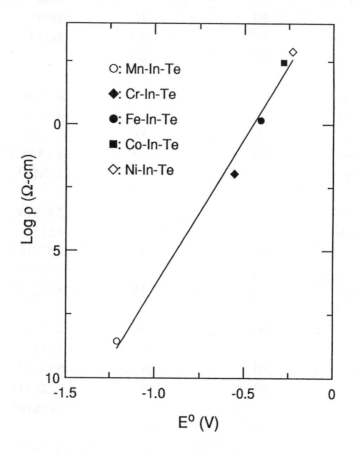

Figure 3.7 The room temperature resistivity of $M_5(InTe_4)_2$ plotted against the standard reduction potential of the $M:M^{+2}$ couple.

used to prepare the amorphous solids. Figure 3.7 illustrates the log of the conductivities for $M_5(InTe_4)_2$ plotted as a function of the standard reduction potential of the precursor divalent transition metal ion. An excellent linear correlation is observed for all five transition metal cations under investigation.

The temperature dependences of the resistivity in the temperature region of 20 to 320 K for Fe, Co, and Ni analogues are shown in Figure 3.8. The Fe and

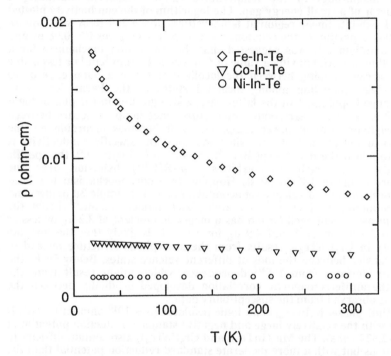

Figure 3.8 Plots of specific resistivity as a function of temperature
for the complexes $M_5(InTe_4)_2$ (M=Fe, Co, and Ni).

Co analogues exhibit a negative temperature coefficient ($d\rho/dT$) throughout the temperature range investigated, indicating semiconductivity. The Ni analogue displays nearly temperature independent resistivity which is at the boundary expected for metallic behavior.

The specific resistivity of $Mn_5(InTe_4)_2$ is significantly larger than that of other complexes in this series, for example, six orders larger than that of $Cr_5(InTe_4)_2$. A linear-least square fit of conductivity data to the relationship, $\sigma = C \exp(-E/k_BT)$ gives the activation energies for conduction, E = 0.51 ev in the temperature region 328-268K and E = 0.20 ev in the region 268-210K. The change in the activation energy may be due to the structural phase change.

Below 210 K, the specific conductivity is essentially temperature independent. This electronic behavior is consistent with that expected for a semiconductor with p- or n- type impurity centers. The onset of temperature independent behavior may then be attributed to a transition from intrinsic to extrinsic semiconductor behavior.

The specific conductivity of $Cr_5(InTe_4)_2$ varies with a range of over 4 orders of magnitudes in the temperature range measured, indicating a semiconductor of a small energy gap. The logarithm of the conductivity plotted against 1/T exhibits linear region at temperatures above 200 K. A least square fit of the linear portion to the relation, $\sigma = C \exp(-E/kT)$, gives E = 0.08 ev and C = 0.23 ohm^{-1}cm^{-1}. It was suggested that the conduction mechanism for a non-crystalline material having a value of C of order 10 ohm^{-1}cm^{-1} or less is due either to carrier hopping between the localized states at a band edge or to phonon-assisted tunneling among localized states at the Fermi level (i.e., variable range hopping).[61] In the latter case, a straight line in the plot of log σ versus 1/T is expected near room temperature since hopping occurs between nearest neighbors and at lower temperatures it becomes favorable for the carriers to tunnel to more distant sites and thus, the specific conductivity is expected to follow the $\sigma = A \exp(-B/T^{1/4})$. As shown in Figure 3.9, the specific conductivity of $Cr_5(InTe_4)_2$ exhibits $\exp(-B/T^{-1/4})$ behavior over the temperature range of 180 to 50 K. Hopping transport mechanism has been observed in many mixed valence semiconductors like the simple 3d oxides, and glasses containing 3d ions,[62,63] and amorphous semiconductor.[64] From the magnetic measurement each Cr ion has a magnetic moment of 3.2 μ_B or less as opposed to 3.8 μ_B for Cr^{3+} and 4.9 μ_B for Cr^{2+}. It is likely that the hopping conductivity in $Cr_5(InTe_4)_2$ results from the amorphous structure instead of electron transfer between the ions of different valence states. Below 50 K, the conductivity deviates from $1/T^{1/4}$ dependence, which might result from the effect of the antiferromagnetic correlation developed gradually between the moments as observed from the susceptibility data.

The Zintl phase, K_5InTe_4, is an ionic insulator ($\rho > 10^{10}$ ohm-cm). This is consistent with the relatively large and negative standard reduction potential of K^+ (E^0 = -2.924 volts). The $Mn_5(InTe_4)_2$ and $Cr_5(InTe_4)_2$ also contain difficult to reduce cations but with a more moderate standard reduction potential than the potassium ion. The $Mn_5(InTe_4)_2$ and $Cr_5(InTe_4)_2$ exhibit a highly temperature dependent semiconductivity. The $Ni_5(InTe_4)_2$ is formed from a more easily reduced metal ion and shows substantially increased conductivity within the boundary expected for metallic behavior; however, an inspection of the data plotted in Figure 3.8 reveals that the temperature dependent slope (dρ/dT) begins to level for the more easily reduced cations, but nevertheless remains positive, indicating that the material is formally a semiconductor. Therefore, it appears that the magnitude of resistivity for the materials of M_5InTe_4 shows a correlation with the standard reduction potential of the parent divalent transition metal ion. This correlation likely arises from the degree of the transfer of electrons from the Zintl polyanion to cation when the product materials are formed.

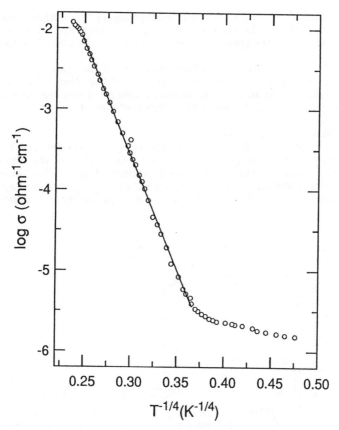

Figure 3.9 The temperature dependence of the conductivity of
$Cr_5(InTe_4)_2$ plotted as $\log\sigma$ versus $1/T^{1/4}$.

4. PHOTO-MAGNETISM OF SPIN GLASS SYSTEMS

The physical properties of a spin glass material make it a good candidate to exhibit unusual photomagnetic effects. In previous reports, we have described the measurement of the photomagnetic response of some spin glass systems. We have observed the photo-generation of magnetic bubbles in the amorphous intermetallic material Fe_2SnTe_4.[66] Ferre and coworkers have also reported a related photomagnetic phenomenon that results from irradiation of frozen spin glasses consisting of metal doped aluminosilicate insulating spin glasses.[67-71] These experiments involve a pulse of radiation that is delivered to the frozen spin glass and the subsequent magnetic response monitored with either

Faraday rotation[67-71] or a SQUID susceptometer.[66]   Other reports of photomagnetic effects observed in spin glass systems deal primarily with radiational heating of the spin glass state resulting in a "melting" of the frozen spins.[72,73]

In this section, the magnetic characterization of the amorphous material, $Ni_3(SbTe_3)_2$, which forms from the reaction of the Zintl phase material, K_3SbTe_3, with $NiBr_2$ in an aqueous solution is described and the observation of a photomagnetic effect in the frozen spin glass is discussed.

4.1 Photo-Magnetic Apparatus

The photo magnetic data are recorded using an experimental technique based on the STEPS technique[74] that permits illumination of the sample with high intensity radiation while very precise magnetic data is being recorded on the SQUID susceptometer. A schematic diagram of the experimental apparatus is illustrated in Figure 4.1. The radiation is supplied by a Lumonics Inc.

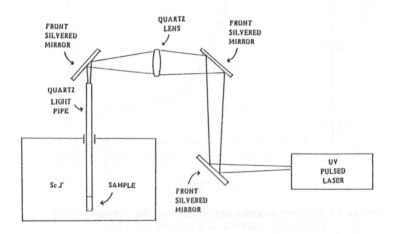

Figure 4.1 Schematic diagram of the experimental apparatus for the irradiation of the spin glass samples with a uv laser.

Eximer-500 laser operating with $He-N_2$ exiting gas (wavelength = 335 nm) and a pulse rate of 10 pulses per second (9 nsec per pulse). A high (optical) purity quartz supracil light pipe with cross section area of 16 mm^2 delivered the radiation to the specimen in the sample chamber of the superconducting susceptometer. Approximately 100mg of $Ni_3(SbTe_3)_2$ was packed into a small quartz tube (5mm i.d.) that was attached to the end of the quartz light pipe. The photomagnetic response of the $Ni_3(SbTe_3)_2$ spin glass was measured at 2.2K and a static magnetic field of 1.0 kG following zero field cooling.

4.2 Magnetic Characterization

The magnetic susceptibility data of $Ni_3(SbTe_3)_2$ were measured over the 2.2 to 60 K temperature region. The measured data are illustrated in Figure 4.2 as

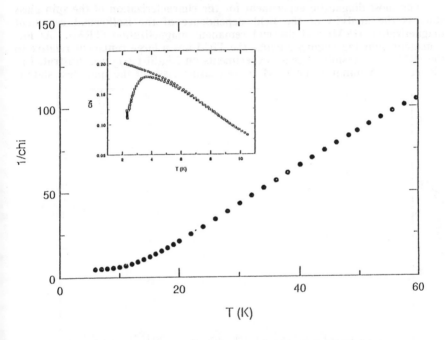

Figure 4.2 Plot of the inverse susceptibility of $Ni_2(SbTe_3)_2$ as a function of temperature. (inset) Plot of the magnetic susceptibility as a function of temperature measured at a field of 1 kG (open circle - ZFC, open square - FC).

inverse susceptibility plotted as a function of temperature. At temperatures above 10K, the inverse magnetic susceptibility data exhibit linear behavior consistent with the Curie Weiss equation [χ = C/(T-θ)]. The data were fit to this equation to give the paramagnetic Curie temperature, θ = 9.9K and the

Curie constant C = 0.47 emu·K/mol. The positive value of θ indicates the presence of ferromagnetic interactions between the unpaired electrons on the nickel atoms.

At lower temperatures the $Ni_3(SbTe_3)_2$ sample exhibits anomalous temperature and condition dependent magnetic behavior. The inset of Figure 4.2 illustrates plots of the temperature dependent magnetic susceptibility of $Ni_3(SbTe_3)_2$ for field cooled and zero field cooled experiments. The maximum that is observed in the zero field cooled magnetic susceptibility data shown in the inset of Figure 4.2 is strong evidence that $Ni_3(SbTe_3)_2$ exists in a spin glass state with a freezing temperature of about 4K.

The most diagnostic experiment for the characterization of the spin glass state is the analysis of the field dependence of the isothermal remanent magnetization (IRM) and thermal remanent magnetization (TRM). At low remanent inducing magnetic fields, the TRM has a large remanent relative to the IRM. The results of these experiments on $Ni_3(SbTe_3)_2$ are illustrated in Figure 4.3. A hump in the TRM curve characteristic of the spin glass state is

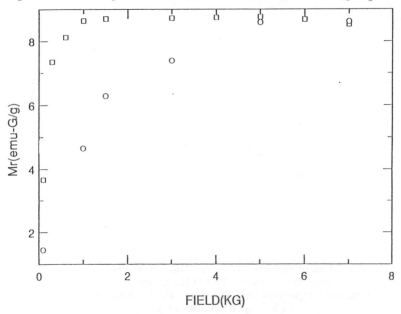

Figure 4.3 The thermal remanent magnetization and isothermal remanent magnetization measured at 2.2 K and plotted as a function of remanent inducing magnetic field.
(open square - TRM, open circle - IRM).

apparent in the field dependent plot of the TRM data in Figure 4.3. Both IRM and TRM data curves converge to the same saturation remanent at higher magnetic fields (*ca.* 7.0 kG) because the increasing importance of short range correlations destroy the spin glass state.

The effects seen for the differences between the IRM and TRM data are greatly enhanced by performing the experiment at low temperatures. As the temperature is decreased the TRM tends to increase in value whereas the IRM is much less affected by the temperature. The TRM and IRM measured as a function of temperature for $Ni_3(SbTe_3)_2$ are illustrated in Figure 4.4. The

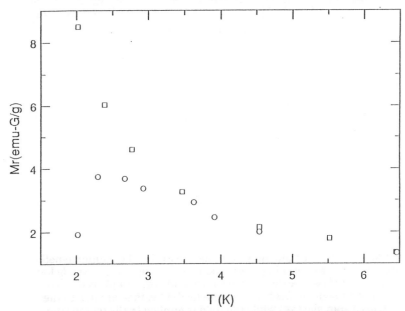

Figure 4.4 The thermal remanent magnetization and isothermal remanent magnetization measured with a magnetization inducing field of 1.0 kG and plotted as a function of temperature. (open square - TRM, open circle - IRM).

TRM and IRM curves merge near 4K, the spin glass freezing point, which also corresponds to the maximum in the susceptibility data for zero field cooling sample. Above the freezing point, the remanent decreases to zero.

The results of resistivity measurements using the van der Pauw four-probe method show that the material exhibits metallic behavior at room temperature, with a specific resistivity of $\rho=5.04\times10^{-4}$ Ω-cm.

4.3 Photo-Magnetism

The unusual magnetic properties of spin glasses make them excellent candidates for an experiment that uses light to generate magnetic bubbles in the material. A process for generating magnetic bubbles in a spin glass material is illustrated schematically in Figure 4.5. The procedure for this

132

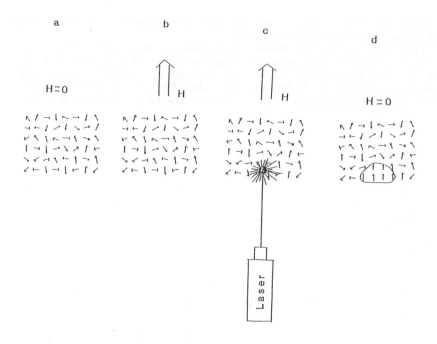

Figure 4.5 A schematic diagram that illustrates the photomagnetic experiment: a) the spin glass is cooled to a temperature well below the spin glass freezing temperature while the sample is in zero applied magnetic field; b) a magnetic field is then applied to the frozen spin glass; c) while the field is applied to the frozen spin glass, the sample is exposed to uv-radiation; d) a domain of magnetization is induced in the frozen spin glass material.

experiment involves cooling the spin glass material to temperatures well below the spin glass freezing temperature while in the sample is not exposed to an applied magnetic field (H = 0kG). This results in the freezing of a random orientation of the spins in the spin glass material. The magnetic field is then applied to the frozen spin glass. Because of the nature of the spin glass state, the magnetization will show a slight increase to a value represented by the zero field cooled magnetic susceptibility data. While the field is applied to the frozen spin glass specimen, the sample is exposed to radiation. The radiation is of sufficient intensity to cause local disruption of the spin glass state and a realignment of the spin glass spins to the direction of the applied magnetic field. The result of this radiation will be the generation of a spin glass moment within the domain walls determined by the boundary of the pulse. If the magnetic field is turned off, the domain of induced magnetization will remain as a remanent domain. In other words, the creation of a magnetic bubble in the material.

In our experiment, the uv laser radiation is used to generate magnetic bubbles on the surface of a material. A schematic diagram of the experimental apparatus is illustrated in Figure 4.1. First, the spin glass material is placed in the superinduction susceptometer (Scχ) at the end of a quartz light pipe. The sample is then cooled to a temperature well below the spin glass temperature ($T_f = 4K$) in zero applied magnetic field. This results in the random freezing of the spins in the spin glass specimen. The magnetic field is then applied to the sample. For a 2 second period, A series of laser pulses (10 pulses per sec, 9 nsec per pulse) are focused through a quartz light pipe to the frozen spin glass. This uv radiation causes a local disruption of the spin glass state and results in the generation of a region of increased magnetization in the spin glass within the

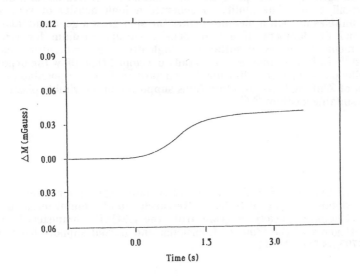

Figure 4.6 The SQUID magnetometer output for a sample of Ni$_2$(SbTe$_2$)$_2$ at 2.2 K plotted as a function of time before, during, and after 30 pulses of uv radiation over a 2 second period.

domain wall determined by the boundary of the radiation delivered to the sample. The effect of this experiment on the magnetization is to cause an increase in the remanent magnetic moment of the sample. Following the pulse, there is an increase in the magnetization to a new steady state value that is larger than the original value. A plot of the SQUID magnetometer output as a function of time before, during and after the frozen spin glass is exposed to the radiation is illustrated in Figure 4.6.

The magnetic data that is obtained from the SQUID magnetometer cannot verify the presence of sharp domain walls, or whether the bubbles that are created are stationary or can easily diffuse through the specimen. However, recent reports on the photomagnetic properties of metal doped aluminosilicate spin glasses show the same effect occurs in insulating amorphous spin glasses.[71,75] By analogy with the observed properties of the manganese doped aluminosilicate spin glass when irradiated in a SQUID magnetometer,[75] and in comparison with the data from the present study, we conclude that the bubbles that are generated in the $Ni_3(SbTe_3)_2$ spin glass** are stationary, at least on the time scales of our experimental investigation.

The magnetic bubbles created by this technique are apparently stationary, and the homogeneity of the spin glass phase allows the generation of a domain of very small area. The ability to generate a high density of very small magnetic bubbles in such a medium makes the spin glass a good candidate for the eventual development of a high density storage medium for erasable magnetic memory devices, if sufficiently high ordering temperatures can be discovered.[76] In fact techniques are already developed that allow the deposit of fine metallic powders onto a disk for this purpose.[77] It is also possible that the deposition of Zintl adducts onto thin films supported on polyimide plastics will provide a suitable medium.[78,79]

ACKNOWLEDGMENTS: C.J.O. wishes to acknowledge support from a grant from the donors of the Petroleum Research Fund administered by the American Chemical Society, a grant from the LEQSF administered by the Board of Regents of the State of Louisiana. and partial support from grant number 0793/88 from NATO.

REFERENCES

1. R.C. Haushalter, C.J. O'Connor, J.P. Haushalter, A.M. Umarji and G.K. Shenoy, *Angewante Chemie 97* (1983) 147.
2. R.C. Haushalter, C.J. O'Connor, A.M. Umarji, G.K. Shenoy, and C.K. Saw, *Solid State Commun. 49* (1084) 929.

3. C.J. O'Connor, J.W. Foise, and R.C. Haushalter, *Proc. Natl. Acad. Sci., Ind. Chem. Sci. 58* (1987) 69.
4. C.J. O'Connor, J.W. Foise, and R.C. Haushalter, *Solid State Commun. 53* (1987) 349.
5. J.H. Zhang, A.J. vanDuyneveldt, J.A. Mydosh, and C.J. O'Connor, *Chem. Mater. 1* (1089) 04-406.
6. J. H. Zhang, B. Wu, and C. J. O'Connor, *Chem. Mater.*, in press.
7. J.-S. Jung, L. Ren, and C. J. O'Connor, *Inorg Chim Acta*, in press.
8. H. Schäfer, *Ann. Rev. Mater. Sci.* 15 (1985) 1.
9. H. Schäfer, B. Eisenmann and W. Muller, *Angew. Chem., Internat. Ed. 12* (1973) 694, and references therein.
10. H.G. von Schnering, *Bol. Soc. Quim. 33* (1988) 41.
11. J.D. Corbett, *Chem. Rev. 85* (1985) 383, and references therein.
12. R.C. Haushalter, D.P. Goshorn, M.G. Sewchok, C.B. Roxlo, *Mater. Res. Bull. 22* (1987) 761.
13. C. J. O'Connor, and J.F. Noonan, *J. Chem. Phys. Solid 48* (1987) 69.
14. J.W. Foise and C.J. O'Connor, *Inorg. Chim. Acta, 162* (1989) 5.
15. K. Moorjani and J. M. D. Coey, *"Magnetic Glasses"*, (Elsevier, 1985).
16. J.A. Mydosh, *J. Mag. and Mag. Mat.*, 7 (1978) 237.
17. S.F. Edwards and P.W. Anderson, *J. Phys. F, Metal Phys.*, 5 (1975) 965.
18. K. H. Fisher, *Phys. Status Solidi B.*, 116 (1983) 353.
19. K. Binder and A.P. Young, *Rev. Mod. Phys.*, 58 (1986) 801.
20. J, Durand, *Glassy metals, Magnetic CHemical and Structural Properties* (Ed., R. Hasegawa, CRC Press, Boca Raton, Florida, 1983) 108.
21. J. J. Rhyne, *Magnetic Pkase Transitions* (Springer Series in Solid State Sciences, vol. 48, Ed. Mausloos and R. J. Elliott, Springer Verlag, Berlin, 1983) 241.
22. J. Durand, *Glassy Metals: Magnetic, Chemial, and Structural Properties* (Ed. R. Hasegawa, CRC Press, Boca Raton, Florida, 1983) 108.
23. J.J. Rhyne, *Magnetic Phase Transitions* (Springer Series in Solid State Sciences, Vol 48,Ed. Mausloos and R.J. Elliott, Springer Verlag, Berlin, 1983) 241.
24. A. R. Frerchmin, S. Kobe, *Amorphous Magnetism and Metallic Magnetic Materials Digest* (Selected Topics in Solid State Physics, Vol. XVII, Ed. E. P. Wolfarth, North Holland, Amsterdam, 1983) 55.
25. S.F. Edwards and P.W. Anderson, *J. Phys. F., Metal Phys.*, 6, (1976) 1927.
26. P.W. Anderson in *"Amorphous Magnetism II"* (Plenum Press, N.Y., 1977) 1.
27. G. Heber, *J. Appl. Phys.*, 10, (1976) 101.
28. K.H. Fisher, *Physica, 86-88B*, (1977) 813.
29. J. Souletie, *Ann. de Physique, 10* (1985) 69.
30. K. Binder, *Adv. Solid St. Phys.* (1977) 55.
31. J.L. Tholence and R. Tournier, *J. Phys. (Paris) 35* (1974) 229.
32. J.L. Tholence and R. Tournier, *Physica 86-88B* (1977) 873.
33. J.O. Thomson and J.R. Thomson, *J.Phys. F 11* (1981) 247.
34. K.H. Fisher, *J. Mag. Magn. Mater. 15-18* (1980).
35. A.F.J. Morgownik and J.A. Mydosh, *Physica 107B+C*, (1981) 305.
36. A.F.J. Morgownik, J.A. Mydosh and L.E. Wenger, *Appl. Phys. 53* (1982) 2211.
37. A.F.J. Morgownik, and J.A. Mydosh *Phys. Rev. B. 24* (1981) 5277.

38. K.V. Rao, M. H.Fahnle, E. Figureroa, O. Beckman, and L. Hendman, *Phys. Rev. B. 27* (1983) 3104.
39. Ali, N. and Woods, S.B., *Soild St. Commun. 45* (1983) 471.
40. V. Cannella, J. A. Mydosh, and J. I. Budnick, J. *Appl. Phys. 42* (1971) 1689..
41. C. N. Guy, *J. Appl. Phys. 50* (1979) 7308, and references there in.
42. L. E. Wenger and P. H. Keesom, *Phys. Rev. B 11* (1975) 3497.
43. L. E. Wenger and P. H. Keesom, *Phys. Rev. B 13* (1976) 4053.
44. P. J. Ford and J. A. Mydosh, *Phys. Rev. B 21* (1980) 1902.
45. O. Laborde and P. Radakrishna, *J. Phys. F 3* (1973) 1731.
46. R. R.Chianelli, *Int. Rev. Phys. Chem. 2* (1982) 127.
47. D. Sherrington *AIP Conference Proceedings 29* (1976) 224
48. R. R. Chianelli and M. B. Dines, *Inorg. Chem. 17* (1978) 418.
49. E. Prouzet, G. Ourvard, R. Brec and P. Seguineau, *Solid State Ionics 31* (1988) 79.
50. P. J. Ford, *Contemp. Phys, 23* (1982) 141;
51. S. Kobe, A. R. Fierchmin, H. Nose, F Stobiecki, Supplement A of53, *J. Magn. Mat. 60* (1986) 1.
52. M. A. Rudderman and C. Kittel, *Phys. Rev. 96* (1954) 99.
53. T. Kasuya, *Prog. Theor. Phys. 15* (1956) 45.
54. K. Yoshida, *Phys Rev. 106* (1957) 893.
55. C. J. O'Connor, *Prog. Inorg. Chem. 29* (1982) 202.
56. J. C. Huffman, J. P. Haushalter, A. M. Umarji, G. K. Shenoy, and R. C. Haushalter, *Inorg. Chem. 23* (1984)312.
57. J.-S. Jung, E. D. Stevens, and C. J. O'Connor, J. *Solid State Chem. 94* (1991) 362.
58. J.W. Foise and C.J. O'Connor, *Inorg. Chim. Acta, 162* (1989) 5.
59. J. Zhang, B. Wu, C. J. O'Connor, and W. B. Simmons, *J. Appl. Phys.*, submitted.
60. J. H. Zhang, B. Wu, and C. J. O'Connor, *Chem. Mater.*, in press.
61. N.F. Mott and E.A. Davis, *Electronic Processes in non-Cryst. Materials* (Oxford 2nd ed., 1979).
62. A. R. West, *Solid State Chemistry and Its Applications* (John Wiley & Sons, 1984) 516.
63. A. Ghosh, *J. Phys.: Condens. Matter 1* (1989) 7819.
64. D. Adler, *Amorphous Semiconductor* (CRC press, 1970).
65. R.C. Haushalter, D.P. Goshorn, M.G. Sewchok, C.B. Roxlo, *Mater. Res. Bull. 22* (1987) 761.
66. C. J. O'Connor and J. F. Noonan, *J. Chem. Phys. Solids 48* (1987) 69.
67. M. Ayadi and J. Ferre, *J. Magn. Magn. Mafs. 54-57* (1986) 91.
68. M. Ayadi and J. Ferre, *International Conference on Magnetism* (an Franciso, 1985, Abstract # 4Pc16).
69. M. Ayadi and J. Ferre, *Phys. Rev. Lett. 50* (1983) 274.
70. M. Ayadi and J. Ferre, *J. Appl. Phys. 55* (1984) 1720.
71. 5. M. Ayadi and J. Ferre, *Phys. Rev. 44* (1991) 10079.
72. S.S. Dindun and E.A. Raitman, *Sov. Phys. Solid State 20* (1978) 1108.
73. K. Kovalenko and I.I. Kondilenko, *Sov. Phys. JETP 47* (1978) 386.
74. C. J. O'Connor, E. Sinn, T.J. Bucelot, and B.S. Deaver, *Chem. Phys. Lett. 74* (1980) 27.

75. L. Ren, J.-S. Jung, J. Ferre, and C. J. O'Connor, manuscript in preparation.
76. G.A.N. Connell, *J. Mag. Magnet. Mats. 54-57* (1986) 1561.
77. H. Fujiwara, K. Mizushima, A. Sakemoto, A. Naganawa, and T. Doi, *J. Mag. Magnet. Mats. 54-57* (1986) 1567.
78. R.C. Haushalter and L. J. Krauss, *Thin Solid Films 102* (1983) 2312;
79. R. C. Haushalter, *Angew. Chemie 95* (1983) 560.

THE MAGNETIC PHASE DIAGRAM OF HIGH-T$_c$ SUPERCONDUCTING OXIDES

L. Krusin-Elbaum

IBM Research, Yorktown Heights NY 10598-0218

1. INTRODUCTION

Among the many exciting and controversial aspects of high temperature superconductivity are the new features discovered in the magnetic phase diagram. The remarkable phenomena, such as the unusual "irreversibility line", the possibly melted vortex liquid above that line, the nearly logarithmic time relaxation of magnetic properties below it, and the collective behavior of vortices in the presence of disorder are still a subject of several competing theories. Many of the magnetic properties are very different from those of conventional superconductors, generating interesting new physical concepts and having an important impact on applications as well. Since the magnetic measurements are so convenient for characterizing superconducting oxides, because they are non-contact and thus avoid the problem of electrical connection through surface oxide layers, the amount of high-T$_c$ magnetic literature over the last few years has been indeed overwhelming. In this brief review the focus will be on some of the recent developments in "macroscopics" and primarily on vortex statics and dynamics, rather than on "microscopics", such as diamagnetic fluctuations, NMR, neutron spectroscopy of local moments or high temperature susceptibility[1]. We also will avoid the discussion of rare earth and transition metal doping with the resulting richness of low-temperature magnetic phases[1]. We will mostly consider here the bulk crystalline material YBa$_2$Cu$_3$O$_7$ (YBaCuO), although Bi$_2$Sr$_2$Ca$_1$Cu$_2$O$_x$ (BiSrCaCuO) and some thin film results will be mentioned.

1.1 Basic Characteristics of Superconductivity

The two fundamental signatures of superconductivity are vanishing electrical resistance below a certain critical temperature T$_c$, and the expulsion of magnetic flux[2] as the superconductor is cooled through T$_c$. This *perfect diamagnetism*, known as Meissner effect[2], is reversible in pure samples and can not be explained by perfect conductivity; it implies that superconductivity disappears beyond a

140

certain critical magnetic field H_c. H_c is related to the energy needed to keep the field out against the magnetic pressure. Empirically, the thermodynamic critical field is well approximated by a parabolic temperature dependence $H_c = H_c(0)[1 - (T/T_c)^2]$. The zero-field transition at T_c into the superconducting state is second order, but in the presence of the field the transition is first order with an associated latent heat[2,3].

1.2 Type II Superconductors

Below H_c magnetic field does not plummet to zero at the normal-superconducting interface, but decays exponentially with the characteristic penetration depth $\lambda = [mc^2/4\pi n_s e^2]^{1/2}$. Here c is the velocity of light, n_s is the density of superconducting electrons (or holes), m is their band mass and e is the electron charge. The magnetic field is screened from the interior of the superconductor by the *supercurrents* flowing within λ. The above is true for so called type I superconductors.

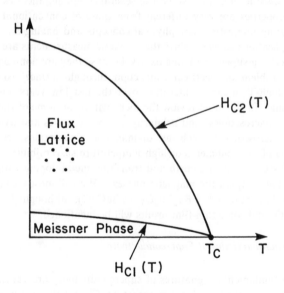

Fig. 1. Phase diagram of a type II superconductor as a function of temperature T and magnetic field H. Effects of thermal fluctuations and pinning defects are ignored.

There is another large class of superconductors, so called type II, in which superconductivity is not destroyed when magnetic field in excess of the lower critical field H_{c1} penetrates the interior. Instead, a *mixed state* is formed in which magnetic flux enters in the form of quantized vortices each containing the flux quantum $\Phi_0 = hc/2e$. The charge is 2e, because in the superconducting state, the electrons are bound into "Cooper pairs"[4], which are described by a coherent quantum-mechanical wavefunction $\psi = |\psi| e^{i\phi}$. In the center of the vortex $|\psi|$ is suppressed. This suppression is on the scale of another basic characteristic superconducting length scale, the coherence length ξ, which defines the core of the vortex. The extent of the vortex is given by λ, which is the scale over which magnetic field varies in the superconductor. The vortices, in the absence of thermal fluctuations or the disorder caused by the defects in the material, self-organize in a regular (hexagonal) lattice (Abrikosov lattice) and in the conventional low-temperature superconductors this line lattice was believed to exist at essentially all temperatures up to the upper critical field H_{c2}; beyond H_{c2} superconductivity was destroyed[2,4]. The distinction between type I and type II superconductors evolves naturally from the phenomenological mean-field theory of Ginzburg and Landau[4], where the important parameter is $\kappa = \lambda/\xi$; the boundary between the two types of superconductors is at $\kappa = 1/\sqrt{2}$.

1.3 Defects, Vortices, and Thermal Fluctuations

In the ideal (pure) conventional superconductor in the Abrikosov phase the vortex lattice is free to slide in response to a current $\mathbf{J}$. The current exerts a force on the vortex line in the direction perpendicular to the current, $\mathbf{F} = \mathbf{J} \times \mathbf{\Phi}_0$, and if the vortices move with the velocity $\mathbf{v}$, there is a finite electric field $\mathbf{E} = \mathbf{J} \times \mathbf{v}$ and hence a non-zero resistance. *Thus, a pure type II material in the mixed phase is not a superconductor in the sense of zero-resistance.*

The presence of defects, if they can pin vortices should make type II material a *true* superconductor. Still the vortices can very slowly creep by thermal activation across the pinning energy barriers which leads to a finite linear (ohmic) resistivity with a thermally activated (Arrhenius) form[5]: $\rho_1 \sim \exp(-U/k_B T)$. For conventional superconductors $U/k_B T$ is large and ρ_1 is practically immeasurable (except very close to H_{c2}). As we will discuss later the situation is very different for high-T_c oxides.

Effective pinning by defects establishes a critical flux gradient in the interior of the superconductor which can be related to a bulk *critical current density* J_c via Maxwell's equation $\nabla \times \mathbf{B} = 4\pi \mathbf{J}/c$. The resulting strongly irreversible magnetic behavior has been interpreted in the framework of the Bean or critical-state

142

model[6], which predicts a more or less rectangular hysteresis loop M(H). As we will see later, there are important new features in M(H) in high-T_c materials which are not predicted by the Bean phenomenology and which have to do with the collective pinning of vortices and large thermal fluctuations. Unlike in the low-T_c materials, thermal fluctuations play a major role in the copper oxides, since the temperatures are high and the energies needed to nucleate vortices and move them around are low. The relevant energy scale is that required to create a segment of the vortex line whose length is one coherence length ξ[Ref.7]. This energy varies as ξ/λ^2, and it is much lower in the high-T_c superconducting oxides, where ξ is very short.

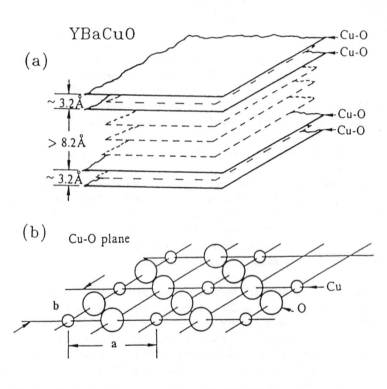

Fig. 2. (a) A schematic representation of the YBaCuO layered structure. Yittrium atoms reside on a sparsely occupied plane between two immediately adjacent Cu-O planes as indicated by a light set of dashes. The three other metal-O planes are shown by heavier dashes. (b) An infinite plane of Cu-O atoms emphasizing the square planar bonding. Superconductivity is thought to reside on Cu-O planes.

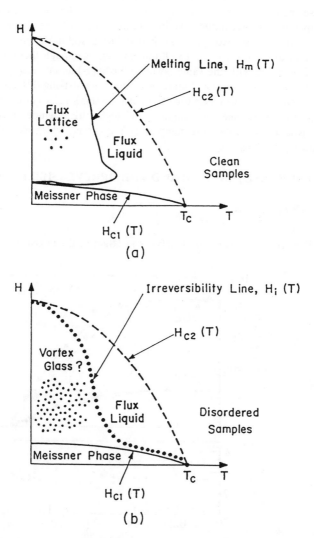

Fig. 3. Phase diagrams for (a) clean and (b) highly disordered high temperature superconductor. The dashed line indicates that H_{c2} is probably not a sharp phase transition.

144

The copper oxide superconductors are extreme type II, with κ values of nearly 100 and in this discussion we will focus on the unusual mixed state properties of these materials. Indeed today, nearly 7 years after the discovery of the high temperature superconductivity in ceramic LaBaCuO by Bednorz and Müller[8], the nature of the mixed phase in these oxides is still in the midst of a raging controversy and certainly is not well understood. Although these materials seem to be describable with the same s-wave Ginzburg-Landau theory as was used by Abrikosov for conventional materials[4], there are striking differences arising from their layered structure (see Fig. 2), unusually small values of interplanar coupling constants, short coherence length ξ, and high critical temperatures. The result of this is a considerably more complex H-T diagram, which we will discuss in the following sections.

2. H-T PHASE DIAGRAM AND THE "IRREVERSIBILITY LINE"

2.1 Critical Fields

The critical fields not only define the extent of the mixed phase, but also very

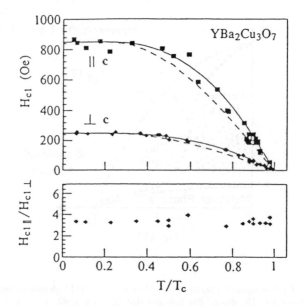

Fig. 4. H_{c1} for H parallel and perpendicular to the c-axis of single crystal YBaCuO (after Ref. 9). Similar experimental results were reported in Ref. 10.

importantly identify the fundamental superconducting length scales: the anisotropic coherence length ξ from the upper critical fields H_{c2} and the anisotropic (and temperature dependent) penetration depth λ from the lower critical fields H_{c1}. In YBaCuO, when the field is along the c-axis $H_{c1} \simeq \Phi_0 \ln \kappa / 4\pi\lambda^2$ is below 1000 Oe; it is 3 - 5 times larger than when the field is in the a-b plane[9,10] (see Fig. 4). The deduced values of $\lambda_{ab} \simeq 1400$Å and $\lambda_c \simeq 4500 - 7000$Å are large and in agreement with the direct measurements of λ[11,12]. In BiSrCaCuO the reported anisotropy ratio $\lambda_c / \lambda_{ab} \sim 15$ is even larger[13]. The $H_{c2} \simeq \Phi_0 / 2\pi\xi^2$ has a mean-field linear behavior near T_c with the slopes dH_{c2}/dT of about 2 and 10 Tesla/K for H∥c and H⊥c respectively[14], implying low-T coherence lengths $\xi_{ab} \simeq 16$Å and $\xi_c \simeq 3$Å. Such small values of ξ compared to λ confirm the strongly type II limit for the high-T_c oxides.

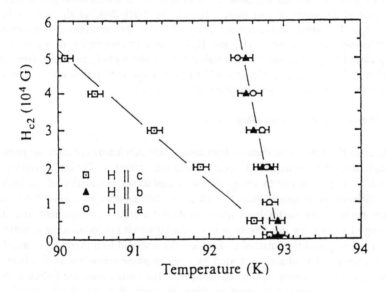

Fig. 5. H_{c2} for H parallel to the a, b and c-axis of single crystal YBaCuO (after Ref. 14). The anisotropy in a-b plane is very small.

2.2 Melting of the Vortex Lattice

A phase diagram for a pure material in an applied field H perpendicular to the copper-oxygen planes (parallel to the c-axis) which was recently proposed by Nelson and co-workers[15,16] is shown in Fig. 3a. In contrast to the conventional superconductor, the Abrikosov lattice occupies now only a portion of the mixed phase. When at high enough temperatures thermal fluctuations displace segments of vortices over distances comparable to the vortex lattice spacing $a_o = \sqrt{\Phi_o/B}$, the lattice is expected to melt (Lindemann criterion of melting is when the displacement reaches $\sim 0.1a_o^{[Ref.15]}$). The possibility of a melting transition *near* H_{c2} has been suggested by Labusch[17] long ago, but for a large melted regime to exist where the vortices can *meander, entangle, and cross*, an extraordinary combination of high T, small ξ and large anisotropy must be present[16]. Nelson et al. [15,16] argued that in these materials thermal fluctuations are being felt far away from the critical regime (i.e. from H_{c2}) and lead to a new *entangled liquid phase* in a magnetic field, due to flux line wandering as vortex filaments traverse the sample. The entropy effects also destabilize the flux lattice relative to an entangled vortex liquid sufficiently close to H_{c1}, where the distances between the vortex lines are of the order of λ or more. Hence, the "back-flow" of the melting line near H_{c1}. For most conventional superconductors, the reentrant melting curve is undistinguishably close to H_{c1} and H_{c2}. The very tiny region of melted flux fluid close to H_{c2} may not be observable even in the high-T_c materials (see discussion in Section 2.7). The mean-field transition line at H_{c2} need not be a sharp transition when fluctuations are taken into account[7].

2.3 Defects and the Vortex-Glass Phase

Even in the absence of thermal fluctuations, the Abrikosov lattice is not robust enough to survive the imperfections present in real materials. The *arbitrarily weak* disorder has been shown to *always* destroy translational correlation of the lattice at sufficiently large length scales[18]. There are two competing effects: the interaction between the vortex lines which promotes a lattice arrangement and the pinning of vortices by the defects in the material which randomizes the structure. Fisher[7,19] proposed that instead of a lattice, there should be a thermodynamically distinct *vortex-glass* phase, separated by a *sharp* phase transition at T_{vg} from a high temperature vortex liquid. In this phase the vortex lines are frozen into a sample-specific compromise configuration insuring that the linear resistivity $\rho_l \equiv \lim_{J \to 0} \{E/J\} = 0$. If the transition at T_{vg} is second order (i.e. continuous), one should observe a critical scaling behavior controlled by a characteristic length scale ξ_{vg} over which the phase ϕ of the superconducting pair wavefunction is correlated. ξ_{vg} will diverge near T_{vg} as a power-law, $|T - T_{vg}|^{-\nu}$, with some characteristic exponent ν. The linear resistivity is predicted to vanish as $\rho_l \sim (T - T_{vg})^s$, when the transition is approached from the high temperature side, with another critical exponent s related to ν. Moreover, Fisher's scaling argument predicts that the functional dependence of the electric field E on the current density J will be independent on temperature just above T_{vg}, provided that the dissipation $\rho(J) \equiv E/J$ is normalized by the linear resistivity ρ_l and the current density is normalized by J_1, a characteristic current density at which the vortex-liquid state is significantly altered[19]. This leads to a power-law relationship, $E \sim J^{(z+1)/2}$, between voltage and current, where z is the critical exponent governing the relaxation of fluctuations in the critical regime. The confirmation of the phenomenological scaling approach of the vortex-glass theory was first reported by Koch et al.[20] (see Fig. 6) in their remarkable fit to resistivity data in thin films of YBaCuO near the transition and later by Worthington et al. in polycrystalline samples[21].

2.4 Irreversibility Line

In addition to transport measurements[20,21], the boundary between vortex-liquid and solid phases has been probed early on with d.c. magnetic measurements on ceramic LaBaCaO by Bednorz and Müller[22] and later observed in d.c. and a.c. susceptibility[23-26], vibrating reed[27] experiments, and torque magnetometry[28]. With the diversity of experimental techniques and theoretical concepts it has been assigned a plethora of names: the depinning line, the melting line, the vortex-glass transition line, and the "irreversibility line". Here we will use the term "irreversibility line" as the most model-neutral. It is not at all clear if the irreversibility line in disordered systems (Fig. 3b) is related to melting in pure systems[28,29]. And it is still controversial whether this is a true phase transition (from a vortex liquid to a vortex solid, be it crystalline or glassy), or just a dynamic crossover (i.e. thermal relaxation becomes so fast that the system reaches equilibrium on the time scale of the experiment).

Empirically, the irreversibility line is an upward curving line which essentially delineates the onset of magnetic irreversibility at lower temperatures. It is well described by a power law

$$H_{irr} = A(1 - T/T_c)^\alpha. \tag{1}$$

There are differences between various experiments in the values of amplitude A as well as in the exponent α; values of α between 4/3 and 2 have been reported[20-28]. The irreversibility line and its nearly logarithmic frequency dependence was quite elegantly related by Yeshurun and Malozemoff[23] to a very large time relaxation of magnetization which they argued was due to the thermal activation of vortices as described by Anderson and Kim[5]. In the Anderson-Kim model, thermal activation of vortices over an energy barrier U is modulated by a Lorentz force on the vortices proportional to the transport current density J,

148

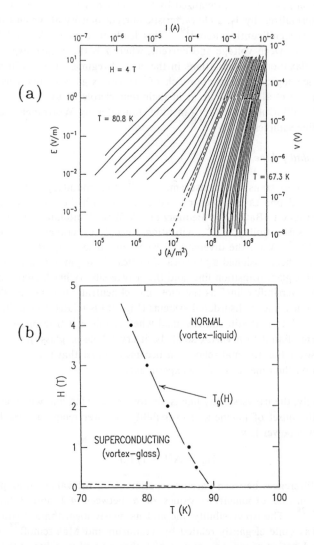

Fig. 6. (a) E-J (voltage-current) curves for a YBaCuO c-axis oriented film in a 4 Tesla magnetic field applied along the c-axis. The scale is logarithmic to cover many decades of voltage and current. The E vs. J isotherms are taken at 0.3 K intervals and the transition into the vortex-glass phase at $T_g \simeq 77.4$ K (where $E \sim J^{(z+1)/2}$ with $z \simeq 4.8 \pm 0.2$) is indicated by the dashed line. (b) The equilibrium boundary $T_g(H)$ between the vortex-glass and vortex-liquid phases (after Ref. 19).

which equals the critical current density J_c in the critical state. The hopping is governed by the Arrhenius equation with an attempt frequency ω_0 ($\simeq 10^{10}$Hz):

$$\omega = \omega_0 \exp\left[\frac{-U}{k_B T}\left(1 - \frac{J}{J_c}\right)\right], \tag{2}$$

which can be solved for J_c:

$$J_c = J_{c0}\left[1 - \left(\frac{k_B T}{U}\right)\ln\frac{\omega_0}{\omega}\right] = J_{c0}\left[1 - \left(\frac{k_B T}{U}\right)\ln\frac{t}{t_0}\right]. \tag{3}$$

The values of U are small; in YBaCuO U $\simeq$ 0.2eV for H$\perp$c and $\simeq$ 0.02eV for H$\|$c[Ref.23]. It is obvious from Eq. 3, that the decay of J in time (and thus of

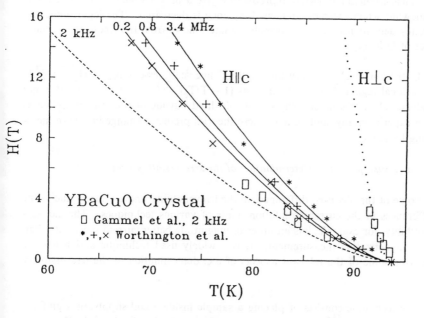

Fig. 7. Irreversibility or melting lines for YBaCuO crystals. The data shown as dots for H$\perp$c and as stars, crosses and x's for H$\|$c are from the a.c. suscepti-bility data of Worthington et al.[25]. For H$\perp$c the frequency dependence is weak, but for H$\|$c the frequency dependence is strong as is seen in data at 0.2, 0.8, and 3.4 MHz. The dashed line is an extrapolation of the data to 2 kHz using flux creep formalism[23]. The open squares are data from Gammel et al.[27] obtained from measurements with a 2 kHz mechanical oscillator. The curves were shifted to adjust for differences in T_c.

magnetization) is logarithmic and large, since the U values are an order of magnitude or two smaller than in low-temperature superconductors, while the temperatures are an order of magnitude higher.

The simple scaling argument for U goes as follows: if U is proportional to the magnetic condensation energy $H_c^2/8\pi$ times a characteristic excitation volume, for sufficiently large fields that volume could be limited laterally by the area which a single flux quantum occupies in the flux line lattice. This area is roughly $a_0^2 = \Phi_0/B$. Along the applied field, the smallest extent of the activation volume is ξ, hence $U \propto (H_c^2/8\pi)a_0^2\xi$. According to Ginzburg-Landau theory $H_c \propto (1 - T/T_c)$ near H_{c2} and ξ scales as $(1 - T/T_c)^{-1/2}$. This gives $U \propto (1 - T/T_c)^{3/2}/B$ or $\alpha = 3/2$. Indeed, $\alpha = 3/2$ is consistently seen in d.c. magnetization experiments, while a.c. measurements at 1 MHz reproducibly give α of 4/3 which seems to increase somewhat when the measurement frequency is decreased. The logarithmic frequency dependence of the irreversibility line follows from the frequency dependence of U in Eq. 2.

In the vortex-glass theory the "melting" line scales essentially as $H_{c2} \propto 1/\xi^2$ in the critical regime[19]. There ξ scales as $(1 - T/T_c)^{-2/3}$, giving $\alpha = 4/3$. It is not understood in this model why 4/3 should hold outside of the critical regime as seen experimentally, and none of these theories predict a change in α with measuring frequency.

2.5 a.c. Susceptibility - Determination of the Irreversibility Line

Some of the observed differences in the irreversibility line are surely due to the differences in the current resolution limits of various techniques. The a.c. susceptibility, for example, has an order of magnitude better current sensitivity than d.c. magnetization measurement. It is a widely used technique and below we describe some of the relevant aspects of a.c. determination of the irreversibility line.

The technique consists of placing a sample inside a coil supplying a uniform a.c. field $h_{ac} = h_0 e^{i\omega t}$, which may be superimposed with the d.c. field H_{dc}. In a normal metal or in a superconductor without pinning characterized by a linear resistivity ρ, the a.c. field will induce eddy currents which will decay in a characteristic length known as the skin-depth $\delta_s = (c^2\rho/2\pi\omega)^{1/2}$. The real or in-phase component of the a.c. susceptibility χ' measures the amount of shielding, and the imaginary or out-of-phase component χ'' measures the energy loss due to the induced currents. χ'' exhibits a maximum when the skin-depth δ_s equals the thickness[30] of the sample. This resistive absorption is a *linear* effect and so δ_s

does not depend on the amplitude of the ac field h_0.[31] In a superconductor with pinning, E-J (I-V) curves are nonlinear in some range of fields and temperatures. The condition "penetration depth $\simeq$ thickness" may now occur[31,32] in the linear or nonlinear portion of the E-J curves depending on J, ω and sample dimensions. In the fully developed critical state[6], Maxwell's law implies that the magnetic induction B will decay linearly from the surface, not exponentially as in the ohmic case. The a.c. field will penetrate to a depth $D_c = \frac{c}{4\pi} h_0/J_c$, beyond which the a.c. field is totally screened. Thus in the nonlinear regime, the penetration depth D_c (and χ) will depend on h_0.

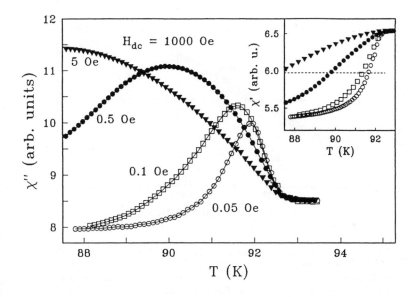

Fig. 8. Out-of-phase component of the a.c. susceptibility χ'' measured in a single crystal of YBaCuO at 1MHz in 1000 Oe dc field for several amplitudes of h_0. Inset: Corresponding χ' curves. Half-screening is indicated by the dashed line.

Fig. 8 shows χ'' for several values of h_0. For a 30μm thick crystal, at $\omega = 1$MHz and small excitation amplitudes h_0 in the range of 0.05 - 0.1 Oe, the position of the maximum roughly separates regions of linear and nonlinear amplitude response. For a crystal of different dimensions, the frequency must be adjusted in order to have the transition into a nonlinear response to be at the maximum of χ''.

152

We *define* the irreversibility line at the *onset of nonlinearity* which, of course, is a result of *pinning*[33]. Fig. 8 shows that the maximum in χ'' roughly coincides with half-screening as determined from χ', shown in the inset. For the 30 μm crystal of Fig. 8, this condition corresponds to a current density of $\simeq$ 20 A/cm². At ω = 1 MHz, the peak in χ'' occurs at a resistivity of $3.6 \times 10^{-7}\Omega$cm. Transport measurements on a similar crystal[21] confirm that under these conditions and for this crystal, the maximum in χ'' will occur very near the temperature where nonlinear behavior is first noted in E-J curves (see Fig. 9).

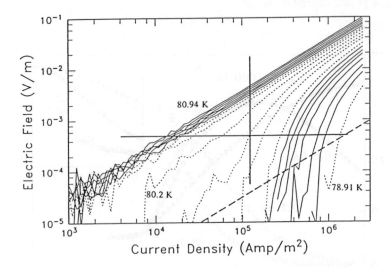

Fig. 9. Nineteen E-J curves measured at 7 Tesla at 0.1 K intervals from 80.94 to 78.91 K. Solid horizontal and vertical lines indicate values of E and J respectively for a typical crystal at half-screening in a.c. susceptibility measurements at ω = 0.5 MHz and h_0 = 0.05 Oe. Dashed inclined line corresponds to ρ_1 of $3 \times 10^{-10}\Omega$/m (from Ref. 21).

Irreversibility lines for $YBa_2Cu_3O_7$ crystal[34] measured at 1 MHz with h_0 = 0.1 Oe and in d.c. fields up to 6 Tesla are shown in Fig. 10(a) for the two extreme alignments of the crystal with respect to the applied d.c. field.

2.6. Sudden Irreversibility Collapse at Low Fields

So far we have discussed the a.c. response in high d.c. fields. Now we will describe a remarkable behavior we have recently discovered in low d.c. fields,

153

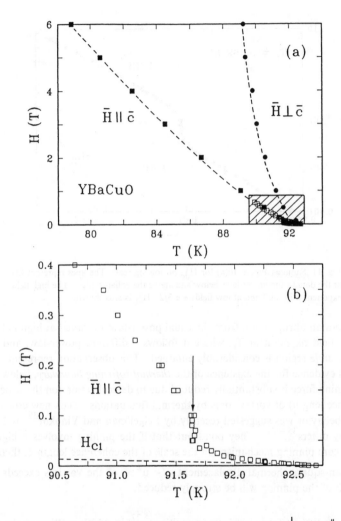

Fig. 10. (a) Irreversibility lines for YBaCuO crystal of Fig. 1 for $H_{dc}^{\perp}$ and $H_{dc}^{\parallel}$ to the c-axis measured at 1MHz with $h_0 = 0.1$ Oe. (b) Details of the irreversibility line at low fields for the $H_{dc}^{\parallel}$ orientation (the shaded region in (a)). $H_{c1}^{\parallel}$ is shown for comparison.

154

namely a *sudden collapse of the irreversibility line*[34] near T_c above the lower critical field H_{c1}. Observed in a.c. susceptibility measurements, the irreversibility line

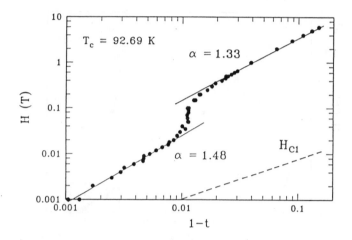

Fig. 11. Same as Figure 10(a) for $H_{dc}^{\parallel}$ on log-log scale. The lines represent fits of the data to the power-laws below and above the collapse edge. The high field exponent α is $\simeq 4/3$ and at low fields $\alpha \simeq 3/2$. $H_{c1}^{\parallel}$ is also shown.

undergoes an abrupt jump from the usual power-law observed at high fields into a lower field regime near T_c, where it follows a different power-law, and where the *reversible* region is considerably enlarged. This observation represents experimental evidence for the existence of the *thermal softening boundary*, below which the pinning force is substantially reduced due to delocalization (on the scale of the coherence length) of vortex cores by thermal fluctuations. Such new crossover in vortex behavior was suggested recently by Feigel'man and Vinokur[35], and elaborated by others[35–37]. They point out that if the pinning involves a highly localized core pinning mechanism on the scale of the coherence length ξ, then when the mean-square thermal displacement $\sqrt{<u^2>}$ of the vortices exceeds ξ, the strength of the pinning will be strongly reduced.

This is distinct from Lindemann melting[16,27], in which melting of the vortex lattice is expected when $\sqrt{<u^2>}$ becomes larger than some fraction of the distance between vortices, and it leads to a qualitatively different prediction, namely of a *thermal softening boundary* which *crosses* the usual irreversibility line. Indeed the prediction of the Feigel'man and Vinokur theory [35] is that this new boundary *increases* in temperature with increasing field. It is important to recognize that this effect coexists with the other line or lines, that is, there can be an irreversibility line above and below the thermal softening boundary. If, however, the position of the irreversibility line depends on the pinning strength (as it does in the thermal-activation and vortex glass models but not in the lattice melting model), then one might expect the irreversibility line to be shifted to lower fields below thermal softening boundary; i.e. an anomaly where the two lines cross.

In Fig. 10 (b) we show in detail the data below 0.6 Tesla and near T_c for dc field parallel to c-axis ($H_{dc}^{\parallel}$), the region shaded in Fig. 10 (a). A sharp (and reproducible on cycling) step in $H_{irr}(T)$ initiates at $H_{dc}^{\parallel} \simeq 0.1$ Tesla and completes, within $\Delta T \sim 0.020$ K, at $H_{dc}^{\parallel} \simeq 200$ Oe. The collapse is even more dramatic when displayed in a log-log plot of Fig. 11. The high-field power-law is described here by the exponent $\alpha = 1.33 \pm 0.05$ ($\sim 4/3$), while below the step (or edge) the power-law exponent is 1.48 ± 0.08 ($\sim 3/2$). This new transition occurs in the same field range in both twinned and fully untwinned crystals having T_c between 92.5 and 93.5 K, very narrow superconducting transitions with ΔT_c of ≤ 400 mK and 100% diamagnetic shielding. The α values are consistently 1.2 to 1.4 at high fields and somewhat larger, between 1.5 and 2, at lower fields.

The idea of Feigel'man and Vinokur[35], is essentially, that if the pinning centers are of the atomic origin (i.e. oxygen vacancies) vortex cores become too large (delocalized on the scale of ξ) for the pinning wells to be fully effective. The critical current J_c is shown to be strongly reduced[35,36] when the thermal displacement of the vortex core $<u^2> \geq (1.4\xi)^2$. In an anisotropic superconductor, the harmonic thermal fluctuation of the vortex line at high temperatures is given (within a logarithmic factor) by[36]:

$$<u^2> \simeq \frac{4\pi^2 k_B T \lambda^2 \sqrt{\Gamma}}{\Phi_0^{3/2} \sqrt{B}}, \tag{4}$$

where Γ is the anisotropy ($\sqrt{\Gamma} = \lambda_c/\lambda_{ab}$). Equating RHS of Eq. 4 to $(1.4\xi)^2$, a parabolic ($\sim T^2$) thermal softening boundary is obtained:

$$B \simeq \frac{16\pi^4 \kappa^4 \Gamma}{1.96 \Phi_0^3} (k_B T)^2. \tag{5}$$

156

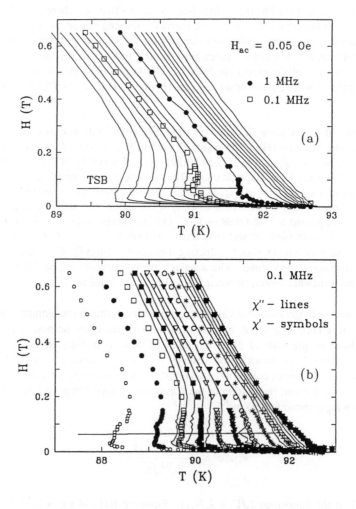

Fig. 12. (a) Lines obtained from the maximum in $\chi''(T,H)$ at 1 MHz (irreversibility line shown in solid dots) and 0.1 MHz (open squares). The low field anomaly is present in both and is consistent with the thermal softening boundary (TSB) B $\propto$ T^2 (see Eq. 5), indicated as a solid line; TSB crosses the irreversibility line. The contours probing nonlinear regime below $H_{irr}(T)$ also show the anomaly, suggesting a change in pinning behavior (after Ref. 34). (b) Contours obtained from the analysis of χ' at ω = 0.1 MHz.

This boundary *crosses* the usual upward-curving irreversibility line. We propose that the collapse edge articulates where the irreversibility and thermal softening lines meet. From Eq. 5 we estimate the magnetic field at which the crossing should occur is $B \simeq 700$ Oe[34], in reasonable agreement with the middle of the observed anomaly.

2.7 Contour Analysis of χ'' - Lines of Constant J_c

As we argued earlier, the position of the maximum in χ'', $\chi''_{max}(T,H)$, is a reasonable description of $H_{irr}(T)$ line for a 30 μm thick crystal at 1 MHz. At lower frequencies we are probing further into nonlinear regime, and if our argument holds, the anomaly we observe in $H_{irr}(T)$ should also appear there. This is clearly evident in Fig. 12 where we also show the line obtained from χ''_{max} measured at 0.1 MHz. Another way to examine the nonlinear regime is by projecting the *entire* peak onto the temperature axis (i.e. the temperature locations of, for example, 90%, 80%, 70% etc. of the peak height on both sides of the maximum) and tracking it as a function of H_{dc}. The contours in H-T plane obtained by such procedure are distinctly different above and below the irreversibility line (Fig. 12(a)). Above, in the linear regime, they can be interpreted as lines of constant linear resistivity. Our anomaly appears at the irreversibility line and propagates into nonlinear regime towards lower temperatures, convincingly indicating the thermal softening boundary (TSB) which crosses the usual irreversibility line. The 0.1 MHz line, of course, coincides with one of the contours in the nonlinear regime, which all show a reentrant behavior below TSB. Similar analysis of χ' shows an identical contour behavior (see Fig. 12(b)).

An alternative explanation is suggested by a possible superfluid entangled vortex liquid ground state[15,16]; indeed, as we have discussed in Section 2.2, a reentrant boundary between entangled and lattice phases is predicted[15] near H_{c1} (see Fig. 3a). In YBaCuO, the estimate of the field range above H_{c1} where the entropic effects due to flux line wondering are dominant is $(H - H_{c1})/H_{c1} \sim 1/\ln\kappa \sim 0.25$, two orders of magnitude smaller than we observe. Thus, unless pinning (which is not included in this theory) modifies such estimate, we argue against it.

Finally, we note that the applied dc field at which the collapse occurs is one-to-two orders of magnitude above the lower critical field $H_{c1}^{\parallel}$. In our crystals, H_{c1} shows a simple linear behavior (with the slope ~ 10 Oe/K[Ref.10]) extrapolating to T_c, as shown in Fig. 10(b) and 4. At these fields the flux lattice spacing is smaller than magnetic penetration depth λ and hence, it is unlikely that this new effect is due to the change in the nature of the vortex-vortex interaction (i.e. from logarithmic at higher fields to exponential near H_{c1}).

3. VORTEX SOLID PHASE - CRITICAL CURRENTS

The behavior of critical currents in type II superconductors is most directly probed in transport measurements. Such measurements, however, are an exceedingly difficult task in high temperature superconductors, because of the many serious problems in electrically contacting these materials through surface oxide layers, especially in small and delicate single crystals. Magnetic measurements, since they are non-contact, are convenient. But to extract the information about the critical currents J_c one pays the price of a more complex analysis and a reliance on a model relating magnetization to J_c. This is commonly done using Bean critical state model[6], which explains many features of the high-field magnetic hysteresis loops. The idea here, as we have mentioned in the Introduction, is that in a *homogenuous* superconductor with defects, the hysteretic magnetization is related to J_c via an Ampere's law $dB/dx = (4\pi/c)J_c$, where B is the locally averaged magnetic induction inside the superconductor.

In such description, the width of the magnetic hysteresis ΔM can be used to derive J_c values and in a slab geometry of thickness d and field applied along the plane of the slab, $J_c \simeq 40\Delta M/d$[Ref.2,6]. The d.c. M(H) measurements also allow determination of the onset of reversible magnetization, and when performed at various temperatures give an estimate of the irreversibility line, $H_{irr}(T)$. It is not widely appreciated, however, that by varying the available parameters in the a.c. technique, such as frequency and the amplitude of the a.c. excitation one can examine the regime not only near and at the irreversibility line, but also of J_c in the fully developed critical state well below the irreversibility line.

3.1 a.c. Response - Determination of J_c

The complicated E-J behavior below the irreversibility line causes a rather complex behavior[31] of χ'', where the high temperature portion of the peak is a measure of the resistivity in the linear regime, and the lower temperature portion a measure of the development of *nonlinear* E-J character.

When the frequency of the a.c. excitation is lowered, the nonlinear E-J behavior should be reflected in the a.c. response. For $\omega = 100$ kHz, χ'' exhibits a maximum well into the nonlinear regime. This is shown in Figures 13 (a) and (b), where $\chi'(T)$ and $\chi''(T)$ respectively are plotted for a d.c. field of 0.5 Tesla and 6 values of h_0, which follow the sequence 1:2:4:8:16:32, which corresponds to h_0

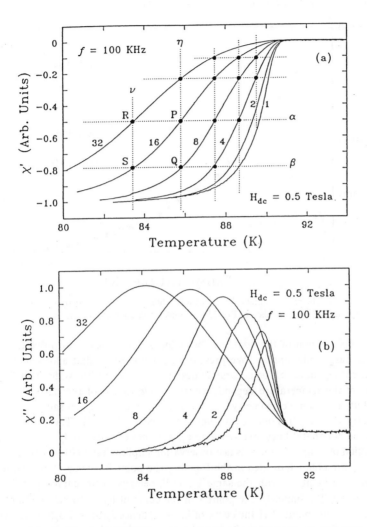

Fig. 13. The a.c susceptibility response ((a) is χ' and (b) is χ'') of a YBaCuO crystal measured with $H_{dc}^{\parallel}$ to the c-axis of 0.5 Tesla using a 100 kHz a.c. field at six different a.c. amplitudes. The geometrical construction in (a) is used to extract $J_c(T)$ (from Ref. 31).

from 0.1 to 3.2 Oe. There is a clear amplitude dependence which is a consequence of the nonlinearity in the electrodynamic response.

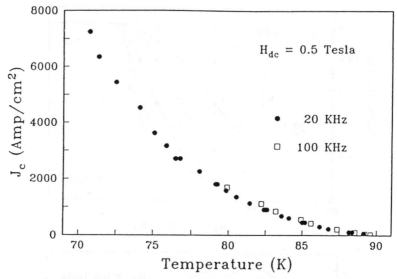

Fig. 14. The critical current density as a function of temperature derived from the a.c. susceptibility data of Fig. 13 and similar data taken at 20 kHz.

The observation of the nonlinear behavior is not sufficient to prove that the critical state model applies. We present now a procedure that allows us to test whether the χ' data can be accurately described by the critical state model[6], and to derive the temperature dependence of J_c *independent* of the particular field distribution or sample geometry[6]. In the critical state for a given sample and coil geometry χ' (and χ'') is only a function of D_c (see Section 2.5). Therefore, in Fig. 13(a) the horizontal lines, marked α and β are lines of constant J_c. But J_c is only a function of temperature. This last observation implies that $J_c(P) = J_c(Q)$, where P and Q are points indicated in the Figure. The point P belongs to the curve h_0 = 16. The point R, with the same L_p as P, belongs to the curve h_0 = 32, so $J_c(R) = 2 J_c(P)$. Similarly, the point S of the curve of h_0 = 16 has the same D_c as the point the point Q of the curve of h_0 = 8, hence $J_c(S) = 2 J_c(Q)$. The consequence, of course, is that $J_c(R) = J_c(S)$, correctly implying that R and S are at the same temperature. That the four points P,Q,R and S form a rectangle stems from the fact that χ' is completely determined by D_c and thus a proof that the critical state model applies. The other points indicated in Fig. 13(a) clearly show that this rectangular construction does not work everywhere; indeed it does not

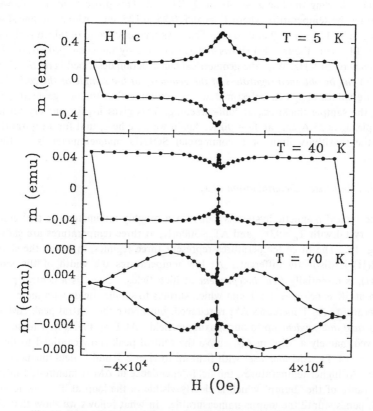

Fig. 15. Magnetic hysteresis loops of a YBaCuO crystal, displaying three characteristic shapes. The width Δm (m = MV, where V is the crystal volume) is field independent (outside of the region dominated by the self-field) at low (5-25K) and intermediate (25-45K) temperatures with a measurable "dip" at low fields above 25K. A "bump" is always present above 45K.

work *at all* at lower a.c. amplitudes as the critical state model *must* fail if h_0 is small enough.

The geometrical construction described above gives us a way to determine the temperature dependence of J_c. Suppose in arbitrary units we define that $J_c(P) = 1$. Then following the line α we obtain $J_c(R) = 2$. This process can be followed to extract the temperature where $J_c = 0.5, 0.25, 0.125$, etc. Along the line β we find $J_c(Q) = 1$, $J_c(S) = 2$ and so on. The J_c vs temperature so obtained is shown in Fig. 14. This Figure also shows the J_c obtained by the same procedure at a frequency of 20 KHz. With the geometrical construction described here, *the shape of $J_c(T)$ can be obtained regardless of the geometry of the sample or the coil*. The absolute values of J_c can be estimated by assuming that for our geometry, $D_c = d$, the sample thickness, at half-screening. This gives us, for our crystal for example, $J_c \simeq 25$ A/cm^2 at T = 90 K. Again we note here, that for a crystal this size, the sensitivity limit of a commercial SQUID magnetometer is ~ 1000 A/cm^2.

3.2 d.c. Response - Determination of J_c

The typical magnetic hysteresis loops for a roughly millimeter size and $20\mu m$ thick crystal with $T_c = 93K$ and $\Delta T_c \sim 300mK$, at three temperatures are shown in Fig. 15. The three temperatures represent three regimes, in which the shape of M(H) is distinctly different. At low temperatures the width of the loop, $\Delta M(H)$, is essentially *field independent* at high fields. There is a central peak, which at 5K is below $\simeq 1.5$ T and which shrinks to smaller fields with increasing temperature. As T increases ΔM is reduced, but above the central peak ΔM *remains field independent* up to our maximum field. At T = 40K a new feature is observed, namely a "dip" in ΔM above the central peak (now confined to fields below $\simeq 0.1$ T): ΔM *increases* with field up to H $\simeq 2$ T and is field independent above. At higher temperatures, the field-dependence of ΔM is manifested in the appearance of the "bump" which is clearly visible in the loop at T = 70 K and which persists until the loop is immeasurable. In what follows we show that the remarkable features of M(H,T) are intimately related to the non-equilibrium crossovers in the vortex-solid phase.

The quantity measured either with a superconducting quantum interference device (SQUID) magnetometer (example is the data of Fig. 15) or a vibrating-sample magnetometer is the total magnetic moment of the sample $\mathbf{m} = \frac{1}{2c} \int_V d^3r' (\mathbf{r}' \times \mathbf{J}(\mathbf{r}'))$, where V is the volume of the sample in which the currents are flowing. The Bean model[6] imposes $|\mathbf{J}(\mathbf{r}')| = J_c$, thus Δm equals J_c times a geometric coefficient. The field-independent ΔM ($= \Delta m/V$) implies a field-independent J_c, indicating that the vortices in the array are in the single-vortex pinning regime[35], i.e. they do not interact.

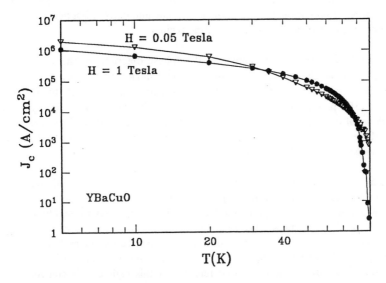

Fig. 16. The critical current density as a function of temperature derived from the d.c. magnetization data of Fig. 15 and similar data taken at other temperatures. At low temperatures the fall-off of $J_c(T)$ is nearly exponential.

4. EVIDENCE FOR COLLECTIVE PINNING EFFECTS IN THE VORTEX SOLID PHASE

4.1 Collective Pinning of Vortices

The behavior of vortices in the diminished solid phase below the melting transition is undoubtedly controlled by the numerous weak randomly distributed defects, such as oxygen vacancies[38], and it has been argued that the collective pinning theory[18], describing the critical current density J_c, should be relevant. As we have mentioned in Section 2.3, random point defects destroy the positional order of the vortex lattice at distances larger than certain correlation lengths R_c (in the direction $\perp H$) and L_c (in the direction $\parallel H$)[18]. If the pinning is weak enough, the vortices can be pinned collectively forming "Larkin domains" of area $\propto R_c^2$ of ordered hexagonal structure, which are disordered only on a larger scale (considered in the vortex-glass approach[7]). Vortex-lattice melting will then occur *inside* the Larkin domains[29].

Feigel'man et al.[35] explored the vortex-solid phase in the framework of collective pinning by weak disorder. Considering thermal *harmonic* fluctuations of vortices, but neglecting thermally activated creep, they proposed a "phase diagram" of *non-equilibrium regimes* in the H-T plane and derived boundaries for the single-vortex and various collective pinning regimes.

Recently we were able to construct the "phase diagram" of the vortex-solid regimes[39] below the melting line for YBaCuO single crystals, by tracing the boundaries for the transitions from the single vortex pinning (1D) to collective pinning (3D) predicted in the theory of collective pinning[35]. We show now how the *shape* of dc magnetization M(H,T) can be associated with various regimes in H-T plane, which are strongly influenced by the *field-dependent* thermal relaxation consistent with collective creep effect[40].

In a "static" picture (i.e. without creep), the relevant length in this regime is L_c, the longitudinal correlation length[18,35]. If L_c is less than the inter-vortex spacing a_0, the vortices are pinned independently and J_c is determined by pinning barriers for single vortices. We refer to this case as one-dimensional (1D) pinning. At high enough temperatures or fields the inter-vortex interaction becomes significant. The two relevant lengths are now L_c and $R_c > a_0$, the longitudinal and transverse size of the correlated region. In this regime J_c is controlled by collective pinning of vortex bundles confined by a correlation volume and thus is field dependent, as observed at T = 70 K. We refer to this as three-dimensional (3D) pinning.

4.2 Non-equilibrium Phase Diagram - Single Vortex Pinning Regime

The connection with the "phase diagram" of Feigel'man and Vinokur[35] is made by translating the M(H) data such as shown in Fig. 15 into a flow pattern of critical current density in the H-T plane[39] with $J_c(H,T)$ = const. calculated using Bean model[6]. The contours shown in Fig. 17 are at temperatures chosen for the visual clarity of the low temperature regime (the high temperature regime is shown later in Fig. 19.

We first explore the transition from the *single-vortex* (1D) to a *collective pinning* (3D) regime. We argue that the essentially *vertical* up to 5.5 Tesla boundary at $T_{sv} \simeq 45K$, indicates a crossover from the 1D regime *below* T_{sv} to the 3D regime *above*. Along this line, $L_c \simeq a_0$. The collective pinning theory[35] predicts that the 1D regime is bounded by $B_{sv}(T) \propto \Upsilon H_{c2}$, where Υ depends on anisotropy Γ and the pinning strength. At H = 0, the crossover from the 1D to 3D region is pre-

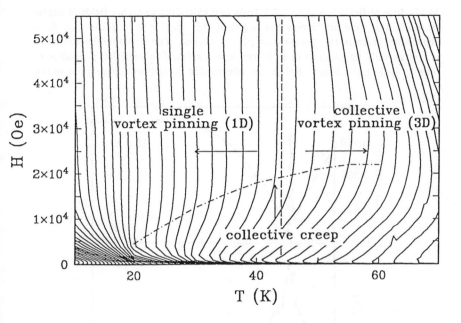

Fig. 17. Contours of constant J_c obtained from the data such as in Fig. 15. The contours are spaced as $\sqrt{J_c}$ for easier visualization. The values of J_c at 5 K, 10 K and 70 K are about 2×10^6 A/cm^2, $\sim 6.5 \times 10^5$ A/cm^2 and $\sim 2.65 \times 10^4$ A/cm^2 respectively. The boundary between the single-vortex and collective pinning regimes is indicated, with the "back-flow" resulting from the transition to a collective creep (after Ref. 39).

dicted for YBaCuO at $T_{sv} \sim 30$ K[Ref.35]. At $T = 0$ we evaluate $B_{sv} \sim 2$ Tesla, although this is an order-of-magnitude estimate and can be as high as 10 Tesla[39]. Experimentally, below T_{sv} there is no curvature in the J_c flow up to the 8-9 Tesla fields[41].

4.3 Collective Creep of Vortices

Next we consider the low field features of M(H) in the single-vortex pinning regime. The central peak is essentially related to the curvature of the vortices at low applied fields due to self-field effects (i.e. generated by the currents flowing in the sample). Here we focus on the field range above the self-field dominated region. It is clear from Fig. 15 and 17 that there is a noticeable "back-flow" of the J_c = const. lines starting at about 20-30 K, corresponding to the "dip" in

166

M(H), first at fields less than ~0.5 Tesla and extending to higher fields at higher temperatures. We argue now the origin of this "dip" is a *collective (i.e. field-dependent) creep in the single-vortex pinning regime*. The dc magnetization gives, of course, only an estimate of J_c; it measures a *relaxed* value of $J(t) \leq J_c$ at some time t_m (here $t_m \sim 10^2$sec). Relaxation effects, due to vortices creeping out of the

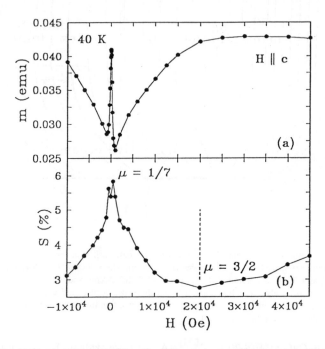

Fig. 18. Top portion of M(H) at T = 40K and the corresponding normalized relaxation rate $S = -d \ln M(t)/d \ln t$. S is a mirror image of M, indicating that the "dip" is caused by a larger relaxation there. Theoretical predictions for the exponent μ in single-vortex creep regime ($\mu = 1/7$), which is followed by a creep of vortex bundles ($\mu = 3/2$) are indicated (see text for the discussion of μ).

dense random pinning wells, affect M(H) at all temperatures[42]. In the theory of collective creep[40], the key new concept is that as J decreases from its initial value J_c, both the size of the correlated volume and the pinning energy U *increase*. The theory gives an expression for the normalized relaxation rate $S = -d \ln M(t)/d \ln(t) = -d \ln J(t)/d \ln(t)$:

$$S = \frac{T}{U_c + \mu T \ln(t/\tau_0)}, \qquad (6)$$

where U_c is the depth of the well, τ_0 is some attempt time, and μ is the exponent which governs the growth of the potential barriers with declining current, $U(J) \propto J^{-\mu}$. In the single-vortex pinning regime, the initial (fast) relaxation($J \lesssim J_c$) is characterized[40] by $\mu = 1/7$. As J decreases with time, the size of the activated loop increases as $L = L_c(J_c/J)^{7/5}$. When L reaches a_0, the creep becomes collective and much slower with $\mu = 3/2$. Below T_{sv} we are in the single-vortex pinning regime in the entire experimental field range, thus initially $\mu = 1/7$ and J relaxes fast at all H. The condition $L(t) = a_0$ (the slow-down of the relaxation) occurs at a current J^* which grows with field as $J^* \propto a_0^{-5/7} \propto B^{5/14}$; i.e. at high fields the slow-down occurs sooner than at low fields. This implies that at lower fields J is smaller because it was relaxing fast (with respect to the field independent J_c) during a longer time interval. If this interpretation is correct, the lower J at low fields should be linked to a faster relaxation rate S. This is confirmed by the data in Fig. 18, showing the top of the hysteresis loop M(H) at 40K and the corresponding normalized relaxation rate as a function of H. At each field, S is measured by increasing H up to 5.5 T, decreasing it to the target H and recording M(t) during approximately one hour (in this short time window the time dependence of S is undetectable). Remarkably, the increase in the relaxation rate at low fields is a *mirror image* of the "dip" in the magnetization. Hence, the "dip" in M(H) in the intermediate temperatures is a result of a *transition to a collective creep in the single-vortex pinning regime*.

4.4 Non-equilibrium Phase Diagram - Collective Pinning of Vortex Bundles

Now we turn to the flow pattern at temperatures above T_{sv} shown in Fig. 19, and throughout the remainder of this Chapter we will present the experimental evidence identifying various regimes of collective pinning. The visualization of Fig. 19 is remarkable on several accounts. First we note that above T_{sv}, the curvature in the current flow pattern becomes more pronounced, reflecting the "fish tail" shape of M(H). While, as we have discussed earlier, at low fields the observed J is dominated by the collective creep, at high fields (the large bundle regime) the pinning energy is large, hence the relaxation is small and the field dependence of J is essentially determined by $J_c(H)$. In the 3D collective pinning regime $J_c = (W/V_c)^{1/2} H^{-1}$, where $W \propto H$ is the mean square value of the random pinning force. The dimensions of the correlated region $V_c = L_c R_c^2$ are $L_c \simeq R_c(C_{44}/C_{66})^{1/2}$ and $R_c \simeq C_{44}^{1/2} C_{66}^{3/2} \xi^2/W$. At high fields $R_c > \lambda$, thus the vortex system is in the *local* regime, where $C_{66} \propto H$ and $C_{44} \propto H^2$. The volume of the correlated region (bundle) in this regime expands as H^5 and J_c falls off as $1/H^3$. We indeed find that the dependence $J_c \propto 1/H^3$ fits remarkably well at high fields[39]. The calculated $J_c(H)$ value[35] at H = 3 Tesla is $\sim 7 \times 10^3$ A/cm^2 as compared with the measured value of $\sim 2 \times 10^3$ A/cm², consistent with the relatively slow relaxation in this regime[39].

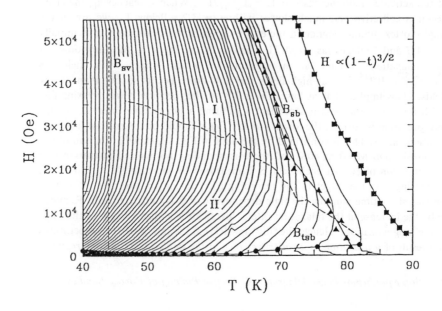

Fig. 19. Same as Figure 17 expanded at higher temperatures with equal contour spacing. The range of the critical current densities is $\sim$2500 A/cm² at 80 K, and $\sim$2×10⁵ A/cm² at 40 K. The collective pinning boundaries are indicated as follows: the vertical line B_{sv} is a transition from the single-vortex to a collective pinning regime. The solid triangles indicate a crossover from the non-local into a local regime $B_{sb}(T)$ (see text) and $B_{tsb} \propto T^2$ is a thermal softening boundary (also see Ref. 34). The dc irreversibility line (solid squares), was determined with a current criterion of 100 A/cm²; it fits well to $H \propto (1-T/T_c)^{3/2}$. A dashed line is a guide to the eye through the "bump"; it separates region I (controlled by $J_c(H)$) and region II (controlled by relaxation).

With decreasing field R_c shrinks until $R_c \lesssim \lambda$, and the *local* description fails. This crossover is predicted to occur at a field B_{sb} which is is about $3B_{sv}$ at $T = 0$. The transition from the non-local into a local regime, taken as the onset of the $1/H^3$ fall-off of J_c, is plotted as $B_{sb}(T)$ in Fig. 19. The "bump" in the *non-local* regime may be understood as follows. On the low-field side of the "bump", similarly to what we have discussed earlier, J is controlled by collective creep. At higher fields near the top of the "bump", the relaxation slows down and J(t) is controlled by $J_c(H)$.

4.5 Thermal Softening of the Core Pinning

So far we have considered the temperature regime, where the harmonic fluctuations of vortices $\sqrt{<u^2>}$ are smaller than the coherence length ξ. A *thermal softening of the core pinning* will occur when $\sqrt{<u^2>}$ becomes comparable to ξ (see Section 2.1) and thus the pinning landscape is smeared out on the same length scale[35]. The thermal softening boundary (TSB) is the *only* known crossover in the phase diagram which *increases* with temperature[35]; i.e. $B_{tsb} \propto T^2$. Note that constant J_c contours turn back and then *reenter* at low fields. The point of reentry, corresponding to a *minimum* in ΔM at low fields also increases as T^2 as shown in Fig. 19, and is suggestive of TSB. The T^2 behavior is clearly seen above 60 K; it is obscured at lower temperatures by the growing prominence of the self-field peak. The value of B_{tsb} is ~0.2 Tesla at 85 K, *above* the lower critical field $H_{c1}^{[Ref.34]}$. As we have seen in Section 2.1, the *same flow pattern* in a.c. response much closer to T_c. There we have proposed that the thermal softening boundary, crosses the irreversibility line H_{irr} shifting it to lower fields. The remnant of the H_{irr} *collapse* in H-T plane is seen below H_{irr} in both a.c. and d.c. response. However, the nature of the "back-flow" and the reentrant regime at high temperatures and low fields still remains to be established. The theory is yet to provide a more complete description of the regime close to H_{c1} where the elastic energy becomes negligible.

5. ANISOTROPY OF a.c. RESPONSE - INTRINSIC PINNING

The strikingly anisotropic nature of high temperature oxide superconductors has received much attention since many interpretations of observed phenomena, such as depinning or flux-lattice melting transitions depend on possible two-dimensional behavior. It has been questioned whether a 3D anisotropic Ginzburg-Landau description is valid[4,43] at low enough temperatures and in the case of weak coupling between Cu-O layers, i.e. when the coherence length in the c-direction, ξ_c, is smaller than the spacing between the layers. A Lawrence-Doniach model[44], incorporating the discrete arrangement of Cu-O layers, with coupling between the planes provided by Josephson tunneling or proximity effect[2-4], has been considered[44-46]. In this scenario one expects an unusual behavior of the vortex core and indeed, two-dimensional pancake vortices have been discussed[46], residing only in Cu-O layers. The vortex cores are drastically

170

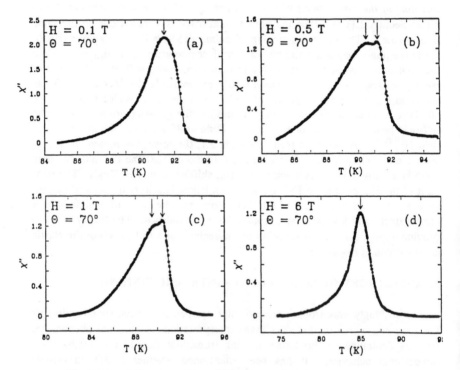

Fig. 20. Out-of-phase component of the a.c. susceptibility χ'' measured in a single crystal of YBaCuO at 1MHz for (a) $H_{dc} = 0.1$ T, (b) 0.5 T, (c) 1 T and (d) 6 T fields applied at 70° angle away from the c-axis. A double peak in χ'' is evident in a finite field range.

affected by the lattice discreteness and may vanish between the layers, where the vortices are Josephson-like[46].

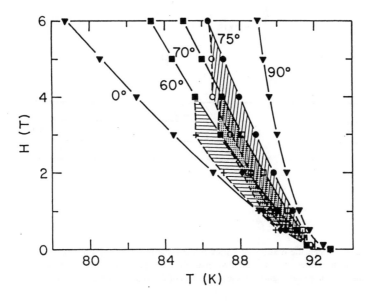

Fig. 21. Temperatures of the peaks in χ'' as a function of the field orientation Θ for different values of H_{dc}. The shaded areas indicate the finite range of fields and angles at which the double dissipation is observed.

The variation in core energy along the c-axis has been considered by Tachiki and Takahashi[47] and Feinberg and Villard[48], leading to an intrinsic barrier, similar to Peierls-Nabarro barrier[49], for the motion of dislocations across the planes. The barrier is against the kink formation in the vortex core; the kinks should appear at some critical orientation of magnetic field Θ_c away from the Cu-O layers.

These ideas were tested by the angle resolved measurements of χ in YBaCuO single crystals, which probe the anisotropy of the irreversibility line [50]. There are two distinct dissipation peaks in χ'' in a range of magnetic fields which is dependent on the orientation of the crystal with respect to the field. We suggest that the appearance of two peaks indicates the presence of two distinct components of critical current density J_c (T), and argue that they arise from vortices conforming to the intrinsic layered crystal structure by kink formation in a "staircase" pattern, as suggested by Feinberg and Villard[48].

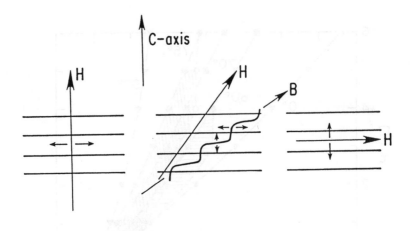

Fig. 22. Schematic drawing of vortex motion due to a.c. excitation. For the intermediate orientation Θ, the two channels of dissipation arise from the motion of the in-plane segment of the vortex and of the kink crossing the a-b plane.

As we discussed earlier, the dissipation peak in χ'' is used as an indicator of the onset of the hysteretic behavior in the principal orientations, i.e. with the applied field either parallel or perpendicular to the a-b plane. However, in the intermediate orientations the behavior is more complex. Figure 20 shows χ'' for several dc magnetic fields at an intermediate angle of $\Theta = 70°$, where Θ is measured from the c-axis. At 0.1 Tesla, χ'' is a single peak, as seen in the principal orientations. As the field is increased, in addition to the shift towards lower temperatures and broadening of the whole feature, which also occurs in the principal orientations, a double peak is emerging. This double peak exists over a range of magnetic fields, which depends on Θ, and disappears again at high enough fields (Fig. 20(d)). The temperature locations of the peaks obtained from both single- and double-peaked χ'' as a function of the field orientation Θ for various fields are shown in Fig. 21. We have already seen in Section 2.5, that with the applied field either parallel ($\Theta = 0°$) or perpendicular ($\Theta = 90°$) to the c-axis, the anisotropy of irreversibility line is ~5 and mimics that of H_{c2}. The shaded areas for each applied field show the extent of the double dissipation peak region and we note again that it occurs in the *finite range* of fields and angles as shown in Fig. 20 and 21.

Double peaks in χ'' have been observed in granular materials, where both intergrain and intragrain currents contribute to dissipation[51]. There, however, the peak associated with the weaker intergranular currents moves very rapidly towards lower temperatures when relatively small d.c. field is applied, because of long Josephson penetration length. We do not see any effects of granularity in our single crystals[41,50] and we argue that the two peaks correspond to two components of J_c; one which is parallel to Cu-O planes and another, crossing the planes as shown schematically in Fig. 22. The two channels of dissipation correspond to losses due to the motion of two (albeit coupled) vortex segments: one in the ab-plane and another one corresponding to a kink crossing the ab-plane. This new vortex structure is present in a finite region of H-Θ space, as shown in Fig. 23, where the data for the intermediate angles are summarized. Feinberg and Villard describe such "staircase" vortices above a critical boundary, below which a perfect lock-in of flux lines parallel to the layers and on their full length is predicted[48]. The region of strong intrinsic pinning is confined by a "lock-in boundary" given by[48]:

$$\cos \Theta_c = \frac{1}{\pi} \left[2\alpha \frac{H^\star}{H} \{ 1 + \Gamma^2 \frac{H^\star}{H} \} \right]^{1/2}, \qquad (7)$$

where, $H^\star$ is of the order of H_{c1}, Γ is the anisotropy parameter and $\alpha \simeq 0.5$. This is represented by boundary 1 in Fig. 23(b), where we take $\Gamma = 5$. At angles above this boundary, kinks in vortex core form and a "staircase vortex" will appear with the average B-field away from the direction of applied field. This boundary is within less than 5 ° of the onset of the shaded region in Fig. 23(b).

Such remarkable effect should be easier to observe in Bi or Tl based copper oxides, which are considered to be more anisotropic[52,53], than in YBaCuO[10,14,54], and perhaps two-dimensional. Indeed, single crystals of BiSrCaCuO show two distinct peaks in the mechanical oscillator response as well as in a.c. susceptibility[55]. There, two separate and independent boundaries in H-T plane are proposed, which are completely determined by the field component along the c-axis, $H\cos\Theta$. We do not find such $H\cos\Theta$ scaling in YBaCuO; the two J_c components appeared to be not decoupled, as it would be in the 2D case at low enough fields. Recently, Bulaevskii[56] proposed a two dimensional description taking into account 2D pancake structure of vortex cores, which also predicts a boundary for intrinsic pinning:

$$\cos \Theta_p = \frac{\Phi_0 \alpha}{4\pi \lambda_{ab}^2 H}. \qquad (8)$$

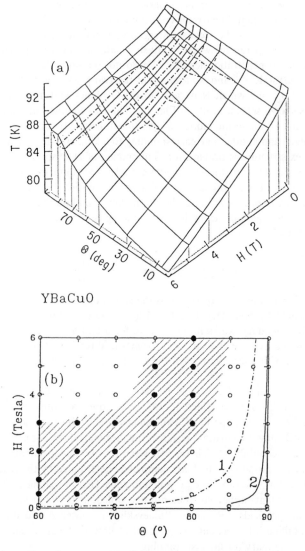

Fig. 23. (a) Phase diagram in H-T-Θ space. The region where the double peak is found is where the dashed lines are drawn underneath the solid lines. (b) In the projection onto a H-Θ plane this region is shaded (pluses are the actual data points). The single peak data are shown by the solid dots. The prediction of Feinberg and Villard (dot-dash line 1) for the lock-in boundary and of Bulaevskii (solid line 2) are indicated.

Here, λ_{ab} is penetration depth for fields in the ab-plane and α is of order unity. This boundary is labeled 2 in Fig. 23(b) and it appears at angles much closer to the ab-plane than "lock-in boundary"[48] and is further separated from the data bounded in the shaded region. Indeed, these results indicate that a 2D description is less likely to be applicable to YBaCuO. At high fields, and closer to the c-axis, the interaction between the vortices will stiffen them, and one expects 3D Ginzburg-Landau description to hold. Indeed, a double peak feature is not observed in that regime.

6. CONCLUDING REMARKS

In summary, a considerable insight has been gained into the macroscopic magnetic behavior of high-T_c superconductors in both linear and nonlinear response regime and we have discussed only some of the interesting new pinning behaviors apparent from the magnetic measurements. We have seen that the effect of large thermal fluctuations combined with the presence of the atomic scale disorder leads to surprising phase diagrams which are drastically different from those of low-T_c type II superconductors. Although there is already a substantial experimental support for the vortex-glass phase and collective pinning and creep of magnetic vortices, this is clearly just a beginning of our unraveling of a complex magnetic phase diagram of the oxide superconductors and exploring many novel phase transitions and pinning crossovers.

REFERENCES

1. for a general review of microscopic properties see **Physical Properties of High Temperature Superconductors**, Volumes I and II, edited by D. Ginsberg (World Scientific, Singapore, 1989).

2. M. Tinkham, **Introduction to Superconductivity** (McGraw-Hill, New York, 1975).

3. P.G. de Gennes, **Superconductivity of Metals and Alloys** (Addison-Wesley, New York, 1989).

4. **Superconductivity**, Volumes I and II, edited by R.D. Parks (Marcel Dekker, New York, 1969).

5. P.W. Anderson and Y.B. Kim, Rev. Mod. Phys. **36**, 39 (1964).

6. C.P. Bean, Phys. Rev. Lett. **8**, 250 (1962).

7. D.S. Fisher, M.P.A. Fisher, and D.A. Huse, Phys. Rev. B **43**, 130 (1991).

8. J.G. Bednorz and K.A. Müller, Z. Phys. B **64**, 189 (1986).

9. D.H. Wu and S. Sridhar, Phys. Rev. Lett. **65**, 2074 (1990).

10. L. Krusin-Elbaum, A.P. Malozemoff, Y. Yeshurun, D.C. Cronemeyer, and F. Holtzberg, Phys. Rev. B **39**, 2936 (1989).

11. L. Krusin-Elbaum, R.L. Greene, A.P. Malozemoff, Y. Yeshurun, and F. Holtzberg, Phys. Rev. Lett. **62**, 217 (1989).

12. D. R. Harshmann et al., Phys. Rev. B **36**, 2386 (1987); Y.J. Uemura et al., Phys. Rev. B **38**, 909 (1988).

13. T.T.M. Palstra, B. Batlogg, L.F. Schneemeyer, R.B. Van Dover, and J.V. Waszczak, Phys. Rev. B **62**, 252 (1988).

14. U. Welp, W.K. Kwok, G.W. Crabtree, K.G. Vandervoort, and J.Z. Liu, Phys. Rev. Lett. **62**, 1908 (1989); Physica C**161**, 1 (1989).

15. D.R. Nelson, Phys. Rev. Lett. **60**, 1973 (1988); D.R. Nelson, J. Stat. Phys. **57**, 511 (1989); D.R. Nelson and H.S. Seung, Phys. Rev. B **39**, 9153 (1989).

16. D.R. Nelson in **Phenomenology and Applications of High Temperature Superconductors**, edited by K. Bedell et al. (Addison-Wesley, New York, 1992).

17. R. Labusch, Phys. Status Solidi **32**, 439 (1969).

18. A.I. Larkin and Yu.N. Ovchinnikov, Zh. Eksp. Teor. Fiz. **65**, 704 (1973); J. Low Temp. Phys. **34**, 409 (1979).

19. M.P.A. Fisher, Phys. Rev. Lett. **63**, 1511 (1989).

20. R.H. Koch, V. Foglietti, W.J. Gallagher, G. Koren, A. Gupta, and M.P.A. Fisher, Phys. Rev. Lett. **63**, 1511 (1989).

21. T.K. Worthington, E. Olsson, C.S. Nichols, T.M. Shaw, and D.R. Clarke, Phys. Rev. B **43**, 10538 (1991); T.K. Worthington, F. Holtzberg, and C.A. Feild, Cryogenics **30**, 417 (1990).

22. K.A. Müller, M. Takashige and J.G. Bednorz, Phys. Rev. Lett. **58**, 408 (1987).

23. Y. Yeshurun and A. P. Malozemoff, Phys. Rev. Lett. **60**, 2202 (1988).

24. A. P. Malozemoff, T.K. Worthington, Y. Yeshurun, F. Holtzberg, and P.H. Kes, Phys. Rev. B **38**(10), 7203 (1988).

25. T.K. Worthington, Y. Yeshurun, A.P. Malozemoff, R. Yandrofski, F. Holtzberg, and T.R. Dinger, J. Phys. Colloque (C8 Suppl.), **12**, 2093 (1989).

26. Y. Yeshurun, A.P. Malozemoff, T.K. Worthington, R. Yandrofski, L. Krusin-Elbaum, T.R. Dinger, F. Holtzberg, and G. V. Chandrasekhar, Cryogenics **29**, 258 (1989).

27. P.L. Gammel, D.J. Bishop, G.I. Dolan, J.R. Kwo, C.A. Murray, L.F. Schneemeyer, and J.V. Waszczak, Phys. Rev. Lett. **59**, 2592 (1987); P.L. Gammel, L.F. Schneemeyer, J.V. Waszczak, and D.J. Bishop, Phys. Rev. Lett. **62**, 1666 (1988).

28. D.E. Farrel, J.P. Rice, and D.M. Ginsberg, Phys. Rev. Lett. **67**, 1165 (1991).

29. T.K. Worthington, M.P.A. Fisher, D.A. Huse, J. Toner, A.D. Marwick, T. Zabel, C.A. Feild, and F. Holtzberg, Phys. Rev. B **46**, 11854 (1992).

30. J.R. Clem, H.R. Kirchner, and S.T. Sekula, Phys. Rev. B **14**, 1893 (1976).

31. L. Civale, T.K. Worthington, L. Krusin-Elbaum, and F. Holtzberg, **Magnetic Susceptibility of Superconductors and other Spin Systems**, edited by R.A. Hein et al., (Plenum, New York, 1992).

32. V.B. Geshkenbein, V.M. Vinokur, and R. Fehrenbacher, Phys. Rev. B **43**, 3748 (1991).

33. Another way is to search for the onset of the 3rd harmonic in a.c. response. See for example A. Shaulov and D. Dorman, Appl. Phys. Lett. **53**, 2680 (1988).

34. L. Krusin-Elbaum, L. Civale, F. Holtzberg, A.P. Malozemoff, and C. Feild, Phys. Rev. Lett. **67**, 3156 (1991).

35. M.V. Feigel'man and V.M. Vinokur, Phys. Rev. B **41**(13), 8986 (1990).

36. A.E. Koshelev and V.M. Vinokur, Physica C, **173**, 465, (1991).

37. A. Houghton, R.A. Pelcovitz, and A. Sudbo, AT&T Techical Memorandum, (1990); Phys. Rev. B **40**, 6763 (1989).

38. T.L. Hylton and M.R. Beasley, Phys. Rev. B **41**, 11669 (1990).

39. L. Krusin-Elbaum, L. Civale, V.M. Vinokur, and F. Holtzberg, Phys. Rev. Lett. **69**, 2280 (1992).

40. M.V. Feigel'man, V.B. Geshkenbein, A.I. Larkin, and V.M. Vinokur, Phys. Rev. Lett. **63**, 2301 (1989); M.V. Feigel'man, V.B. Geshkenbein, and V.M. Vinokur, Phys. Rev. B **43**, 6263 (1991).

41. T. Tamegai, L. Krusin-Elbaum, L. Civale, P. Santhanam, M.J. Brady, W.T. Masselink, F. Holtzberg, and C. Feild, Phys. Rev. B **45**, 8201 (1991).

42. A.P. Malozemoff and M.P.A. Fisher, Phys. Rev. B **42**, 6784 (1990); M.P. Maley, J.O. Willis, H. Lessure, and M.E. McHenry, Phys. Rev. B **42**, 2639 (1990).

43. V.G. Kogan, Phys. Rev. B **24**, 1572 (1981).

44. W.E. Lawrence and S. Doniach, in **Proceedings of the 12th International Conference on Low Temperature Physics**, edited by J.G. Daunt, D.V.

Edwards, F.J. Milford, and M. Jaqub (Plenum, New York, 1965), Part A, p. 566.

45. L.N. Bulaevskii, Zh. Eksp. Teor. Fiz. **64**, 2241 (1973) [Sov. Phys. JETP **37**, 1133 (1973)]

46. J.R. Clem, Phys. Rev. B **43**, 7837 (1991).

47. M. Tachiki and S. Takahashi, Solid State Comm. **70**, 291 (1989).

48. F. Feinberg and C. Villard, Phys. Rev. Lett. **65**, 919 (1990).

49. J. Friedel, in **Dislocations** (Pergamon Press, 1964).

50. L. Krusin-Elbaum, L. Civale, T.K. Worthington, and F. Holtzberg, Physica C **185**, 2337 (1991).

51. Q.H. Lam, Y. Kim, and C.D. Jeffries, Phys. Rev. B **42**, 4846 (1990).

52. S. Martin, A.T. Fiory, R.M. Flemming, L.F. Schneemeyer, and J.V. Waszczak, Phys. Rev. Lett. **60**, 2194 (1988).

53. D.E. Farrell, S. Bonham, J. Foster, Y.C. Chang, P.Z. Ziang, K.G. Vandervoort, D.J. Lam, and V.G. Kogan, Phys. Rev. Lett. **63**, 782 (1989).

54. G.J. Dolan, F. Holtzberg, C. Feild and T. Dinger, Phys. Rev. Lett. **62**, 2184 (1989).

55. C. Durán, J. Yazyi, F. de la Cruz, D.J. Bishop, D.B. Mitzi, and A. Kapitulnik, Phys. Rev. B **44**, 7737 (1991).

56. L. N. Bulaevskii, J. Mod. Phys. B **4**, 1849 (1990).

MAGNETIC PROPERTIES OF MOLECULAR COMPOUNDS CONTAINING LANTHANIDE(III) AND COPPER(II) IONS.

Olivier KAHN and Olivier GUILLOU

Laboratoire de Chimie Inorganique, URA n° 420, Université de Paris Sud, 91405 Orsay, France

INTRODUCTION

Since two decades or so, a large number of heteropolymetallic compounds have been described. The studies concerning these compounds have often been performed, either in relation with the modelling of some metalloenzymes containing several kinds of metal ions, or with the perspective of designing novel molecular materials, in particular molecular-based magnets. A particular emphasis has been brought to the magnetic properties[1]. The main idea emerging from those studies is that the interaction between two nonequivalent magnetic centers may lead to situations wich cannot be encountered with species containing a unique kind of spin carriers. In fact, the investigation of the magnetic properties of heteropolymetallic compounds in the last period has represented quite an important contribution to the development of molecular magnetism as a whole[2], and has allowed to introduce several important new concepts. Let us briefly recall some of them: (i) the importance of the relative symmetries of the interacting magnetic orbitals[3,4]; (ii) the strict orthogonality of the magnetic orbitals favoring the stabilization of the state of highest spin[3,5-7]; (iii) the irregular spin state structure leading to molecular systems with a high spin in the ground state despite antiferromagnetic interactions between nearest neighbor ions[8]; (iv) the one-dimensional ferrimagnetism[9-15]; (v) the design of molecular-based magnets through crystal engineering of ferrimagnetic chain compounds[16,17].

All the results mentioned above have been obtained with compounds involving only 3d metal ions. In the meantime some studies concerning heteropolymetallic species containing both lanthanide and 3d ions appeared. The pionnering work along this line was performed by Gatteschi and coworkers who found that in a series of Cu(II)Gd(III)Cu(II) trinuclear compounds the Gd(III)-Cu(II) interaction was ferromagnetic, irrespective of the details of the molecular structure[18-20]. The same situation holds when Cu(II) is replaced by nitronyl nitroxide radicals[21,22]. The ferromagnetic nature of the Gd(III)-Cu(II) pair has been subsequently confirmed by Matsumoto, Okawa and coworkers on a binuclear compound[23,24], then by us in one- and two-dimensional molecular materials[25-27].

This short review is devoted to the magnetic properties of compounds containing both lanthanide and 3d ions. The heart of the paper deals with the Gd(III)-Cu(II) pair; it is clearly the simplest situation since the $^8S_{7/2}$ free-ion ground state of Gd(III) has no first-order angular momentum. The case where Gd(III) is replaced by another lanthanide(III) ion with an angular momentum in the ground state will be also approached.

SOME SELECTED EXPERIMENTAL RESULTS

The Gd(III)Cu(II) dinuclear compound [GdCu(fsa(en)$_2$(H$_2$O)$_4$](ClO$_4$)

To our knowledge only one compound containing discrete Gd(III)Cu(II) binuclear units has been reported. The formula of the unit is [GdCufsa(en)$_2$(H$_2$O)$_4$]$^+$ where (fsa)$_2$en^{4-} is the binucleating ligand deriving from N,N'-(2-hydroxy-3-carboxybenzilidene)ethylenediamine. Matsumoto et al. investigated the nitrate derivative[23,24] while we investigated the perchlorate. In spite of many efforts single crystals suitable for X-ray work have not been obtained yet. The infrared data unambiguously indicate that the Cu(II) ion is located in the N$_2$O$_2$- inside site and the Gd(III) in the outside O$_4$- site. However due to its size this Gd(III) ion is most probably pulled out of the plane of the (fsa)$_2$en^{4-} ligand, as schematized below, and completes its coordination sphere with water molecules:

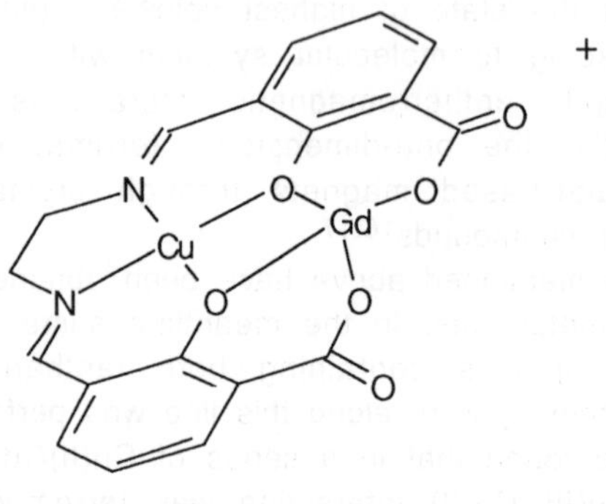

The magnetic behavior of the perchlorate derivative is represented in Figure 1 in the form of the $\chi_M T$ versus T plot, χ_M being the molar magnetic susceptibility and T the temperature. When the temperature is lowered, $\chi_M T$ first remains constant and equal to 8.3 cm^3 K mol^{-1}, then increases below ca. 50 K, and eventually reaches a plateau with $\chi_M T$ = 10.0 cm^3 K mol^{-1} in the low-temperature range. This behavior is typical of a Gd(III)-Cu(II) ferromagnetic interaction. As a matter of fact, the interaction between the S_{Gd} = 7/2 and S_{Cu} = 1/2 local ground states gives rise to two molecular states, S = 4 and S = 3, with a septet-nonet energy gap of 4J if the interaction Hamitonian is $H = -J\mathbf{S}_{Gd}.\mathbf{S}_{Cu}$. Upon cooling down the excited S = 3 state is progressively depopulated. The low-temperature plateau of $\chi_M T$ corresponds to the temperature range where only the S = 4 ground state is thermally populated.

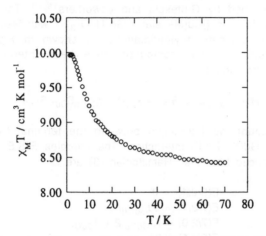

Figure 1: $\chi_M T$ versus T plot for [GdCufsa(en)$_2$(H$_2$O)$_4$](ClO$_4$)

The fitting of the experimental data by the theoretical expression of $\chi_M T$:

$$\chi_M T = \frac{4N\beta^2}{k} \frac{15g_4^2 + 7g_3^2\exp(-4J/kT)}{9 + 7\exp(-4J/kT)} \tag{1}$$

in which g_3 and g_4 are the Zeeman factors for the S = 3 and S = 4 states, respectively, leads to J = 2.1 cm^{-1} K mol^{-1}. The molecular Zeeman factors, g_3 and g_4, are related to the local Zeeman factors through[28]:

$$g_3 = (9g_{Gd} - g_{Cu})/8$$
$$g_4 = (7g_{Gd} + g_{Cu})/8 \tag{2}$$

The spectrum of the low-lying state is then as:

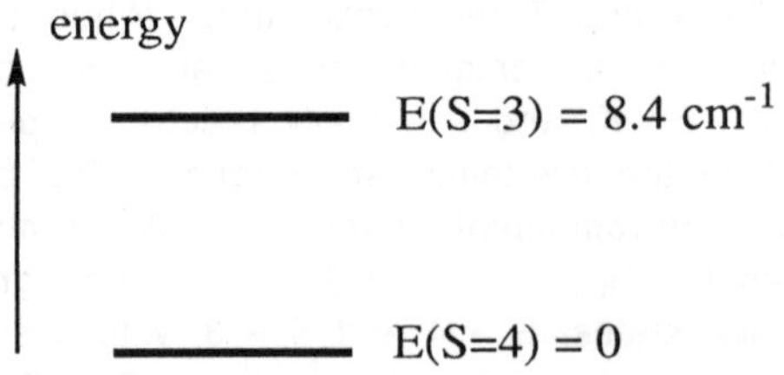

energy

$E(S=3) = 8.4$ cm^{-1}

$E(S=4) = 0$

Compounds containing GdCu$_2$ triads

The first trinuclear species containing a Gd(III) and two Cu(II) ions have been described by Gatteschi and coworkers[18,19]. The structure of one of them, of formula $\{[Cu(Mesalen)]_2Gd(H_2O)_3\}^{3+}$ with Mesalen = N,N'-bis(methylsalicylaldiminato)ethylenediamine, is shown in Figure 2. The magnetic properties of this compound have been interpreted with a Hamiltonian of the form :

$$H = -J_{GdCu}\mathbf{S}_{Gd}\cdot(\mathbf{S}_{Cu1} + \mathbf{S}_{Cu2}) + J_{CuCu}\mathbf{S}_{Cu1}\cdot\mathbf{S}_{Cu2} \qquad (3)$$

taking into account the interaction between the terminal Cu(II) ions in addition to the Gd(III)-Cu(II) interaction. The energies E(S,S') of the low-lying states deduced from the Hamiltonian (3) are :

$$E(9/2,1) = 0$$
$$E(7/2,1) = 9J_{GdCu}/2$$
$$E(7/2,0) = 7J_{GdCu}/2 + J_{CuCu}$$
$$E(5/2,1) = 8J_{GdCu} \qquad (4)$$

S ans S' being the spin quantum numbers associated with the spin operators:

$$\mathbf{S} = \mathbf{S'} + \mathbf{S}_{Gd}$$
$$\mathbf{S'} = \mathbf{S}_{Cu1} + \mathbf{S}_{Cu2} \qquad (5)$$

The theoretical expression of the magnetic susceptibility is :

$$\chi_M T = (N\beta^2/4k) [165g_{9/2,1}{}^2 + 84g_{7/2,1}{}^2\exp(-9J_{GdCu}/2kT) +$$
$$84g_{7/2,0}{}^2\exp(-7J_{GdCu}/2kT)\exp(-J_{CuCu}/kT) + 35g_{5/2,1}{}^2\exp(-8J_{GdCu}/kT)]$$
$$/ [5 + 4\exp(-9J_{GdCu}/2kT) + 4\exp(-7J_{GdCu}/2kT)\exp(-J_{CuCu}/kT) +$$
$$3\exp(-8J_{GdCu}/kT)] \qquad (6)$$

where the $g_{S,S'}$ molecular Zeeman factors are related to the local Zeeman factors through[29] :

$$g_{9/2,1} = (7g_{Gd} + 2g_{Cu})/9$$
$$g_{7/2,1} = (59g_{Gd} + 4g_{Cu})/63$$
$$g_{7/2,0} = g_{Gd}$$
$$g_{5/2,1} = (9g_{Gd} - 2g_{Cu})/7 \tag{7}$$

The $\chi_M T$ versus T plot for this compound again increases upon cooling down to 1.2 K, then reaches a value correponding to a S = 9/2 ground state in which all the local spins are aligned in a parallel fashion. The fitting of the experimental data is complicated by the fact that the two interaction parameters J_{GdCu} and J_{CuCu} are strongly correlated. The values $J_{GdCu} = 5.5$ cm^{-1} and $J_{GdCu} = -4.2$ cm^{-1} have been proposed.

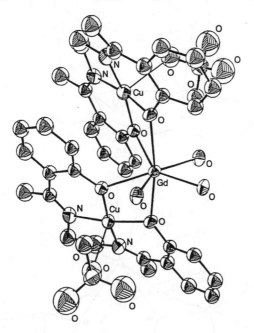

Figure 2 : Molecular structure of $\{[Cu(Mesalen)]_2Gd(H_2O)_3\}^{3+}$ (from ref. 18).

Let us consider now the hexanuclear clusters $[Ln_2Cu_4(fsaaep)_4(NO_3)_6].CH_3OH.H_2O$ abbreviated as $[Ln_2Cu_4]$; Ln is a trivalent lanthanide and (fsaaep)$^{2-}$ is the ligand deriving from 3-(N-2-pyridylethylformimidoyl)salicylic acid. The crystal structure of this cluster has been solved for Ln = Pr, and is isomorphous with the La and

184

Gd derivatives[30]. The structure consists of $[Ln_2Cu_4]$ entities in which the metal ions form a chair-shaped hexagon. The two lanthanide atoms are located on both sides of a double layer containing the four copper atoms. The interlayer Cu(II)-Cu(II) interactions are negligible such that from a magnetic point of view the hexanuclear cluster $[Ln_2Cu_4]$ may be viewed as two independent $LnCu_2$ units. The stucture of the $[Pr_2Cu_4]$ entity along with the detail of a $PrCu_2$ unit are shown in Figure 3.

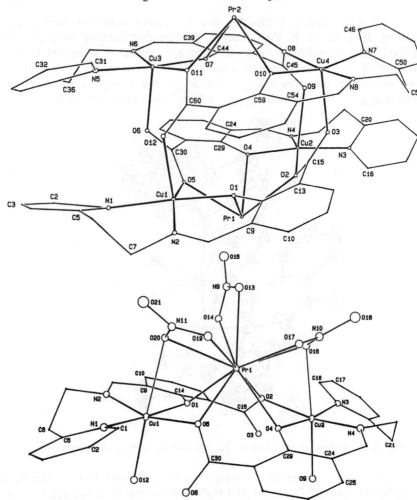

Figure 3 : (Top) Molecular structure of the hexanuclear cluster $[Pr_2Cu_4]$. For a sake of clarity the nitrato ions chelated to the lanthanide atoms are not represented.
(Bottom) View of one of the $PrCu_2$ units showing the chelation of the nitrato ions.

The $\chi_M T$ versus T plot for [Gd$_2$Cu$_4$] is displayed in Figure 4 (left). $\chi_M T$ is constant and equal to 16.6(1) cm^3 K mol^{-1} down to ca. 60 K, then increases as the temperature is lowered further, and eventually reaches a plateau with $\chi_M T$ = 24.4(1) cm^{-1} K mol^{-1}. Since the Cu(II)-Cu(II) interlayer interaction is negligible, the magnetic behavior of [Gd$_2$Cu$_4$] is that of two independent Cu(II)Gd(III)Cu(II) triads, and the theoretical expression of $\chi_M T$ is given by Eq. (6) with a factor two. J_{CuCu} is determined independently from the magnetic properties of the [La$_2$Cu$_4$] cluster in which the Cu(II) ions interact through the closed-shell La(III) ion, and found equal to -3.13 cm^{-1}. The fitting of the susceptibility data for [Gd$_2$Cu$_4$] leads to J_{GdCu} = 6.0 cm^{-1}. The spectrum of the low-lying states is then as:

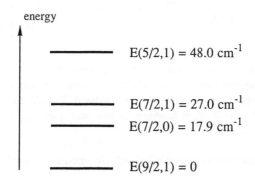

To confirm the nature of the ground state, the variation of the magnetization M versus the field H was measured at both 2 K and 30 K. At 2 K, only the ground state is significantly populated, and, as expected, the M = f(H) curve closely follows the Brillouin function for two independent S = 9/2 spins. On the other hand, at 30 K, the excited states are significantly populated as well, and the magnetization curve is just between the Brillouin function for the six independent local spins, and the Brillouin function for two independent S = 9/2 triad spins, as shown in Figure 4 (right) where M is expressed in Nβ units and H in Tesla.

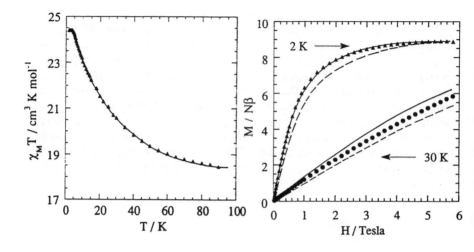

Figure 4: (Left) $\chi_M T$ versus T plot for [Gd$_2$Cu$_4$] within a field of 10 G. (Right) Field dependence of the magnetization for [Gd$_2$Cu$_4$] at both 2 K and 30 K. The full lines represent the Brillouin functions for two uncorrelated S = 9/2 spins with g = 1.99, and the dotted lines the Brillouin functions for two local spins S_{Gd} = 7/2 and four local spins S_{Cu} = 1/2.

One and two-dimensional Gd(III)Cu(II) molecular materials

Two low-dimensional compounds containing Gd(III) and Cu(II) ions have been reported[25,26]. Both result from the reaction of Gd(III) with the Cu(II) precursor [Cu(pba)]$^{2-}$ shown below :

where pba stands for 1,3-N,N'-propylenebis(oxamato). These compounds may be described as arrangements built from ladder-like motifs:

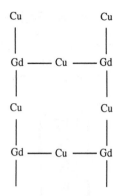

where the horizontal and vertical units — Cu — stand for $[Cu(pba)]^{2-}$. The former compound, of formula $Gd_2(ox)[Cu(pba)]_3[Cu(H_2O)_5].20H_2O$ (hereafter abbreviated as Gd_2Cu_3Cu), is constructed by a combination of ladder-like motifs and oxalato bridges promoting a two-dimensional pattern of association of the ladders, with a double-sheet thick honeycomb network based on hexagonal rings as shown in Figure 5. In addition $[Cu(H_2O)_5]^{2+}$ dications are interspersed in the gap between the double-sheets, anchoring the slabs to each other.

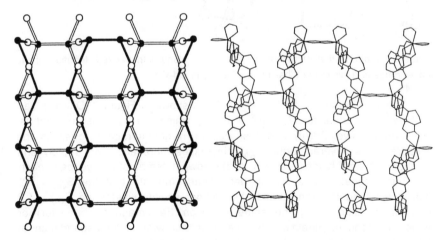

Figure 5: (Left) Schematic representation of four ladder-like motifs linked together by oxalato goups along the b direction for Gd_2Cu_3Cu. The solid (open) lines symbolize linkages within the upper (lower) layer. (Right) Detail of the upper layer of the two-dimensional pattern. This Figure emphasizes the honeycomb pattern of Gd_6 rings.

188

The later compound, of formula $Gd_2[Cu(pba)]_3 \cdot 23H_2O$ (hereafter abbreviated as Gd_2Cu_3), can be seen as resulting from the condensation of two ladders together with a redistribution of the rungs of the ladders, which achieves an infinite tube-like structure of an essentially square section and quasi four-fold symmetry as shown in Figure 6.

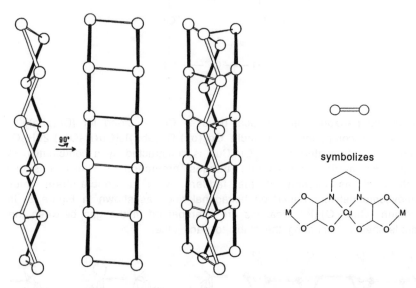

Figure 6: Condensation of two ladder-like motifs along with the redistribution of the ladder rungs to achieve the tube-like motif found in Gd_2Cu_3.

The $\chi_M T$ versus T plot for Gd_2Cu_3Cu is shown in Figure 7. Down to ca. 30 K, $\chi_M T$ is equal to 17.2 cm^3 K mol^{-1}, which corresponds to what is expected for two Gd(III) and four copper(II) noninteracting ions. As the temperature is lowered further, $\chi_M T$ increases more and more rapidly and reaches 60 cm^3 K mol^{-1} at 1.3 K, the lowest temperature we can reach with our magnetometers. Those data clearly indicate that the $S_{Gd} = 7/2$ local spins tend to align parallel, but do not specify wheter the $S_{Cu} = 1/2$ local spins align along the same direction (ferromagnetic behavior) or along the opposite direction (ferrimagnetic behavior). The dependence of the molar magnetization M as a function of the field H was recorded at various temperatures below 5 K both in the low-field regime (0 ≤ H ≤ 200 G) and in the high-field regime (0 ≤ H ≤ 20 Tesla). Up to 200 G, the magnetization is strictly linear versus the field, the slope being in perfect agreement with the magnetic susceptibility value measured independently. Those low-field measurements do not show any indication of spin decoupling, i.e. a change of sign of d^2M/dH^2. The experimental data in the high-field regime at 1.6 and 4.2 K are shown in Figure 8 together with the theoretical curves for two Gd(III) and four Cu(II) noncoupled ions, i.e. Brillouin's functions. At both temperatures the magnetization increases faster than expected for noninteracting ions. The saturation magnetization M_S is equal to 18 Nβ, which exactly corresponds to the value anticipated when all local spins are aligned parallel. Those data unambiguously show that even in zero-field all local spins tend to align parallel. The Gd(III)-Cu(II) interaction through the oxamato bridge is ferromagnetic. In the 2-20 K temperature range the magnetic susceptibility data could be fitted with a Curie-Weiss law $\chi = C/(T-\theta)$ and a Weiss constant $\theta = 1.25(3)$ K, which in the mean-field approximation leads to $J_{GdCu} = 0.15$ cm^{-1} for the interaction between Gd(III) and Cu(II) ions separated by ca. 5.7 Å across the oxamato bridge. The magnetic properties of Gd_2Cu_3 are quite similar to those of Gd_2Cu_3Cu. They also reveal a Gd(III)-Cu(II) ferromagnetic interaction.

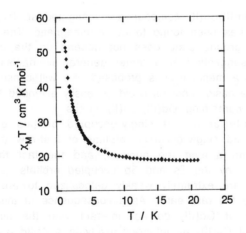

Figure 7: $\chi_M T$ versus T plot for Gd_2Cu_3Cu.

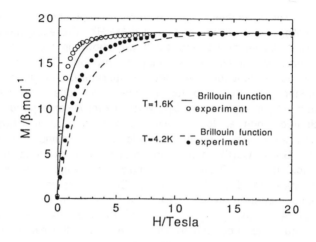

Figure 8 : Magnetization M versus magnetic field H curves for Gd_2Cu_3Cu at 1.6 and 4.2K.

WHY IS THE Gd(III)-Cu(II) INTERACTION FERROMAGNETIC ?

In all the Gd(III)Cu(II) compounds investigated so far, the Gd(III)-Cu(II) interaction has been found to be ferromagnetic. The parallel spin alignment in the ground state does not depend on the details of the structure, which suggests that a rather general mechanism applies. In this Section such a mechanism is proposed. We will consider a Gd(III)-Cu(II) pair, but the basic concepts could be easily extended to systems of higher nuclearity containing Gd(III)-Cu(II) motifs.

The peculiarity of the 4f singly-occupied orbitals of Gd(III) as compared to the 3d singly-occupied orbitals of first-row transition ions is their contraction around the nucleus, and the fact that they are efficiently shielded by the 5s and 5p occupied orbitals. It follows that these 4f orbitals are extremely weakly delocalized toward the oxygen atoms surrounding the rare earth. As a consequence of this, the 4f-type magnetic orbitals of Gd(III) do not interact with the single 3d-type magnetic orbital of Cu(II). All integrals involving a 4f-3d overlap density vanish; the Heitler-London type interaction[2,38] between the local ground states of Gd(III) and Cu(II) is therefore zero.

An alternative mechanism lies in the coupling between the 4f-3d ground configuration (GC) and the metal-metal charge-transfer configurations (CTC). The CTC's of lowest energy correspond to either the 3d → 4f or the 4f → 3d process; both lead to a S = 3 excited pair state.

The transfer integrals β_{4f-3d}, however, are zero, such that the GC-CTC interaction cannot stabilize the low-lying $S = 3$ state. We are in the case where both Anderson's mechanism[31,32] ($\beta_{4f-3d}{}^2/U = 0$) and Kahn's mechanism[33,34] ($\beta_{4f-3d}S_{4f-3d} = 0$) are inoperative.

The CTC of energy immediately above is associated with the 3d → 5d process; an electron is transferred from the singly-occupied orbital centered on copper toward an empty orbital centered on gadolinium. Two excited states, $S = 3$ and $S = 4$, arise from this CTC. Due to Hund's rule, the latter is lower in energy than the former. The energy gap Δ between these two excited states is easily calculated as :

$$\Delta = E(S=3)_{CTC} - E(S=4)_{CTC} = 8\,k^0{}_{4f-5d} + \text{two-site ionic integrals} \qquad (8)$$

where $k^0{}_{4f-5d}$ is a mean one-site exchange integral which may be expressed as:

$$k^0{}_{4f-5d} = (1/7)\sum_{i=1}^{7} <4f_i(1)5d(2)|1/r_{12}|4f_i(2)5d(1)> \qquad (9)$$

Let us define by U' the energy gap between the barycenters of the $S = 3$ and $S = 4$ pair states arising from the CTC on the one hand, and the ground configuration on the other hand. U' represents the energy cost associated with the 3d → 5d electron transfer. Neglecting the two-site ionic integrals occurring in Eq. (8), the two coupling matrix elements $<(S=3)_{GC}|H|(S=3)_{CTC}>$ and $<(S=4)_{GC}|H|(S=4)_{CTC}>$ are equal to β_{5d-3d}. Thus, the GC-CTC mixing stabilizes the $S = 3$ and $S = 4$ low-lying pair states of - $\beta_{5d-3d}{}^2/(U'+\Delta/2)$ and - $\beta_{5d-3d}{}^2/(U'-\Delta/2)$, respectively. β_{5d-3d} is the transfer integral between a 5d-type orbital for gadolinium and the 3d-type magnetic orbital for copper. In contrast with β_{4f-3d}, β_{5d-3d} is far from being negligible. Indeed, the 5d-type gadolinium orbitals are very diffuse, and may be delocalized toward the oxygen atoms surrounding the rare earth. The J_{GdCu} interaction parameter occurring in the spin Hamiltonian $-J_{GdCu}S_{Gd}\cdot S_{Cu}$ is equal to $[E(S=3) - E(S=4)]/4$, which leads to:

$$J_{GdCu} = \sum_{i=1}^{5} [\beta_{5d-3d}{}^2 \, \Delta/(4U'^2-\Delta^2)]_i \qquad (10)$$

where the index $i = 1$ to 5 arises from the fact that the five 5d orbitals may be involved in the electron transfer. Eq. (10) accounts for a ferromagnetic interaction.

A rough estimation of J_{GdCu} given in Eq. (10) has been carried out in the case of $[Ln_2Cu_4(fsaaep)_4(NO_3)_6].CH_3OH.H_2O$. In principle the index i in Eq. (10) is attached to the transfer integrals β_{5d-3d} as well as Δ and U'. In fact Δ and U' have been deduced from atomic values, and therefore can be viewed as mean values for the five 5d-type orbitals. Δ may be estimated from the ^7D - ^9D energy gap between the two terms arising from the $4f^75d^1$ configuration of gadolinium(II). This gap has been found as 8488 cm^{-1} by Callahan[35]. A very rough estimation of U' is given by the energy difference between the ionization potentials of Cu(II) and Gd(III), namely 120,000 cm^{-1}. The transfer integrals β_{5d-3d} are much more difficult to estimate. We used an Extended Hückel approach, and found values in the range 1,411 - 3,838 cm^{-1}, which leads to J_{GdCu} = 4.8 cm^{-1}. The rusticity of the Extended Hückel method is well known, and the fairly good agreement with the experimentaly determined J value is obviously somewhat fortuitous. Our calculation, however, suggests that Eq. (10) not only gives the good sign but also the good order of magnitude for J_{GdCu}.

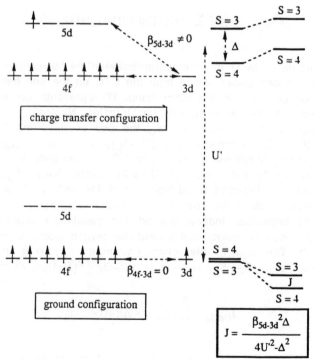

Figure 9: Schematic representation of the orbital mechanism explaining the ferromagnetic nature of the Gd(III)-Cu(II) interaction.

Our mechanism, schematized in Figure 9, is in no way novel. It has been introduced by Goodenough[36] as early as 1963, and recently invoked by Tchougreeff[37] to justify the ferromagnetic ordering of decamethylferrocenium tetracyanoethenide, by Kinoshita and coworkers to explain the intermolecular ferromagnetic coupling in paranitrophenyl nitronyl nitroxide[38,39], and by Wieghardt, Girerd and coworkers[40] to interpret the magnetic properties of μ-oxo manganese(III) compounds. The situation encountered in the Gd(III)Cu(II) compounds, however, has something of specific. Let us show this specificity in quite a simple fashion. For that we consider a dissymmetric AB pair with the orbital pattern shown below:

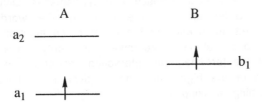

where a_1, a_2 and b_1 are natural orbitals[2,34] (they are not orthogonalized). The relative energies of the low-lying singlet and triplet states depend on three contributions arising from different mechanisms, namely: (i) the Heitler-London interaction within the ground configuration $(a_1)^1(b_1)^1$, which leads to a singlet-triplet energy gap :

$$J_i \approx 2k_{11} + 4\beta_{11}S_{11} \qquad (11)$$

where k_{11}, β_{11} and S_{11} are the two-electron exchange, transfer and overlap integrals, respectively, involving the a_1b_1 overlap density. In most cases the $4\beta_{11}S_{11}$ negative term dominates and the singlet state is the lowest. The triplet state becomes the lowest when the a_1 and b_1 orbitals are (quasi-) orthogonal; (ii) the interaction between GC and the first excited charge-transfer configurations $(a_1)^2$ and $(b_1)^2$. This contribution gives a further stabilization of the singlet state J_{ii}:

$$J_{ii} \approx -2\beta_{11}^2(1/U_A+1/U_B) \qquad (12)$$

where U_A and U_B are the energy costs associated with the $b_1 \rightarrow a_1$ and $a_1 \rightarrow b_1$ electron transfers, respectively. J_{ii} vanishes when a_1 and b_1 are orthogonal; (iii) the interaction between GC and the more excited charge-transfer configuration $(a_1)^1(a_2)^1$, which stabilizes the triplet state of a J_{iii} energy :

$$J_{iii} \approx 2\beta_{12}{}^2 k^0{}_{12}/U'^2 \qquad (13)$$

U' being the energy cost associated with the $b_1 \longrightarrow a_2$ electron transfer, and $k^0{}_{12}$ the one-site exchange integral $<a_1(1)a_2(2)|1/r_{12}|a_1(2)a_2(1)>$. J_{iii}, in the general case, is one order of magnitude smaller (in absolute value) than J_i and J_{ii}, and is therefore masked by those two contributions. It is only when a_1 and b_1 are (quasi-) orthogonal that J_{iii} can be detected. But, if it is so, the stabilization of the triplet state may be explained by the $2k_{11}$ term occurring in the Heitler-London contribution J_i given by Eq. (11). There is no need to invoke the J_{iii} term. The situation is different if a_1 is a rare earth 4f orbital. In this case the overlap density a_1b_1 is negligibly small in any point of space, and k_{11} vanishes. Only the J_{iii} term may account for a ferromagnetic interaction. In other words the Gd(III)-Cu(II) case is ideal to investigate the effect of an electron transfer from a singly occupied orbital on a site toward an empty orbital on the other site. Of course the ferromagnetic interaction cannot be very large since the stabilization of the high-spin state is given by a third order term, but there is nothing masking it.

Ln(III)-Cu(II) INTERACTION WITH Ln(III) CARRYING AN ORBITAL MOMENTUM

Up to now we focused on the Gd(III)-Cu(II) interaction between spin carriers without first-order orbital momentum. In most of the compounds Gd(III) can be replaced by any other trivalent lanthanide, and actually we investigated the magnetic properties of quite of few compounds containing Ln(III) ions carrying an angular momentum. To our knowledge there are very few studies dealing with this problem, the most important one being probably that performed by Levy during the sixties[41]. Levy was concerned by the iron garnets, and his results seem to be difficult to transfer to the molecular compound area. Therefore we can say that almost everything remains to be done as far as the magnetic properties of such Ln(III)-Cu(II) species are concerned. We are working on this problem, but several years will be probably necessary to obtain a consistent theoretical frame allowing to simulate quantitatively the magnetic properties.

In this Section we will restrict ourselves in proposing a preliminary qualitative approach which, we hope, will be of interest for the increasing number of colleagues synthesizing and investigating molecular compounds involving 4f and 3d ions.

Let us consider a lanthanide(III) ion in spheric environment. The spin-orbit coupling partially removes the degeneracy of the $^{2S+1}\Gamma$ free-ion ground term to give $^{2S+1}\Gamma_J$ states. For the $4f^1$-$4f^6$ configurations orbital and spin momenta prefer to align antiparallel, and the state of lowest energy has the smallest J. For the $4f^8$-$4f^{13}$ configurations, on the contrary, orbital and spin momenta prefer to align parallel; the state of lowest energy has the highest J. For all trivalent lanthanides, except Sm(III) and Eu(III), the first free-ion excited state is high enough in energy to be fully depopulated at room temperature and below[42-44]. In the following we will ignore the Sm(III) and Eu(III) cases. The complexity of the problem is due to the fact that the free-ion states of the same J arising from different $^{2S+1}\Gamma$ terms may mix through spin-orbit coupling, the resulting states being further split and mixed by the ligand field. A first approximation, however, consists of neglecting the ligand field effect as well as the mixing of the free-ion states. If we do so, we are faced with a Ln(III)-Cu(II) problem in which Ln(III) possesses an angular momentum **L** in addition to a spin momentum **S**; **L** and **S** are aligned antiparallel or parallel according to whether the number of 4f electrons is smaller or larger than seven.

The mechanism of the spin-spin interaction for Gd(III)-Cu(II) remains valid for the other Ln(III) ions. The mixing between the Ln(III)Cu(II) ground configuration and the Ln(II)Cu(III) charge transfer configuration resulting from the transfer of an unpaired electron from a 3d-type copper orbital toward a 5d-type lanthanide orbital stabilizes the state of highest spin. In other words the Ln(III) and Cu(II) spin momenta tend to align parallel. When the number of 4f electrons for Ln(III) is smaller than seven, the resulting spin momentum for the Ln(III)-Cu(II) pair tends to align antiparallel with the Ln(III) orbital momentum, affording an overall antiferromagnetic interaction. On the other hand, when the number of 4f electrons is larger than 7, the pair spin momentum tends to align parallel with the Ln(III) orbital momentum affording an overall ferromagnetic interaction. This mechanism is schematized in Figure 10.

196

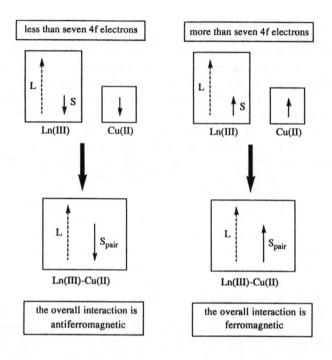

Figure 10: Schematic representation of the angular momentum coupling in a Ln(III)-Cu(II) pair, where Ln(III) is a trivalent lanthanide ion carrying an orbital momentum.

It is possible to go a bit further in our model, and to look for the high- and low-temperature limits of the product $\chi_M T$ for a Ln(III)-Cu(II) pair. Let us define by $(\chi_M T)_{HT}$ and $(\chi_M T)_{LT}$, respectively, those limits.

We consider a Ln(III)-Cu(II) pair where the free-ion ground state for the lanthanide ion Ln(III) is characterized by the quantum numbers S, L and J. We ignore the symmetry lowering around Ln(III), which is the essence of our model. Furthermore, we assume that the g-factor for the Cu(II) ion is isotropic and equal to 2. The high-temperature limit for $\chi_M T$ corresponds to the sum of the uncoupled Ln(III) and Cu(II) contributions, i.e.:

$$(\chi_M T)_{HT} = (N\beta^2/3k)[g_J^2 J(J+1) + 3] \tag{14}$$

with:

$$g_J = 3/2 + \frac{S(S+1) - L(L+1)}{2J(J+1)} \tag{15}$$

In the low-temperature limit only the pair ground state schematized in Figure 11 is thermally populated. This state is characterized by the spin quantum number $S_{pair} = S + S_{Cu} = S + 1/2$ since the local spins tend to align parallel, the orbital quantum number J since the Cu(II) ion has no first-order angular momentum, and the angular quantum number $J_{pair} = L \pm S_{pair}$; the sign + holds when the number of 4f electrons for Ln(III) is larger than seven, and the sign - when this number is smaller than seven. In this low-temperature limit $\chi_M T$ is then given by:

$$(\chi_M T)_{LT} = (N\beta^2/3k)g_{J_{pair}}{}^2 J_{pair}(J_{pair}+1) \tag{16}$$

where $g_{J_{pair}}$ is again given by an expression like Eq. (15). The results for all the Ln(III)-Cu(II) pairs, but those with Ln = Sm and Eu, are given in Table I. For the lanthanide ions with less than seven 4f electrons the ratio $(\chi_M T)_{HT}/(\chi_M T)_{LT}$ is larger than unity; $\chi_M T$ decreases as T is lowered, which corresponds to an overall antiferromagnetic interaction. For those with more than seven 4f electrons this ratio is smaller than unity; $\chi_M T$ increases as T is lowered, which corresponds to an overall ferromagnetic interaction.

Table I: Values of high- and low-temperature limits of $\chi_M T$ in cm^3 K mol^{-1}, $(\chi_M T)_{HT}$ and $(\chi_M T)_{LT}$, for Ln(III)-Cu(II) pairs in the free-ion model developed in this paper.

Ln(III)-Cu(II)	$(\chi_M T)_{HT}$	$(\chi_M T)_{LT}$
Ce(III)-Cu(II)	1.18	0.33
Pr(III)-Cu(II)	1.98	0.88
Nd(III)-Cu(II)	2.02	0.90
Gd(III)-Cu(II)	8.26	10.00
Tb(III)-Cu(II)	12.10	14.43
Dy(III)-Cu(II)	14.55	17.02
Ho(III)-Cu(II)	14.45	16.91
Er(III)-Cu(II)	11.86	14.07
Tm(III)-Cu(II)	7.53	9.23
YB(III)-Cu(II)	2.95	3.91

We can note that the variation of $\chi_M T$ versus T is expected to be more pronounced for n < 7 than for n ≥ 7. In this latter case the free-ion Ln(III)-Cu(II) interaction effect may be masked by the ligand-field effect we have not taken into account. This remark leads us to stress again that our approach is quite preliminary. Only the coupling between the lanthanide(III) and copper(II) angular momenta has been considered. For all Ln(III) ions, but Gd(III), the splitting of the free-ion ground state due to the ligand-field effect is another important factor we have not introduced yet in our approach.

CONCLUSION

Through a series of examples we have confirmed that the Gd(III)-Cu(II) interaction was ferromagnetic, irrespective of the details of the geometry. An interpretation of this behavior has been proposed, based on the coupling between the Gd(III)Cu(II) ground configuration and the Gd(II)Cu(III) excited configuration resulting from the transfer of an unpaired electron from a 3d-type copper orbital toward a 5d-type gadolinium orbital. This mechanism is not novel. What is remarkable in the Gd(III)-Cu(II) case, however, it is that the other interaction mechanisms involving 4f-3d overlap densities are inoperative owing to the nature of the 4f orbitals.

In a second part we briefly approached the very complicated problem of the Ln(III)-Cu(II) interaction, where Ln(III) is now a lanthanide ion carrying both a spin and an orbital momentum. We developed a free-ion model taking into account two factors, namely: (i) the parallel spin alignment of the local spins, as found in the Gd(III)-Cu(II) case; (ii) the antiparallel or parallel alignment of the lanthanide spin and orbital momenta according to whether the 4f shell is less or more than half filled. According to this model the product $\chi_M T$ decreases as T is lowered when the 4f shell is less than half filled, and increases when the 4f shell is more than half filled. Quite a few experimental results concerning both discrete units and one- or two-dimensional networks are qualitatively in line with this expectation. In spite of this agreement, we must now introduce in our model the ligand-field effect which splits the free-ion ground state of the rare earth. We intend to work along this line in the near future.

1. O. Kahn, *Struct. Bonding (Berlin)* **68** (1987) 89.
2. O. Kahn, *Molecular Magnetism* (Verlag-Chemie, New York) in press.
3. O. Kahn, J. Galy, Y. Journaux, J. Jaud, I. Morgenstern-Badarau, *J. Am. Chem. Soc.* **104** (1982) 2165.
4. Y. Journaux, O. Kahn, J. Zarembowitch, J. Galy, J. Jaud, *J. Am. Chem. Soc.* **105** (1983) 7585.
5. O. Kahn, R. Prins, J. Reedijk, J.S. Thompson, *Inorg. Chem.* **26** (1987) 3557.
6. C. Benelli, A. Dei, D. Gatteschi, L. Pardi, *Inorg. Chem.* **27** (1988) 2831.
7. A. Caneschi, A. Dei, D. Gatteschi, *J. Chem. Soc., Chem. Commun.* (1992) 630.
8. Y. Pei, Y. Journaux, O. Kahn, *Inorg. Chem.* **27** (1988) 399.
9. M. Verdaguer, M. Julve, A. Michalowicz, O. Kahn, *Inorg. Chem.* **22** (1983) 2624.
10. A. Gleizes, M. Verdaguer, *J. Am. Chem. Soc.* **106** (1984) 3727.
11. Y. Pei, J. Sletten, O. Kahn, *J. Am. Chem. Soc.* **108** (1986) 3143.
12. Y. Pei, M. Verdaguer, O. Kahn, J. Sletten, J.P. Renard, *Inorg. Chem.* **26** (1987) 138.
13. D. Beltran, E. Escriva, M. Drillon, *J. Chem. Soc., Faraday Trans.* **78** (1982) 1773.
14. M. Drillon, E. Coronado, D. Beltran, R. Georges, *J. Appl. Phys.* **57** (1984) 3353.
15. M. Drillon, E. Coronado, R. Georges, J.C. Ganduzzo, J. Curely, *Phys. Rev. B* **40** (1989) 10992.
16. O. Kahn, Y. Pei, M. Verdaguer, J.P. Renard, J. Sletten, *J. Am. Chem. Soc.* **110** (1988) 782.
17. K. Nakatani, J.Y. Carriat, Y. Journaux, O. Kahn, F. Lioret, J.P. Renard, Y. Pei, J. Sletten, M. Verdaguer, *J. Am. Chem. Soc.* **111** (1989) 5739.
18. A. Bencini, C. Benelli, A. Caneschi, R.L. Carlin, A. Dei, D. Gatteschi, *J. Am. Chem. Soc.* **107** (1985) 8128.
19. A. Bencini, C. Benelli, A. Caneschi, A. Dei, D. Gatteschi, *Inorg. Chem.* **25** (1986) 572.
20. C. Benelli, A. Caneschi, D. Gatteschi, O. Guillou, L. Pardi, *Inorg. Chem.* **29** (1990) 1751.
21. C. Benelli, A. Caneschi, D. Gatteschi, L. Pardi, P. Rey, D.P. Shum, R.L. Carlin, *Inorg. Chem.* **28** (1989) 272.
22. C. Benelli, A. Caneschi, A.C. Fabretti, D. Gatteschi, L. Pardi, *Inorg. Chem.* **29** (1990) 4223.
23. N. Matsumoto, M. Sakamoto, H. Tamaki, H. Okawa, S. Kida, *Chem. Letters* (1989) 853.
24. M. Sakamoto, M. Hashimura, K. Matsuki, N. Matsumoto, K. Inoue, H. Okawa, *Bull. Chem. Soc. Jpn.* 64 (1991) 3639.

25. O. Guillou, P. Bergerat, O. Kahn, E. Bakalbassis, K. Boubekeur, P. Batail, M. Guillot, *Inorg. Chem.* **31** (1992) 110.
26. O. Guillou, R.L. Oushoorn, O. Kahn, K. Boubekeur, P. Batail, *Angew. Chem. Int. Ed. Engl.* 31 (1992) 626.
27. O. Guillou, O. Kahn, R.L. Oushoorn, K. Boubekeur, P. Batail, *Inorg. Chim. Acta* **198-200** (1992) 119.
28. R.P. Scaringe, D. Hodgson and W.E. Hatfield, *Mol. Phys.* **35** (1978) 701.
29. D. Gatteschi and A. Bencini, in *Magneto-Structural Correlations in Exchange Coupled Systems*, eds. R.D. Willett, D. Gatteschi and O. Kahn (D. Reidel, Dordrecht, 1983).
30. M. Andruh, I. Ramade, E. Codjovi, O. Guillou, O. Kahn and J.C. Trombe, *to appear.*
31. P.W. Anderson, *Phys. Rev.* **115** (1956) 2.
32. P.W. Anderson, in *Magnetism*, ed. G.T. Rado and H. Suhl, (Academic Press, New York, 1963) Vol. 1, p. 25.
33. B. Briat, O. Kahn, *J. Chem. Soc., Trans. Faraday* **72** (1976) 268.
34. J.J. Girerd, Y. Journaux, O. Kahn, Chem. Phys. Letters **82** (1981) 534.
35. W.R. Callahan, *J. Opt. Soc. Am.* **55** (1963) 695.
36. J.B. Goodenough, *Magnetism and Chemical Bond*, (Interscience, New York, 1963) p. 165.
37. A.L. Tchougreeff, *J. Chem. Phys.* **96** (1992) 6026.
38 P. Turek, K. Nozawa, D. Shiomi, K. Awaga, T. Inabe, Y. Maruyama, and M. Kinoshita, *Chem. Phys. Lett.* **180** (1991) 327.
39. M. Tamura, Y. Nakazawa, D. Shiomi, K. Nozawa, Y. Hosokoshi, M. Ishikawa, M. Takahashi, and M. Kinoshita, *Chem. Phys. Lett.* **186** (1991) 401.
40. R. Hotzelmann, K. Wieghardt, U. Flörke, H.J. Haupt, D.C. Weatherburn, J. Bonvoisin, G. Blondin, J.J. Girerd, *J. Am. Chem. Soc.* **114** (1992) 1681.
41. P.M. Levy, *Phys. Rev.* **135** (1964) A155; **147** (1966) 147.
42. T.H. Siddall, in *Theory and Applications of Molecular Paramagnetism*, ed. E.A. Boudreaux and L.N. Mulay (Interscience, New-York, 1976) p 257.
43. A.T. Casey, S. Mitra, in *Theory and Applications of Molecular Paramagnetism*, ed. E.A. Boudreaux and L.N. Mulay (Interscience, New-York, 1976) p 271.
44. A.T. Casey, in *Theory and Applications of Molecular Paramagnetism*, ed. E.A. Boudreaux and L.N. Mulay (Interscience, New-York, 1976) p 27.

MAGNETOCHEMISTRY OF SEVERAL IRON(III) COMPOUNDS

Richard L. Carlin
Department of Chemistry
University of Illinois at Chicago
Chicago, Illinois 60680

I.   Introduction

The paramagnetic properties of mononuclear compounds of iron(III) are straightforward. This is because most of the compounds are (quasi-)octahedral in geometry, and have five unpaired electrons in a weak crystalline field. The spin-free compounds are found to have a well-isolated $^6A_1$ ground state, with S = 5/2, and generally a very small zero-field splitting (ZFS) as well. The spin-only moment is calculated as 5.98 μ_B, and this value is observed frequently. The moments are independent of temperature in the absence of cooperative phenomena. Epr spectra of crystals containing iron(III) are readily obtained at room temperature or below. The g-values are generally about 2, for spin-orbit coupling is not strong, and essentially isotropic. Spin-paired compounds have been studied less extensively. We have reviewed elsewhere some of the features of the paramagnetic properties of iron compounds (1). We concentrate here on several compounds which undergo magnetic ordering. Surprisingly few of the latter have been reported. Two series of compounds are of particular interest: the $A_2[FeX_5(H_2O)]$ series of octahedral materials and the $A_3Fe_2X_9$ group of tetrahedral substances.

We have already reviewed the $A_2[FeX_5(H_2O)]$ series of antiferromagnets (2) and we shall discuss here only the recent work on these substances. Many bimetallic ferrimagnets contain iron(III) moieties but these have been reviewed elsewhere (3). A general monograph on magnetochemistry has been published (4); further reference to many topics discussed here may be found there. A general review of the coordination chemistry of iron(III) has been published (5). The magnetochemistry of isoelectronic manganese(II), on the other hand, has been much more extensively investigated.

The five-coordinate ferromagnet $[FeCl(dtc)_2]$, where dtc is an abbreviation for the diethyldithiocarbamate ion, has the unusual (for iron) spin-3/2, $^4A_2$ ground state, with a relatively large zero-field splitting (4). The spin-crossover compounds related to $[Fe(dtc)_3]$ have been reviewed elsewhere and need not be mentioned further (4).

II.   Paramagnetism

Paramagnetic behavior is described by the Curie law. The magnetic susceptibility χ varies inversely with temperature according to the Curie law. This may be written as $\chi = C/T$, where C is the Curie constant; in the common case of spin-only magnetism, C =

$Ng^2\mu_B^2S(S+1)/3k_B$, where N is Avogadro's number, S is the spin quantum number and k_B is the Boltzmann constant. Measurements of the moment of iron compounds at room temperature have been used for years in order to distinguish spin-free from spin-paired electronic configurations. Octahedral complexes of iron(III) may be found which belong to either one of these cases; to date, on the other hand, all known tetrahedral compounds are spin-free.

Deviations from the Curie law are found as the temperature is decreased sufficiently for the magnetic ions to begin to interact with one another. This is described in a high-temperature approximation by the Curie-Weiss law, $\chi = C/(T - \theta)$ where the non-zero θ suggests that such interactions are present. This is a molecular field approximation; more detailed models at the molecular level are available.

The Hamiltonian in use to describe paramagnetism is

$$\mathcal{H} = g\mu_B H \cdot S + D[S_z^2 - 1/3\ S(S+1)] + E(S_x^2 - S_y^2) \qquad [1]$$

where the spin S is 5/2 and D is the axial zero-field splitting parameter in energy units. The term in E is usually small and measures the rhombic distortions. The action of the ZFS is to split the six-fold degeneracy of the $^6A_1$ state into the $|\pm 1/2\rangle$, $|\pm 3/2\rangle$ and $|\pm 5/2\rangle$ doublets. (There are some smaller terms added to Eq. 1 when epr spectra are being analyzed. These are not important for our purposes here.)

The susceptibility of iron(III) compounds has been reported in the ^{4}He region for only a handful of compounds, amazing as that may seem. Even fewer specific heat data are available. Many of the Weiss constants (θ) in the literature have been obtained by the extrapolation of data taken above 77 K, and it is our experience that many of these extrapolations are unreliable. Iron is largely an isotropic ion, has no low-lying energy levels whose population change with temperature, and thus obeys the Curie law (as distinct from the Curie-Weiss law) over a wide temperature interval.

Among the halo complexes, the most stable complexes are formed with the small non-polarizable fluoride ion. Chloro complexes are less stable, bromo complexes even less stable and few iodides are known. Consistent with the stability order $F^- > Cl^- > Br^-$ is the occurrence of higher coordination number Fe(III) complexes with fluoride than with bromide. Thus while F^- readily forms hexacoordinate $[FeF_6]^{3-}$ complex ions and no tetrafluoro complex ions, Cl^- forms both $[FeCl_6]^{3-}$ and $[FeCl_4]^-$ while Br^- apparently does not form a stable hexabromo complex, $[FeBr_6]^{3-}$ (5).

The number and arrangement of halide ligands coordinated to a transition metal in a complex anion is often surprisingly dependent on the nature of the cationic counterions present (6). There are reports in the older literature of a number of chloro-compounds of the ferric ion, but the stoichiometry of many of the materials is unusual and does not hint at the molecular composition. For example, we note the report of $(C_5H_5NH)_5Fe_2Cl_{11}$, $(C_5H_5NH)_2FeCl_5$ and $(C_5H_5NH)_4Fe_2Cl_{10}$ (7). It has been found that $(MeNH_3)_2FeBr_5$ must be formulated as $(MeNH_3)_2[FeBr_4]Br$ (6).

There are relatively few reports of substances which contain the tetrahedral tetrahaloferrate(III) ions, and most of them seem to concern only chloride ligands. The crystal structure reports of these materials are collected in Table 1, where *simple* is used in the title to exclude the $A_3Fe_2X_9$ materials discussed in Section V below. The tetrahaloferrate(III) ions are generally somewhat distorted from true tetrahedral symmetry; this is most likely due to crystal lattice packing forces as well as to hydrogen-bonding effects. As one example, the anion in $[FeCl_2(OPPh_3)_4][FeCl_4]$ displays Fe-Cl distances from 2.111 to 2.217 Å and Cl-Fe-Cl angles from 104.3 to 113.8° (8). In some cases, synthesis has been difficult: as apparently simple a material as $(EtNH_3)FeCl_4$ has proved to be very difficult for us to prepare reproducibly. Discussion of the unusual behavior of the tetrahedral molecule $[NEt_4][FeCl_4]$ will be reserved until Section VI, below.

The epr spectrum of $[PPh_4][FeCl_4]$, diluted by its gallium analog, yields a zero-field splitting parameter D which ranges from -0.0109 cm^{-1} at room temperature to -0.0419 cm^{-1} at 4.2 K (9).

Most of the octahedral hexahaloferrate(III) ions have been found in bimetallic compounds, such as $[Cr(NH_3)_6][FeCl_6]$, and will not concern us further here (3). The generation of hexahalo compounds with such large cations is the rule. The formation in several cases of the hexachloroferrate(III) anion with small cations such as methylammonium or ethylammonium is therefore remarkable. As an example of this phenomenon, crystal structure analysis shows (10) that the methylammonium compound, $(MeNH_3)_4FeCl_7$, contains $[FeCl_6]^{3-}$ ions arranged in a planar lattice. It is more significant, however, that one has been able to stabilize the large hexachloroferrate(III) ion with only the simple, singly charged methylammonium cation (11).

In a strong crystalline field, the ferric ion has but one unpaired electron and a $^2T_2$ ground state. Three orbital states are low-lying, and spin-orbit coupling effects become very important. Low temperatures are required to detect the epr spectra, and the g-values deviate greatly from 2. One well-known compound with this electronic configuration is $K_3[Fe(CN)_6]$; few other such compounds have been studied recently.

Table 1.   Simple Tetrahedral Tetrahaloferrates

Compound	Space Group	Z	Reference
Na[FeCl$_4$]	P2$_1$2$_1$2$_1$	4	a
Cs[FeCl$_4$]	Pbnm	4	b
(CH$_3$NH$_3$)$_2$[FeBr$_4$]Br	Pna2$_1$	8	c
[bettf]$_2$[FeCl$_4$]	P$\bar{1}$	2	d
[bettf][FeBr$_4$]	P$\bar{1}$	2	d
[PCl$_4$][FeCl$_4$]	Pbcm (or Pbc2$_1$)	4	e
[ϕ_4As][FeCl$_4$]	I$\bar{4}$	2	f
[NEt$_4$][FeCl$_4$]	P6$_3$mc	2	g
[4-Et(py)H][FeBr$_4$]	P2$_1$/c	4	h

a.   R. R. Richards and N. W. Gregory, J. Phys. Chem. **69**, 239 (1965).

b.   G. Meyer, Z. Anorg. Allg. Chem. **436**, 87 (1977).

c.   G. D. Sproul and G. D. Stucky, Inorg. Chem. **11**, 1647 (1972).

d.   T. Mallah, C. Hollis, S. Bott, M. Kurmoo, P. Day, M. Allan and R. H. Friend, J. Chem. Soc. Dalton Trans. 859 (1990). The cation is bis(ethylenedithio)tetrathiafulvalene.

e.   T. J. Kistenmacher and G. D. Stucky, Inorg. Chem. **7**, 2150 (1968).

f.   F. A. Cotton and C. A. Murillo, Inorg. Chem. **14**, 2467 (1975). The phosphorus analog appears to be isomorphous.

g.   D. J. Evans, A. Hills, D. L. Hughes and G. J. Leigh, Acta Cryst. **C46**, 1818 (1990).

h.   M. L. Hackert and R. A. Jacobson, Acta Cryst. B**27**, 1658 (1971).

III. Magnetic Interactions (4)

At high temperatures the spins act independently, but cooperative magnetic behavior sets in as the temperature is decreased to a certain value. The transition from the paramagnetic state to the antiferromagnetic one in zero applied field is a second-order phase transition. The principal mechanism of magnetic interaction is the superexchange interaction. This requires a polarization of the filled electron shells of the ligands between metal ions by the spins of the metal ions, and allows spin information to be transmitted throughout the lattice.

The metal ion appears anomalous to a coordination chemist, for few superexchange-coupled, $S = 5/2$ iron(III) antiferromagnets are known. We distinguish this long-range ordered state from the short-range order which is found in the much better known dimers and clusters. The interaction operator used to describe formally this type of interaction is the usual Hamiltonian

$$\mathcal{H}' = -2J\Sigma S_i \cdot S_j \tag{2}$$

in which S_i and S_j are the spin operators of the interacting ions i and j, and J is called the exchange constant. As written, this is an isotropic interaction, which should be generally applicable to iron(III).

A phase transition to long-range order can occur when the interactions (Eq. 2) extend over the lattice and the temperature of the sample is reduced sufficiently. Exchange effects may be accounted for in the usual molecular field approximation by modifying a susceptibility equation as, for example,

$$\chi_i' = \chi_i[1 - (2zJ/Ng_i^2\mu_B^2)\chi_i]^{-1} \tag{3}$$

where $i = \parallel$ or $\perp$.

IV. The $A_2[FeX_5(H_2O)]$ Series

These salts, which are good examples of the Heisenberg magnetic model, continue to be of interest. They are part of a very large series of isomorphous systems, which are listed in Table 2. Despite the fact that these materials were reviewed some time ago (2), much more has been done on these salts; furthermore, we shall point out that there is still much to be done. Most of the compounds have the virtue of being easily prepared, even in single crystal form.

One of the most interesting salts in this series continues to be $(NH_4)_2[FeCl_5(H_2O)]$, which was reported (13) not to exhibit an easy axis at the transition temperature indicated by a λ-anomaly in the specific heat. The authors suggested that the substance was exhibiting canting, while others (2) suggested that perhaps a crystal phase transition occurs as the substance is cooled to cryogenic temperatures. The magnetic results would then require an interpretation in terms of the low temperature crystallographic structure. Recent work has not yet resolved the problem but makes the crystallographic transition explanation unlikely.

Thus, the specific heats of the isomorphous $(NH_4)_2[InCl_5(H_2O)]$ and $(NH_4)_2[InBr_5(H_2O)]$ have been reported over the respective temperature intervals of 8.5 to 349 K and 7.8 to 348 K (14,15). Neither compound exhibits a λ-anomaly in the measured temperature region, indicating that no such crystallographic phase transition occurs. A similar conclusion arises from an examination of the X-ray powder patterns of $(NH_4)_2[FeCl_5(H_2O)]$, which show no evidence for such a phase transition in the 90-298 K temperature range (16).

Nevertheless the specific heat of $(NH_4)_2[InCl_5(H_2O)]$ does exhibit a small, rounded anomaly at $90 < T/K < 125$ (14). (We note that the specific heat of the iron salt, which is the material of prime interest here, has not been reported above cryogenic temperatures.) Though rotation of the ammonium ion usually causes peaks in the specific heat, such rotation cannot be ruled out as a contributor. The analogous bromide does not exhibit such an anomaly (15).

Much of the recent work on these salts has relied on Mössbauer investigations (17). The first such study on $(NH_4)_2[FeCl_5(H_2O)]$, in the 78-300 K temperature range, revealed a strong temperature dependence of the Fe quadrupole splitting Δ in the salt (18). These data were taken as evidence that a structural phase transition indeed occurred at about 240 K. The quadrupole splitting of the potassium analog does not exhibit a significant temperature dependence (18). Despite the above, it appears that the anomalous behavior is associated with reorientation of the ammonium ion (16). It is well-known that NH_4^+ is a hindered rotator in solids, especially low-symmetry solids. The lines are broadened in the antiferromagnetic phase, and the sublattice magnetization angles deduced from the quadrupolar interaction are inconsistent with a simple canted spin structure. It thus appears that the magnetic structure of $(NH_4)_2[FeCl_5(H_2O)]$ is rather complex, in such a way that magnetic equivalence between all iron sites is destroyed. Similar results have been obtained independently (19). Evidence was presented for the existence of two phase transitions between 7-9 K, as well as for unusually high thermal hysteresis. Arguments in favor of a canted antiferromagnetic structure were presented, but these have been questioned (20).

The epr spectrum of Fe^{3+} in $(NH_4)_2[InCl_5(H_2O)]$ between 100 K and room temperature has also been reported (21). The parameter D, which

Table 2. Unit Cell Parameters of Isomorphic Compounds
of the Series $A_2[MX_5(H_2O)]$ (12).[*]

A	M	X	a(Å)	b(Å)	c(Å)
NH_4	In	Cl	14.10	10.17	7.16
K	In	Cl	13.905	9.952	7.185
Rb	In	Cl	14.050	10.087	7.215
Cs	In	Cl	14.410	10.382	7.416
H_3O	Fe	Cl	13.720	9.926	7.038
NH_4	Fe	Cl	13.706	9.924	7.024
K	Fe	Cl	13.75	9.92	6.93
Rb	Fe	Cl	13.825	9.918	7.100
Rb	Fe	Br	14.2	10.4	7.4
Cs	Fe	Br	14.7	10.7	7.6
NH_4	Mo	Cl	13.860	9.944	7.133
NH_4	Mo	Br	14.423	10.373	7.516
K	Ru	Cl	13.53	9.55	6.96

[*]All belong to space group Pnma and have $Z = 4$.

measures the axial distortion around the ferric ion, becomes more negative at decreasing temperatures. A plot of the spin-Hamiltonian parameter D vs. temperature showed a change in slope at about 240 K, which once again indicated not a phase transition but a slowing down of the ammonium ion ionic reorientation. But this work has recently been criticized (22), for a continuous behavior of the spin-Hamiltonian parameters with temperature was observed. Data down to liquid helium temperatures were reported.

The Mössbauer spectra of $Cs_2[FeCl_5(H_2O)]$ have also been investigated (23,24). The phase diagram, reported earlier (2), was confirmed and a crystallographic phase transition at 151.5 K was discovered. Hysteresis at the spin-flop transition has been investigated in $K_2[Fe(Cl_{1-x}Br_x)_5(H_2O)]$ (x = 0.1, 0.2, 0.3) (25). The pure material, $K_2[FeCl_5(H_2O)]$, orders at 14.0 K and the transition temperature at zero-applied field increases to 15.2 K for the sample with x = 0.3. There are as yet no reported measurements on pure $K_2[FeBr_5(H_2O)]$. As we shall see, there are many other mixed chloro/bromo compounds of iron(III) which have been characterized.

The salt $(H_3O)_2[FeCl_5(H_2O)]$ was first reported by Søtofte and Nielsen (26). The ordering temperature is of the order of 5 K (private communication, W. M. Reiff). We have been unable to prepare the substance despite repeated attempts. We note that there are also reports in the older literature (27) of compounds of stoichiometries such as $Rb_2FeBr_3Cl_2 \cdot H_2O$ and $Rb_2FeCl_3Br_2 \cdot H_2O$. Nothing else is known about these materials.

Little is known about the fluoride analogs. Magnetic data on $K_2[FeF_5(H_2O)]$, which unlike any of the other materials acts as a magnetic linear chain, have been reviewed (2). Recent work on this substance involves a Mössbauer investigation of the long-range ordering at 0.8 K, as well as a determination of part of the phase diagram (28). Otherwise, the crystal structure of $Rb_2[FeF_5(H_2O)]$ has been reported (29). Although there is an orthorhombic phase consisting of discrete octahedra, it belongs to the space group Cmcm, with $Z = 4$, $a = 9.730$, $b = 7.905$ and $c = 8.292$ Å. The cesium and ammonium salts have also been reported. There are as yet no magnetic measurements on these substances.

Further studies of the susceptibilities of $Cs_2[FeBr_5(H_2O)]$ and its rubidium isomorph have been carried out (30). The spin-flop phase boundary of the cesium compound was observed and $H_{SF}(0)$, the spin-flop field extrapolated to 0 K, was found to be 50 kOe. Using the mean field results along with the estimated experimental $\chi_\perp(0) = 0.072$ emu/mol, the anisotropy field was estimated as 6.5 kOe and the exchange field as 196 kOe. Then the ratio $\alpha = H_A/H_E = 3.3 \times 10^{-2}$. No evidence was obtained to confirm the previously-reported anisotropy in the two perpendicular susceptibilities. The new data on $Cs_2[FeBr_5(H_2O)]$ show that χ_b and χ_c are equal, within experimental error. This result also throws into question the earlier report of two λ-features in the specific heat of the material.

The addition of small concentrations of the indium isomorph causes dramatic changes in the ac susceptibility of the $K_2[FeCl_5(H_2O)]$ system (31). That is, $K_2[Fe_pIn_{(1-p)}Cl_5(H_2O)]$ with p as large as 0.87 exhibits a rather sharp peak in the dispersion, χ'. The out-of-phase or absorptive component, χ'', behaves similarly. The susceptibilities of the diluted antiferro-magnetic materials are therefore similar to those of a ferro-magnet. This is caused by local fluctuations in the total magnetic moment which arise because, in the neighborhood of a nonmagnetic impurity, the balance of the two antiferromagnetic sublattices is destroyed. The field-dependent magnetizations measured to date are also characteristic of a ferromagnet.

V. The Tetrahaloferrates(III)

We now describe a set of new compounds, of stoichiometry $A_3Fe_2X_9$, which offer a variety of interesting phenomena. These $S = 5/2$ compounds are part of a large series of tetrahaloferrates which are under extensive investigation. The substances are found to be canted antiferromagnets in which the character of the long-range magnetic ordering is dependent on both the nature of the anion $[FeX_4]^-$ (X = Cl, Br or a mixture of the two) and on chemical substitution in the cation, $[4-X(py)H]^+$ (X = H, Cl, or Br). The ability to control both the degree of long-range ordering and its critical temperature by chemical substitution in this group of compounds may provide an insight into the nature of the canted antiferromagnet. This appears to be the first observation of canted antiferromagnetism in coordination compounds of iron(III). The substances are listed in Table 3. [Incidentally, the alternative dimeric structure, so common with so many other metals, *has* been reported in two cases when the counterion is cesium and the halide is either chloride (32) or fluoride (33).]

The structure of $[(4\text{-chloropyridinium})FeCl_4]_2 \cdot 4\text{-chloropyridinium}$ chloride is illustrated in Fig. 1, showing the eight $[FeCl_4]^-$ anions, twelve $[4\text{-Cl}(py)H]^+$ cations and four chloride ions. The unit cell consists of pairs of discrete tetrachloroferrate(III) anions with 4-chloropyridinium cations. This confirms the nature of the compound as a double salt (34).

The ac susceptibility at zero-applied field was measured on a polycrystalline sample of the 4-chloro compound, as well as on an oriented single crystal. The single-crystal data are presented in Fig. 2. The susceptibility of a powdered sample of the chloride exhibits

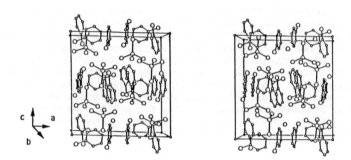

Fig. 1.   The unit cell of $[(4\text{-chloropyridinium})FeCl_4]_2 \cdot 4\text{-}$
chloropyridinium chloride.

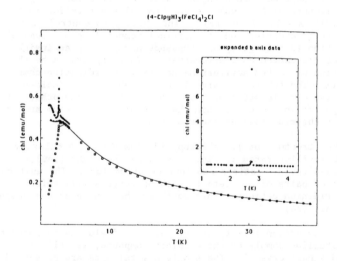

Fig. 2.   The single crystal susceptibilities of $[(4\text{-}$
chloropyridinium$)FeCl_4]_2 \cdot 4\text{-}$chloropyridinium chloride.

Table 3. Magnetic Ordering Temperatures[*]

Compound	T_c, K	Ref
$2[(4-Cl(py)H)FeCl_4] \cdot$		
$[4-Cl(py)H]Cl$	2.73 (χ)	34
	2.685 (c)	35
$2[(4-Br(py)H)FeCl_4] \cdot$		
$[4-Br(py)H]Cl$	2.34 (χ)	34
	2.28 (c)	37
$[4-Br(py)H]_3Fe_2Cl_{1.3}Br_{7.7}$	4.40, 5.67 (χ)	38
$[4-Cl(py)H]_3Fe_2Br_9$	7.96 (χ)	38
	5.50, 7.92 (c)	39
$[4-Cl(py)H]_{1.1}[4-Br(py)H]_{1.9}Fe_2Br_9$		
	7.52, 7.80 (c)	39
	7.73, 7.78 (χ)	39
$[4-Br(py)H]_3Fe_2Br_9$	5.23, 7.64 (c)	39
	7.7 (χ)	
$[(pyH)_3Cl][FeCl_4]_2$	2.212 (c)	40

[*] χ, susceptibility; c, specific heat

a sharp peak with a maximum at T_c = 2.725 K. This is behavior
characteristic usually of either a ferromagnet or a canted
antiferromagnet. The single crystal measurements show that the material
is actually a weak ferromagnet. The susceptibility in the c-direction,
Fig. 2, is the normal parallel susceptibility χ' of an antiferromagnet.
A sharp spike is observed only in data parallel to the b-axis.

Strong absorption (χ'') of the ac signal over the region of the
sharp peak in χ' suggests indeed that the material is a canted
antiferromagnet. The observation of canted antiferromagnetism in this
material is of especial interest because of the isotropic nature of the
ferric ion. Canting usually requires a degree of anisotropy in the
magnetic ion, and is perhaps commonest in cobalt salts, for cobalt
usually exhibits very large g-value anisotropy. On the other hand,
iron(III) is isoelectronic with manganese(II), and there are many
examples of weak ferromagnets of manganese(II). The source of canting
in manganese systems has usually been ascribed to single-ion anisotropy
(zero-field splitting), small as it is with manganese.

The specific heat of $[(4\text{-chloropyridinium})FeCl_4]_2 \cdot 4$-
chloropyridinium chloride has recently been measured (35) in a new
adiabatic calorimetric cryostat; the data are plotted in Fig. 3. A
typical λ-anomaly is observed at T_c = 2.685 K. The lattice specific
heat was determined by means of the modified Komada/Westrum phonon
distribution model (36), which approximates the phonon density of states
based on the experimental specific heat and several physical parameters
related to the mass and structure of the compound in question. The
calculated density of states function is in turn represented by the
single output parameter---the apparent characteristic temperature---θ_{KW}.
Phase transitions typically produce a minimum in the apparent
characteristic temperature. The calculated quantity θ_{KW} for [4-
Cl(py)H]$_3$Fe$_2$Cl$_9$ is presented in Figure 4. Once θ_{KW} is calculated the
lattice specific heat is easily obtained. The lattice specific heat of
[4-Cl(py)H]$_3$Fe$_2$Cl$_9$ is also presented in Figure 3.

Once the lattice specific heat is determined the magnetic specific
heat may be evaluated and analyzed. The specific heat in the transition
region is presented in Figure 3. Four functions are displayed in the
figure. These are the experimental and the lattice specific heats, the
magnetic specific heat and the curve used to extrapolate the magnetic
specific heat to 0 K.

About two-thirds of the disorder process is completed below the
critical temperature as indicated by the entropy values. By comparison
only about 50% of the disorder process in [4-Br(py)H]$_3$Fe$_2$Cl$_9$ takes place
below the transition temperature. This illustrates the fact that short-
range order effects play a much smaller role in the magnetic
interactions of [4-Cl(py)H]$_3$Fe$_2$Cl$_9$ relative to those in [4-
Br(py)H]$_3$Fe$_2$Cl$_9$.

Notes

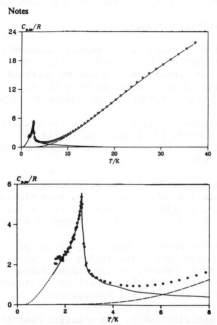

Fig. 3. a) Top: Molar specific heat of $[4\text{-Cl(py)H}]_3Fe_2Cl_9$: o, experimental points; - - -: lattice contribution; ----, excess (magnetic) specific heat; -.-.-., extrapolation of the data to 0 K.   b) Bottom: The transition region on a magnified scale.

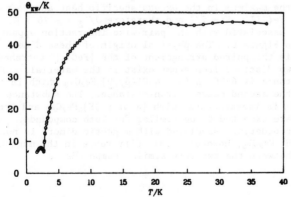

Fig. 4.   Apparent characteristic temperature θ_{KW} for $[4\text{-Cl(py)H}]_3Fe_2Cl_9$

The exchange constant J/k_B may be obtained from the high-temperature tail of the specific heat which follows the relation

$$C_{\underline{p},ex}/R = 2[S(S+1)]^2zJ^2/3k_B^2T^2$$

where z is the magnetic coordination number, assumed to be six for the present system. A plot of the magnetic specific heat against T^{-2} should, therefore, be linear and J/k_B may be obtained from its slope. The exchange constant J/k_B is found to be 0.110 K, in excellent agreement with the values obtained from susceptibility experiments.

The phase diagram of [(4-chloropyridinium)FeCl$_4$]$_2$·4-chloropyridinium chloride is illustrated in Fig. 5. The bicritical point is at 2.40 K and 7 kOe, and $H_{SF}(0) = 5.5$ kOe. With $H_c(0) = 56$ kOe, $\alpha = H_A/H_E = 1.9 \times 10^{-2}$.

We have also prepared [(4-bromopyridinium)FeCl$_4$]$_2$·4-bromopyridinium chloride, which appears to have similar crystallographic and magnetic properties (34). The susceptibility data are similar to those in Fig. 2. Magnetic ordering is found to occur at 2.34 K.

The experimental specific heat of [4-Br(py)H]$_3$Fe$_2$Cl$_9$ is plotted in Figure 6 (37). A λ-type phase transition associated with the onset of long-range ordering is observed at low temperatures. The maximum specific heat associated with the transition was found at 2.280 K and this is identified as the critical temperature.

The lattice contribution to the specific heat of [4-Br(py)H]$_3$Fe$_2$Cl$_9$ was also determined by means of the Komada-Westrum model. A broad minimum in θ_{KW} is observed in the vicinity of 21 K indicating additional excess specific heat around that temperature. This excess specific heat contribution is believed to be that of magnetic dimers which results in a Schottky-like specific heat. For such an interaction the magnetic exchange constant $J/k_{\underline{B}}$ turns out to be dependent on the maximum in the excess specific heat and on the total spin S. For Fe(III), S = 5/2 and therefore $J/k_{\underline{B}} = -26.7$ K. The specific heat associated with the pair-wise interaction appears as the dashed line in Figure 6. The physical origin of these dimer interactions is the paired arrangement of the [FeCl$_4$]$^-$ tetrahedra in the unit cell. Two distinct iron atoms exist in the material. The nearest iron-iron distance is 6.567 Å for [4-Cl(py)H]$_3$Fe$_2$Cl$_9$ which is 1.069 Å shorter then the second nearest iron-distance. The substance [4-Br(py)H]$_3$Fe$_2$Cl$_9$ is isostructural with [4-Cl(py)H]$_3$Fe$_2$Cl$_9$ and iron-iron interactions are expected to be similar for both compounds. Short-range magnetic ordering associated with magnetic dimers is not observed for [4-Cl(py)H]$_3$Fe$_2$Cl$_9$, however. This difference in the magnetic interactions between the two very similar compounds is

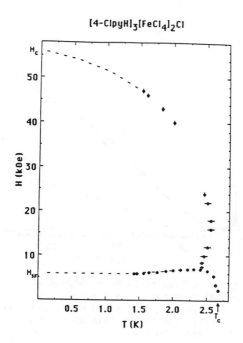

Fig. 5.   The phase diagram in the H-T plane of [(4-chloropyridinium)FeCl₄]₂·4-chloropyridinium chloride.

216

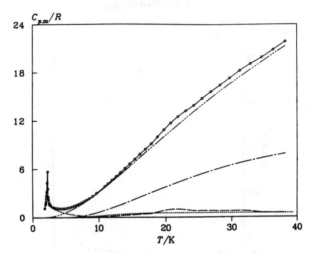

. 6. Molar specific heat of [4-Br(py)H]$_3$Fe$_2$Cl$_9$. o-o-o,
experimental results; --- - - - ---, lattice specific
heat; -.-., specific heat of internal modes as represented
by two Einstein functions (this specific heat is a part of
the lattice contribution); - - - - - -, specific heat of
magnetic dimers; ___ ___ ___ , excess specific heat (see
text).

surprising. It is apparent that the change in cation size allows for long-range exchange which is now the dominant interaction at the expense of short-range pair-wise ordering. Evidently, the dimer interactions are controlled by the nature of the cation and thus chemical substitution may provide a switch able to turn these interactions on and off.

Once the lattice specific heat is resolved, the magnetic specific heat is readily obtained and is also shown in Fig. 6. Two maxima are clearly seen. The broad maximum at higher temperatures appears to agree well with the theoretical prediction for magnetic dimers. A long-range order-disorder transition is identified at lower temperatures. The entropy of the transition between 0 and 9.0 K, is found to be $1.887R$. The theoretical value for the entropy of disorder corresponding to two $S = 5/2$ ions is $2R\ln(2S+1) = 2R\ln(6) = 3.58R$. This value is about twice as large as the experimental entropy of transition. The rest of the entropy has to be included in the Schottky-like anomaly due to the stronger pair exchange constant. Less than half of the disorder process takes place below the peak temperature, 2.280 K. The tail of the transition extends up to 9.0 K with corresponding post-transition entropy of $1.036R$. This indicates that short-range interactions play an important role in the magnetic ordering process. Previously, the compound has been interpreted as an Heisenberg antiferromagnet based upon its magnetic susceptibility and the exchange constant J/k_B was determined to be between -0.111 and -0.094 K. From the critical point, 2.280 K, we find that $J/k_B = -0.114$ K.

In the case of $[(4\text{-bromopyridinium})FeCl_4]_2 \cdot 4\text{-bromopyridinium}$ chloride, the phase diagram yields the results that $H_{SF}(0) = 3$ kOe, $H_c(0) = 42$ kOe, $\alpha = 1.0 \times 10^{-2}$, and the bicritical point is at 2.24 K and 4.2 kOe (34).

Another salt of interest is $[4\text{-Cl(py)H}]_3Fe_2Br_9$, whose structure has been determined (38) and whose magnetic properties and specific heat have been reported (39). A related salt is $[4\text{-Br(py)H}]_3Fe_2Br_9$ (39). Other salts of interest are $[4\text{-Br(py)H}]_3Fe_2Cl_{1.3}Br_{7.7}$ (38), $[4\text{-Cl(py)H}]_{1.1}[4\text{-Br(py)H}]_{1.9}Fe_2Br_9$ (39) and $[(pyH)_3Cl][FeCl_4]_2$ (40) and the analogous bromide (41).

As mentioned above, many stoichiometric substitutions can be made with this series of compounds. The crystal structures of several compositions are known. Thus, $[4\text{-Cl(py)H}]_3Fe_2Br_9$, $[4\text{-Cl(py)H}]_3Fe_2Cl_9$, $[4\text{-Br(py)H}]_3Fe_2Cl_9$, and $[4\text{-Br(py)H}]_3Fe_2Cl_{1.3}Br_{7.7}$ are all monoclinic and isomorphous (34,38). The structures of $[4\text{-Br(py)H}]_3Fe_2Br_9$ and $[4\text{-Cl(py)H}]_{1.1}[4\text{-Br(py)H}]_{1.9}Fe_2Br_9$ are likely to be monoclinic as well and isomorphous with the four compositions.

The magnetic interactions in these materials depend greatly on the chemical composition and in particular on the halide anion. Materials whose halide anion is the smaller chloride tend to order as canted antiferromagnets below 3 K, while the bromide analogs display a more complex ordering behavior at higher temperatures. A strong effect on T_c is observed when bromide replaces chloride at the anionic sites producing $[FeBr_4]^-$, $[FeCl_xBr_{4-x}]^-$ and Br^-. An explanation for the enhancement in superexchange is that there is increased electron delocalization onto the more polarizable bromide, enabling more efficient overlap. The increase in the exchange constant is roughly a factor of 4 on going from the 4-chloropyridinium-containing tetrachloride to the analogous tetrabromide (38).

We assume that $[4-Cl(py)H]_{1.1}[4-Br(py)H]_{1.9}Fe_2Br_9$ is isomorphous to both $[4-Cl(py)H]_3Fe_2Br_9$ and the related salts. The magnetic susceptibility data of a single crystal sample given in Fig. 7 indicate that this compound is likewise a canted antiferromagnet. Under higher resolution, the ferromagnetic spike may be resolved into two peaks with maxima occurring at 7.73 and 7.78 K. Strong absorption (or out-of-phase signal, χ'') was observed in the transition region. These results are confirmed by the specific heat data presented in Fig. 8, which exhibit two sharp peaks at 7.52 K and 7.80 K.

Like the other $[4-X(py)H]_3Fe_2X_9$ compositions the manifestation of magnetic ordering is associated with a sharp peak in the magnetic susceptibility which is accompanied with a strong out-of-phase signal. Such behavior is typical of spin canting. However, unlike most other materials in this chemical system the ferromagnetic peak in the magnetic susceptibility of $[4-Cl(py)H]_{1.1}[4-Br(py)H]_{1.9}Fe_2Br_9$ (Fig. 9) can clearly be resolved into two separate peaks indicating two ordering temperatures and thus two canted phases. The same phenomena are mirrored in the specific heat data.

The entropy distribution and in particular the relatively large degree of magnetic ordering at the second critical temperature (71%) gives strong support to the notion that long-range three-dimensional exchange effects are the prime interactions in this material. By contrast, evidence of substantial short-range order is seen in the compounds $(pyH)_3Fe_2Br_9$, $[4-Br(py)H]_3Fe_2Cl_9$ and $(pyH)_3Fe_2Cl_9$ but not in $[4-Cl(py)H]_3Fe_2Cl_9$. None of these other materials (with the possible exception of the first one) displays two transitions as do the three substances described here. Two critical temperatures, and somewhat more complicated exchange interactions, are also observed in the magnetic susceptibilities of single crystals of $[4-Br(py)H]_3Fe_2Cl_{1.3}Br_{7.7}$. Large, homogeneous, single phase single crystals readily form in all these stoichiometries. It is not yet clear whether in $[4-Cl(py)H]_{1.1}[4-Br(py)H]_{1.9}Fe_2Br_9$ the $[4-X(py)H]^+$ cations are randomly distributed among lattice sites, as in a solid solution or take specific sites to form a

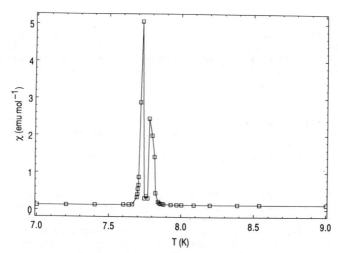

Fig. 7.   Susceptibility measurements for $[4\text{-}Cl(py)H]_{1.1}[4\text{-}Br(py)H]_{1.9}Fe_2Br_9$

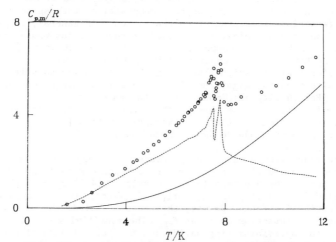

Fig. 8.   Specific heat of $[4\text{-}Cl(py)H]_{1.1}[4\text{-}Br(py)H]_{1.9}Fe_2Br_9$. The solid curve is the calculated lattice specific heat and the dashed curve is the magnetic contribution. (The chopiness of the latter curve is an artifact.)

unique compound. The shape and width of the transitions, and the lack of entropic evidence for short-range magnetic order support the notion that the tetrahaloferrates(III) ions are in a uniform environment, however.

The specific heat of $[4\text{-}Cl(py)H]_3Fe_2Br_9$ is presented in Fig. 9, and is observed to exhibit two λ-type peaks, at 5.50 and 7.92 K. The transition at 7.92 K is in good agreement with the weak ferromagnetic peak observed in the ac susceptibility data. *The susceptibility data provided no indication concerning the existence of the 5.50 K transition.*

The lattice specific heats were again determined using the Komada-Westrum phonon distribution model. The lattice contribution to the specific heat of $[4\text{-}Cl(py)H]_3Fe_2Br_9$ was then determined using $\theta_{KW} =$ 31.9 K, which is the constant value of the apparent characteristic temperature above 19 K.

As in $[4\text{-}Cl(py)H]_{1.1}[4\text{-}Br(py)H]_{1.9}Fe_2Br_9$, but even more so, most of the magnetic ordering is completed at the second critical temperature (81%) which implies that the magnetic interactions are governed by long-range three-dimensional exchange. Short-range exchange interaction plays a smaller role in this material than in any other composition yet examined in this chemical system. These observations are consistent with the magnetic susceptibility data which are found to be reproducible by the simple cubic Heisenberg AF model with $J/k_B = -0.47$ K and g = 1.98. We find that among isomorphous compositions the degree of short-range interactions is dependent on the cation. In general, a smaller degree of short-range order is observed in compositions whose cation is $[4\text{-}Cl(py)H]^+$ relative to compositions in which the cation is $[4\text{-}Br(py)H]^+$. This trend is apparently independent of the tetrahaloferrate(III) anion (bromo, or chloro). Anion substitution has a large effect on the magnitude of the magnetic exchange constant J/k_B, however. For $[4\text{-}Cl(py)H]_3Fe_2Br_9$, where the anion is tetrabromoferrate(III), we observe that long-range order is the dominant exchange interaction and that the exchange constant $|J|/k_B = 0.32$ K (based on T_{c2}) and is the highest in the system.

The transition at lower temperatures, T_{c1}, appears to be associated with a different phenomenon for each composition. Further canting towards the b crystallographic axis in $[4\text{-}Cl(py)H]_3Fe_2Br_9$, which was concluded in $[4\text{-}Cl(py)H]_{1.1}[4\text{-}Br(py)H]_{1.9}Fe_2Br_9$, is not supported by the susceptibility data. We may, therefore, conclude that a new channel for magnetic interaction opens at T_{c1}. The sensitivity of T_{c1} to the nature of the pyridinium ring halide and the predominance of short-range order in $(pyH)_3Fe_2Br_9$ lead us to conclude that Fe-X...X...X-Fe interactions through the pyridinium ring halide may be of importance at lower temperatures in spite of the long exchange pathway. The ordering

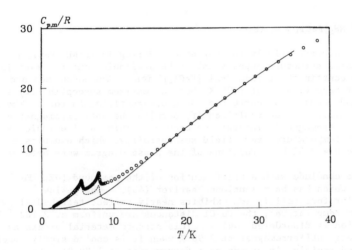

Fig. 9. Specific heat of $[4\text{-}Cl(py)H]_3Fe_2Br_9$. The solid curve is
the calculated lattice contribution and the dashed curve is
the derived magnetic specific heat.

at higher temperatures (i.e., at T_{c2}) in all likelihood originates from a Fe-X...X-Fe exchange pathway, whose sensitivity to the nature of the $[4-X(py)H]^+$ cation is much smaller. Based on the value $T_{c1} = 5.50$ K we calculate a magnetic exchange constant, $J/k_B = -0.22$ K.

Most fascinating, in light of the magnetic behavior displayed so far by these materials, are the properties of $[4-Br(py)H]_3Fe_2Br_9$. This composition yields two λ-type transitions at ca. 5.25 and 7.65 K (39). It is interesting to point out that both T_{c1} and T_{c2} are lower in $[4-Br(py)H]_3Fe_2Br_9$ than in $[4-Cl(py)H]_3Fe_2Br_9$.

There are as yet no Mössbauer spectra reported on any of these compounds.

VI. $[EtNH_3]FeCl_4$ and TEIC

We mention briefly the magnetic ordering in $[EtNH_3]FeCl_4$ (42). The crystal structure appears not to be available, but the material does seem to contain the tetrahedral $[FeCl_4]^-$ ion. The substance orders antiferromagnetically at 1.89 K, but the maximum susceptibility is quite large and needs to be corrected for a demagnetizing factor. The large susceptibility can be explained if, despite the antiferromagnetic order, a large ferromagnetic interaction occurs. This is also indicated by the high-temperature, zero-field susceptibilty, which exhibits a positive $\theta = 2.52$ K. Portions of the phase diagram were determined.

We conclude with a brief mention of the compound $[NEt_4]FeCl_4$ (TEIC), which has been mentioned earlier (43). The yellow tetrachloroferrate(III) ion exhibits nearly perfect tetrahedral symmetry (at room temperature), the Fe-Cl distances are uniform at 2.182 Å but the cation is disordered (44). This strange material orders as a Heisenberg antiferromagnet at 2.98 K when it is cooled slowly; when it is cooled rapidly, it orders as a ferromagnet at 1.89 K! Several crystallographic phase transitions are observed in the specific heat when the sample is cooled and this fact undoubtedly leads to the observed behavior. The analogous bromide, $[NEt_4]FeBr_4$, likewise undergoes crystallographic phase transitions as it is cooled. The substance orders antiferromagnetically at 3.9 K, but a ferromagnetic phase has not been observed (45). The crystal structure of orange $[NEt_4]FeBrCl_3$, which is isostructural to the tetrachloride, has also been reported (44). The bromide ion replaces chloride in either of two sites, giving rise to disorder. There are no magnetic measurements on this substance.

VII. ACKNOWLEDGEMENTS

The author's research has been supported by a succession of grants from the Solid State Chemistry Program of the Division of Materials Research of the National Science Foundation, most recently under DMR-8815798. We thank Roey Shaviv and Fernando Palacio for their comments on the manuscript.

References

1. R. L. Carlin, C. B. Lowe and F. Palacio, An. Quim. **87**, 5 (1991).

2. R. L. Carlin and F. Palacio, Coord. Chem. Revs. **65**, 141 (1985).

3. R. L. Carlin, Comments Inorg. Chem. **11**, 215 (1991).

4. R. L. Carlin, "Magnetochemistry," Springer-Verlag, Berlin, Heidelberg, New York, Tokyo, 1986.

5. S. M. Nelson, in "Comprehensive Coordination Chemistry," Ed. G. Wilkinson, Vol 4., Pergamon Press, Oxford, 1987, p. 217. and references therein.

6. G. D. Sproul and G. D. Stucky, Inorg. Chem. **11**, 1647 (1972).

7. H. Remy, Chem. Ber. **58**, 1565 (1925).

8. E. Durcanska, T. Glowiak, J. Kozisek, I. Ondrejkovicova and G. Ondrejovic, Acta Cryst. C45, 410 (1989).

9. J. C. Deaton, M. S. Gebhard and E. I. Solomon, Inorg. Chem. **28**, 877 (1989).

10. C. B. Lowe, R. L. Carlin, J. A. Zora and K. R. Seddon, unpublished.

11. C. A. Clausen III and M. L. Good, Inorg. Chem. **7**, 2662 (1968).

12. X. Solans, M. C. Moron and F. Palacio, Acta Cryst. C44, 965 (1988).

13. J. N. McElearney and S. Merchant, Inorg. Chem. **17**, 1207 (1978).

14. R. J. C. Brown, J. E. Callanan, R. D. Weir and E. F. Westrum, Jr., J. Chem. Thermo. **20**, 847 (1988).

15. S. M. Acosta, E. F. Westrum, Jr., R. J. C. Brown, J. E. Callanan, and R. D. Weir, J. Chem. Thermo. **20**, 1321 (1988).

16. C. S. M. Partiti, H. R. Rechenberg and J. P. Sanchez, J. Phys. C 21, 5825 (1988).

17. C. E. Johnson, Hyperf. Inter. 49, 19 (1989).

18. C. S. M. Partiti, A. Piccini and H. R. Rechenberg, Sol. St. Comm. 56, 687 (1985).

19. Y. Calage, J. L. Dormann, M. C. Moron and F. Palacio, Hyperf. Inter. 54, 483 (1990).

20. S. R. Brown and I. Hall, J. Mag. Mag. Mat. 104-107, 921 (1992).

21. C. S. M. Partiti and H. R. Rechenberg, J. Phys. Chem. Solids 50, 1023 (1989).

22. S. K. Misra and X. Li, Phys. Rev. B 45, 12883 (1992).

23. J. A. Johnson, C. E. Johnson and M. F. Thomas, J. Phys. C 20, 91 (1987).

24. J. Chadwick and M. F. Thomas, J. Phys. C 20, 3979 (1987).

25. A. Paduan-Filho, C. C. Becerra and F. Palacio, Phys. Rev. B 43, 11107 (1991).

26. I. Søtofte and K. Nielsen, Acta Chem. Scand. A35, 821 (1981).

27. F. Krauss and T. v. Heidlberg, J. Prakt. Chem. 121, 364 (1929).

28. L. Takacs and W. M. Reiff, J. Phys. Chem. Sol. 50, 33 (1989).

29. B. Wallis, U. Bentrup and G. Reck, Eur. J. Sol. St. Inorg. Chem. 27, 681 (1990).

30. R. L. Carlin, D. P. Shum and F. Palacio, unpublished.

31. R. L. Carlin, D. P. Shum, A. Paduan Filho, H. Claus and L. J. de Jongh, unpublished; see also A. Paduan-Filho, V. B. Barbetta, C. C. Becerra, M. Gabas and F. Palacio, to be published. The rubidium salt behaves similarly: M. Gabas, F. Palacio, M. C. Moron and A. Paduan-Filho, J. Mag. Mag. Mat. 114, 246 (1992).

32. M. T. Kovsarnechan, J. Roziere and D. Mascherpa-Corral, J. Inorg. Nucl. Chem. 40, 2009 (1978).

33. J. M. Dance, J. Mur, J. Darriet, P. Hagenmuller, W. Massa, S. Kummer and D. Babel, J. Sol. St. Chem. 63, 446 (1986).

34. J. A. Zora, K. R. Seddon, P. B. Hitchcock, C. B. Lowe, D. P. Shum and R. L. Carlin, Inorg. Chem. **29**, 3302 (1990).

35. R. Shaviv and R. L. Carlin, Inorg. Chem. **31**, 710 (1992).

36. N. Komada and E. F. Westrum, Thermochim. Acta **109**, 11 (1988).

37. R. Shaviv, K. E. Merabet, D. P. Shum, C. B. Lowe, D. Gonzalez, R. Burriel and R. L. Carlin, Inorg. Chem. **31**, 1724 (1992).

38. C. B. Lowe, R. L. Carlin, C.-K. Loong and A. J. Schultz, Inorg. Chem. **29**, 3308 (1990).

39. R. Shaviv, C. B. Lowe and R. L. Carlin, Inorg. Chem., submitted.

40. R. Shaviv, C. B. Lowe, J. A. Zora, C. B. Aakeröy, P. B. Hitchcock, K. R. Seddon and R. L. Carlin, Inorg. Chim. Acta **198-200**, 613 (1992).

41. C. B. Lowe, A. J. Schultz, R. Shaviv and R. L. Carlin, unpublished.

42. E. Lammers, J. C. Verstelle, A. J. van Duyneveldt, C. Lowe and R. L. Carlin, J. Phys. (Paris) Suppl. 12, **49**, C8-1465 (1988).

43. J. A. Puertolas, R. Navarro, F. Palacio, D. Gonzalez, R. L. Carlin and A. J. van Duyneveldt, J. Mag. Mag. Mat. **31-34**, 1067 (1983); J. A. Puertolas, V. M. Orera, F. Palacio and A. J. van Duyneveldt, Phys. Lett. **A98**, 374 (1983).

44. D. J. Evans, A. Hills, D. L. Hughes and G. J. Leigh, Acta Cryst. **C46**, 1818 (1990).

45. R. Navarro, J. A. Puertolas, F. Palacio and D. Gonzalez, J. Chem. Thermo. **20**, 373 (1988).

MAGNETO-STRUCTURAL CORRELATIONS IN

Mn(III) FLUORIDES

Fernando Palacio and M. Carmen Morón

Instituto de Ciencia de Materiales de Aragón.C.S.I.C. - Universidad de Zaragoza.
E–50009 Zaragoza. Spain

1. Introduction

There has been in the recent past much activity in the study of the structural and magnetic properties of Mn(III) fluorides. It is well established that Mn(III) has a firm tendency to produce $[MnF_6]^{3-}$ distorted coordination octahedra mainly due to the influence of a strong Jahn-Teller effect. This favours a tendency to form chains or layers of corner-sharing $[MnF_6]^{3-}$ groups packed in crystal lattices of fairly low symmetry. Another effect of the Jahn-Teller distortion is to increase the single-ion anisotropy of the Mn^{3+} ion, therefore largely affecting the magnetic properties of its derivatives. The physical consequence of low crystal symmetry and magnetic anisotropy is a canting of the magnetic moments observed in the magnetic properties of most of these compounds. This makes the Magnetochemistry of the Mn(III) derivatives far more complex and yet even more fascinating than that of other trivalent transition metal ions [1-4].

Within the rather large number of Mn(III) fluorides it is possible to select some series of structurally related compounds which magnetic properties have been studied. This provides a good opportunity to investigate the relationship between their structural and magnetic properties. In the case of Mn(III) derivatives this is by no means a simple task. The structural distortions always present in these compounds tend to vary within the same series from one compound to other. It is then of foremost importance to carefully select the series in such a way that a minimum number of structural parameters has to be related with the modifications affecting the magnetic properties.

Such is the aim with which we focus this review. In addition, we are not only interested in short range interactions but also in the onset of long range order. For this reason, we are not including in this work many Mn(III) fluorides that either behave as paramagnets in the temperature range they have been studied, as it is the case of several A_3MnF_6 (A = alkaline ion) compounds, or are unique examples within the set of Mn(III) derivatives. We have also left aside more complex bimetallic fluorides where the Mn(III) ion is accompaigned by a divalent transition metal ion, as in the case of several $M^{II}MnF_5$ derivatives or of $Cu_3Mn_2F_{12}·12H_2O$. Compounds within these last series differ in the divalent (magnetic) transition metal, thus introducing many more parameters in the correlations.

The structure of the paper is as follows. We start studying the single ion characteristics of Mn(III) fluorides, since this is a first step necessary to understand their magnetic properties. We then review the properties of some series of one-dimensional and two-dimensional systems. This will be made in sections §3 and §4. Most of the systems here reviewed undergo magnetic ordering below a given temperature, therefore, attention has been paid to the studies concerning the determination of their magnetic structures. Within each section, the structural and magnetic properties of the compounds selected are presented prior to the discussion of the magneto-structural correlations that can be established within each particular series. Some few compounds, chemically related to the series presented in §3 and §4, seem to lack lower dimensional character. Their properties are reviewed in section 5, where the possibilities of a controlled dehydratation to prepare new low dimensional compounds are also discussed. Finally, since this area of research in Magnetochemistry is just at its beginnings, we discuss a variety of perspectives and problems demanding to be solved in the next future.

2. Single ion characteristics

The electronic configuration of Mn(III) ion possess four unpaired electrons in the $3d$ orbitals thus giving place to a ^5D term in the free ion ground state. This ion is isoelectronic with Cr(II). In a weak crystal field of octahedral symmetry the ^5D term splits in a ^5E_g ground state and in a $^5T_{2g}$ excited state. The presence of an orbital doublet as a ground state favors strong Jahn-Teller distortions which confer specific peculiarities to the Mn(III) ion as compared with other trivalent ions of the first raw. The reduction in symmetry caused by the Jahn-Teller distortion splits the ^5E_g state into two singlets, $^5A_{1g}$ and $^5B_{1g}$, which relative energies strongly depend of the form of the distortion. Thus, a compression of the octahedra favors $^5A_{1g}$ as the ground state whereas an elongation favors $^5B_{1g}$ as the lowest energy level [5]. As it will be explained in detail in the next section, a general feature of Mn(III) fluorides is to present elongated octahedra of tetragonal or orthorhombic symmetry.

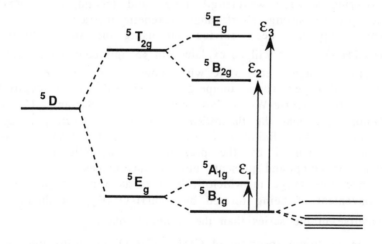

Fig. 1 Schematic energy level diagram of Mn(III)

Spin-orbit coupling splits the ground state furthermore into a singlet and two doublets of lower energy as illustrated in Fig. 1. This zero field splitting is the only that may affect the magnetic properties of the Mn(III) ions since the $^5A_{1g}$ level is at several thousands of cm^{-1}, not reachable by thermally excited electrons. If the zero field splitting is of the order of magnitude of thermal energy then lowering of the temperature will cause depopulation of the energy levels. The consequence is a reduction of the effective spin value with the temperature and the appearance of single-ion anisotropy which, in turns, represents a major contribution to the weak ferromagnetism observed in Mn(III) fluorides, as it will be repeatedly outlined in this review. It is therefore of basic importance to know the order of magnitude of the zero field splitting in these compounds.

Up to the best of our knowledge no EPR experiments have been reported on Mn(III) fluorides besides the fact that the theory has been extensively developed for this ion [6, 7]. In the absence of direct determinations the zero field splitting has been calculated by fitting the susceptibility curve to the Van Vleck equation [8, 9]. This method requires an accurate knowledge of crystal field parameters as calculated from the electronic spectra of these compounds and good susceptibility measurements from a compound possessing as weak magnetic interactions as possible. Both requirements entitle difficulties since first, there is a lack of high quality spectroscopic data covering a wide wavelength range and second, most Mn(III) fluorides possess strong, short range, magnetic interactions.

The fitting has been carried out using the susceptibility of $CsMnF_4 \cdot 2H_2O$. As it will be explained in §5, this compound orders as a 3-d antiferromagnet at 1.53 K with some canting in the magnetic moments. Between room temperature and 8 K the susceptibility follows well a Curie-Weiss law with $\theta = -2.3$ K and C = 3.16 emu.K/mol, very near the theoretical value of 3.06 corresponding to S = 2 and g = 2.02. Below 8 K deviation of the data from the Curie-Weiss law is caused by the presence of magnetic interactions. Although this compound presents obvious advantages because of the wide temperature range in which it behaves as a paramagnet, it has the inconvenient of being formed by $[Mn(H_2O)_2F_4]^-$ octahedra, less representative in the series than the $[MnF_6]^{3-}$ ones.

In the electronic spectrum of $CsMnF_4 \cdot 2H_2O$, the following energy transitions have been assigned to transitions between crystal field states [10][1]:

---

1. The authors actually refer the spectrum to the monohydrate, $CsMnF_4 \cdot H_2O$; however, as it will be evidenced in §5.3, this compound does not seem to be stable. It is therefore reasonable to consider that, in stead, the spectral data refers to $CsMnF_4 \cdot 2H_2O$.

$$\varepsilon_1: {}^5B_{1g} \rightarrow {}^5A_{1g} = 12000 \text{ cm}^{-1}$$
$$\varepsilon_2: {}^5B_{1g} \rightarrow {}^5B_{2g} = 18850 \text{ cm}^{-1}$$
$$\varepsilon_3: {}^5B_{1g} \rightarrow {}^5E_g = 22800 \text{ cm}^{-1}$$

The same assignation of energy transitions has been customarily followed in the interpretation of the electronic spectra of other Mn(III) fluorides as it is summarized in Table I.

TABLE I

Assignments of electronic transitions in spectra of Mn(III) fluorides

Compound	${}^5B_{1g} \rightarrow {}^5A_{1g}$	${}^5B_{1g} \rightarrow {}^5B_{2g}$	${}^5B_{1g} \rightarrow {}^5E_g$	Ref.
Na_3MnF_6	8400	17600	19200	[11]
K_3MnF_6	9000	17500	21000	[12]
Cs_3MnF_6	9200	18000	19500	[11]
Na_2MnF_5	12500	18000	21000	[12]
$(NH_4)_2MnF_5$	12750	18200	21000	[10]
K_2MnF_5	12100	18750	21400	[13]
$K_2MnF_5 \cdot H_2O$	12000	18000	21250	[10]
NH_4MnF_4	14700	18400	22200	[11]
$KMnF_4$	15400	18300	22300	[11]
$RbMnF_4$	15500	18500	22500	[11]
$CsMnF_4$	15500	18000	22000	[11]
$RbMnF_4 \cdot H_2O$	12100	18400	21700	[14]
$CsMnF_4 \cdot 2H_2O$	12000	18850	22800	[14]

Although there is no discussion on the assignation of ε_2 and ε_3 as the transitions from the ground state to the ${}^5T_{2g}$ manifold, the assignation of ε_1 has received more controversy. In addition to the interpretation given above, this energy transition has also been assigned to a spin forbidden transition to the ${}^3T_{1g}$ state [15, 16] and it has also been associated to a charge transfer process between the metal and the fluorine ions [17, 18]. Moreover, when the values of the energy transitions of $CsMnF_4.2H_2O$, as given by Bhattacharjee, are introduce into the Van Vleck equation together with the Slater and spin-orbit coupling parameters, as corresponding to the free ion values of Mn(III) [19], discrepancies between the calculated and the experimental susceptibility curves are observed at temperatures below 30 K. Introducing a covalence correcting factor, $\beta = 0.9$, [20, 21] does not improve the fitting any better.

The splitting of the ^5E_g level has been theoretically estimated in about 4000 cm^{-1} for the MnF_3 [22]. The crystal structure of the MnF_3 consists of $[MnF_6]^{3-}$ distorted octahedra, the octahedron axial deformation being of the same magnitude as in other Mn(III) fluorides. The plausibility of this theoretical prediction does not seem to have been experimentally explored up to the present. As a matter of fact the experimental wavelength range in most spectral studies reported so far on these compounds does not extend below 7000 cm^{-1}. However, preliminary results on spectral studies of these compounds in the wavelength range between 4000 cm^{-1} and 30000 cm^{-1} show that, indeed, a transition at about 5000 cm^{-1} can be observed in these fluorides [8, 9]. The absorption spectrum of $CsMnF_4.2H_2O$ at room temperature is exemplified in Fig. 2.

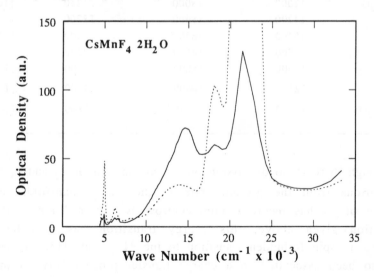

Fig. 2. Polarized crystal spectrum at room temperature of $CsMnF_4.2H_2O$ taken at two perpendicular directions of the electric vector (after [9]).

According to the spectrum of Fig. 2 an alternative assignation of the energy transitions can be proposed for the $CsMnF_4.2H_2O$. Thus, $\varepsilon_1 = 5000$ cm^{-1}, $\varepsilon_2 = 18200$ cm^{-1} and $\varepsilon_3 = 21600$ cm^{-1} would be in better agreement with theoretical estimations [22]. Important enough, when these energy values are introduced into the Van Vleck equation, the experimental susceptibility can be fitted in all the temperature range down to 8 K as illustrated in Fig. 3. Below this temperature magnetic interactions start becoming important in $CsMnF_4.2H_2O$ and the susceptibility separates from the Curie-Weiss law.

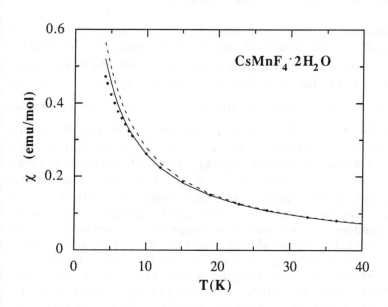

Fig. 3. Magnetic susceptibility of $CsMnF_4.2H_2O$ as a function of the temperature. The broken line is the fitting using the assignation given by Bhattacharjee [10]. The solid line is the fitting using the assignation proposed by Andrés [8].

The fitting of the susceptibility gives 5.5 cm^{-1} and 22.5 cm^{-1} for the energy transitions between the doublet ground state and, respectively, the doublet and the singlet states in which spin orbit coupling splits the $^5B_{1g}$ state. Since the first excited state above the $^5B_{1g}$ manifold is at 5000 cm^{-1}, the magnetic properties of Mn(III) depending on the single ion characteristics are governed by the quintet ground state, $^5B_{1g}$. At high temperature all the energy levels in the quintet are populated and the spin value of the magnetic system is S = 2. At low temperatures excited energy levels of the quintet depopulate and effective spin value decreases down to S = 1/2 at temperatures below 5 K.

3. One-dimensional magnetic systems

One dimensional systems have received most of the attention given to the study of the structural and magnetic properties of Mn(III) fluorides. Two series of compounds are of interest: $A_2MnF_5 \cdot xH_2O$, (x = 0 for A = Li, Na, NH$_4$ and x = 1 for A = K, Rb, Cs, Tl), and $AMnF_4 \cdot H_2O$, (A = NH$_4$, K, Rb and Tl), where the compounds $BMnF_5 \cdot H_2O$, (B = Sr, Ba), should also be included as part of the first series. From an structural point of view, the compounds in both series are formed by infinite zig-zag chains of *trans*-connected octahedra. Within each chain Mn^{3+} ions are linked by common apical F$^-$ ions, the bridge angle Mn-F-Mn being different for each compound. The extend of orbital overlapping throughout the bridges depends therefore of each particular compound and obviously affects to the intrachain superexchange mechanism. The rather large variety of compounds possessing structurally related exchange pathways where only the bridging angle varies as a parameter, renders these series quasi-ideal to investigate how small structural modifications affect the magnetic system. The structural and magnetic properties of the compounds of these series will be reviewed separating in two sub-sections pentafluoromanganate and tetreflouromanganate derivatives, prior to discuss their magneto-structural correlations.

3.1. Structural properties

3.1.1. $A_2MnF_5 \cdot xH_2O$, (x = 0 for A = Li, Na, NH_4 and x = 1 for A = K, Rb, Cs, Tl), and $BMnF_5 \cdot H_2O$, (B = Sr, Ba)

The crystal structures of these pentafluoromanganates(III) have been study by x-ray single crystal diffraction [23-30] and consist of infinite chains of trans–connected $[MnF_4F_{2/2}]^{2-}$ octahedra (Fig. 4). The crystal symmetry of the Li, Na, K, Sr and Ba derivatives is monoclinic while that of the Rb, Cs, Tl and NH_4 is orthorhombic (Table II).

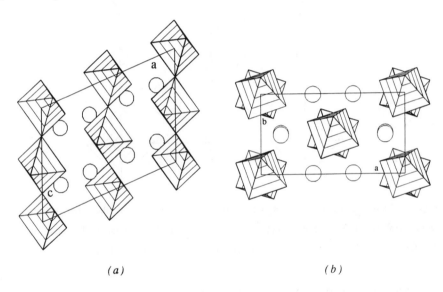

(a)   (b)

Fig. 4. View of the unit cell of Li_2MnF_5: (a) along [010] and (b).along [001]. $[MnF_4F_{2/2}]^{2-}$ units are represented by octahedra and Li^+ ions by circles.

The octahedra are distorted mainly due to the strong Jahn–Teller effect associated to the d^4 high–spin configuration of Mn^{3+}, as stated in §2. The octahedra exhibit three different Mn-F distances (see Table VI in §3.4.1.), the longest lying along the bridging direction following a ferrodistortive ordering, as depicted in Fig. 4a. The lengthening of the Mn-F_{bridge} results from a lower electron density

236

between Mn and F_{bridge} than between Mn and F_{eq}. The geometry of the octahedra remains approximately constant along this series of pentafluoromanganates(III) but the bridge angle Mn–F–Mn varies with the size of the alkaline cation from 121.5° for Li_2MnF_5 to 180° for $Cs_2MnF_5.H_2O$ (see Table VI in §3.4.1.).

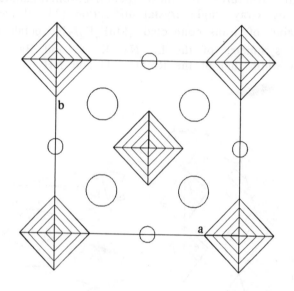

Fig. 5. Crystal structure of $Cs_2MnF_5.H_2O$ projected along the chain direction. Octahedra stand for $[MnF_4F_{2/2}]^{2-}$ units while circles represent the positions of the Cs^+ ions (large circles) and water molecules (small circles).

The A and B cations are located between the anionic chains as do the water molecules in the monoaquo derivatives (Figs. 4b and 5). The water molecules connect two neighboring octahedra belonging to the same chain through, respectively, two O - H ⋯ F hydrogen bonds. Neutron powder diffraction data indicate that $Tl_2MnF_5.H_2O$ exhibits no structural phase transition between room temperature and 2 K [30]. To the authors' knowledge, no low temperature structural data have been reported for the rest of the members of the series.

TABLE II

Crystallographic data of Mn(III) pentafluoromanganates

Compound	space group	a (Å)	b (Å)	c (Å)	β (°)	Z	Ref
Li_2MnF_5	C2/c	10.016(1)	4.948(1)	7.408(1)	112.19(1)	4	23
Na_2MnF_5	$P2_1/c$	7.719(1)	5.236(1)	10.862(2)	108.99(1)	4	24
$(NH_4)_2MnF_5$	Pnma	6.20±.03	7.94±.01	10.72±.01	-	4	25
$SrMnF_5.H_2O$	$P2_1/m$	5.108(1)	7.920(2)	6.106(1)	110.24(1)	2	26
$BaMnF_5.H_2O$	$P2_1/m$	5.370(3)	8.172(2)	6.280(4)	111.17(5)	2	26
$K_2MnF_5.H_2O$	$P2_1/m$	6.04±.01	8.20±.01	5.94±.01	96.5±2	2	27
$Rb_2MnF_5.H_2O$	Cmcm	9.383(2)	8.214(3)	8.348(2)	-	4	28
$Cs_2MnF_5.H_2O$	Cmmm	9.727(8)	8.686(11)	4.254(2)	-	2	29
$Tl_2MnF_5.H_2O$	Cmcm	9.688(2)	8.002(1)	8.339(1)	-	4	30

3.1.2. $AMnF_4 \cdot H_2O$ (A = K, Rb, Tl)

The crystal structure of the isomorphic $AMnF_4 \cdot H_2O$ (A = K, Rb, Tl) compounds consists of chains where alternate trans–$[MnF_4F_{2/2}]^{2-}$ and trans–$[MnF_4(H_2O)_2]^-$ units are connected with each other by sharing apical fluorine atoms, as shown in Fig. 6. X–ray single crystal diffraction experiments at room temperature show that the three compounds crystallize in the monoclinic space group C2/c (Table III) [31-34]. There are two inequivalent Mn^{3+} ions in the unit cell which are related to the $[MnF_4F_{2/2}]^{2-}$ and $[MnF_4(H_2O)_2]^-$ octahedra. The crystal symmetry has also been confirmed by neutron powder diffraction data [34, 35].

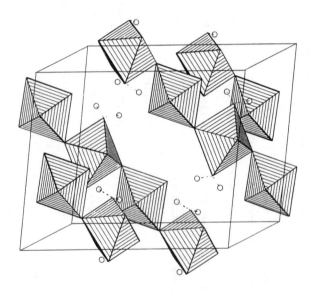

Fig. 6. Crystal structure of $KMnF_4 \cdot H_2O$. Dotted lines indicate $O \cdots H$ bonds. The positions of the K^+ ions are not shown for simplicity.

The essential difference with the series of aquopentafluoromanganates(III) studied in §3.1.1. is that in the $AMnF_4 \cdot H_2O$ family the water molecule is not interstitial but enters the coordination polyhedra of the Mn^{3+} ion. However, the A cations

TABLE III

Crystallographic data of linear chain Mn(III) tetrafluoromanganates

Compound	space group	a (Å)	b (Å)	c (Å)	β (°)	Z	Ref
$KMnF_4 \cdot H_2O$	C2/c	13.907(1)	6.2136(2)	10.492(1)	104.69(1)	8	31,32
	C2/c	13.781(3)[a]	6.160(1)[a]	10.376(3)[a]	104.42(1)[a]	8	35
	C2/c	13.7546(14)[b]	6.1406(5)[b]	10.3343(12)[b]	104.230(5)[b]	8	35
$RbMnF_4 \cdot H_2O$	C2/c	13.932(2)	6.471(1)	10.635(1)	105.54(1)	8	33
	C2/c	13.859(2)[a]	6.446(1)[a]	10.492(1)[a]	104.54(1)[a]	8	35
	C2/c	13.8323(4)[b]	6.4285(2)[b]	10.4837(3)[b]	103.980(2)[b]	8	35
$TlMnF_4 \cdot H_2O$	C2/c	13.784(1)	6.631(1)	10.5377(1)	103.66(1)	8	34
	C2/c	13.6775(4)[c]	6.5862(2)[c]	10.3890(3)[c]	103.176(1)[c]	8	34
	C2/c	13.677(2)[d]	6.5841(8)[d]	10.386(1)[d]	103.12(1)[d]	8	34

a) T = 100 K. b) T = 1.5 K. c) T = 20 K. d) T = 1.3 K.

are interstitial in both series. The $[MnF_4F_{2/2}]^{2-}$ octahedra exhibit three different Mn–F distances (see Table VII in §3.4.2.) the longest lying along the chain direction as in the case of the pentafluoromanganates. On the other hand, the $[MnF_4(H_2O)_2]^-$ anion is elongated in the oxygen direction, the Mn–O bond not following the chain direction. The hydrogen atoms of the water molecule are involved in two hydrogen bonds connecting octahedra belonging to the same and neighboring chains. In addition, neutron powder diffraction experiments performed between room temperature and 1.5 K indicate that no structural phase transition occurs below room temperature [34, 35].

3.2. Magnetic properties

3.2.1. $A_2MnF_5 \cdot xH_2O$, (x = 0 for A = Li, Na, NH₄ and x = 1 for A = K, Rb, Cs, Tl), and $BMnF_5 \cdot H_2O$, (B = Sr, Ba)

The magnetic properties of these compounds have been extensively characterized [8, 23, 30, 36-40]. In the high temperature region the susceptibility follows well the Curie-Weiss law with large negative values of θ (see Table IV), consequence of the presence of rather strong antiferromagnetic interactions in the system.

In the temperature region between about 25 and 120 K the experimental points show a broad maximum indicative of the presence of short range interactions within the chain. The data have been fitted [41] to the isotropic Heisenberg linear chain model following two distinct approaches: the Fisher [42] expression normalized for finite spin values [43, 44]:

$$\chi(T) = \frac{Ng^2\mu_B^2 S(S + 1)}{3kT} + \frac{(1 + u)}{(1 - u)} \qquad (1)$$

where $u = \coth\left[2JS(S + 1)/kT\right] - kT/2JS(S + 1)$ and the high temperature series expansion (h.t.s.) of the reciprocal susceptibility, given by the expression

$$\frac{1}{\chi K} = \frac{3}{S(S + 1)} + \sum_{n = 1}^{\infty} (-1)^n \frac{b_n}{K^n} \qquad (2)$$

where $\bar{\chi}$ is the reduced susceptibility, given by $\bar{\chi} = J\chi/Ng^2\mu_B^2$, K is the reduced temperature, given by $K = kT/J$, g is the Landee factor, N the number of spins in the lattice and b_n are the series expansion coefficients as calculated by Rushbrook and Wood [45]. Either approach leads to fairly similar values of the parameters J/k and g. In Table IV we have summarized the values of J/k selecting those determined from h.t.s. when available. The calculated values of the g factor oscillate between 1.88 and 2.09 in good agreement with the expected one for a fully populated S = 2 orbital ground state. In Fig. 7. the temperature dependence of the susceptibility in the region of the broad maximum is exemplified for the case of $Rb_2MnF_5 \cdot H_2O$. The solid curve is the fit to Eq. (1). In Fig. 8. the reciprocal susceptibility of a selection of $A_2MnF_5 \cdot xH_2O$ and $BMnF_5 \cdot H_2O$ compounds is depicted as a function of the temperature.

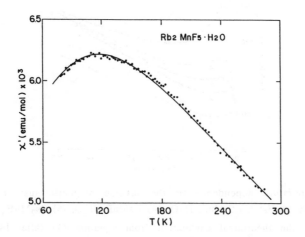

Fig. 7. Temperature dependence of the susceptibility of $Rb_2MnF_5 \cdot H_2O$ in the region where the linear chain phenomena manifest. The full curve is the theoretical calculation from equation (1) (after [8]).

Magnetic ordering has been observed for $(NH_4)_2MnF_5$, $K_2MnF_5 \cdot H_2O$, $Tl_2MnF_5 \cdot H_2O$ and $Rb_2MnF_5 \cdot H_2O$ [8, 36-39]. In addition, an increase observed in the susceptibility curve of Li_2MnF_5 at the lowest measured temperatures might be indicative of the magnetic

242

ordering of the compound at below 1.5 K [36]. The magnetic ordering must be attributed to the existence of interchain magnetic interaction, J', since the absence of long range order is a thermodynamic property intrinsic of linear chain systems.

Two different values of T_c have been reported for $(NH_4)_2MnF_5$. Emori et al. [36] observe a sharp peak in the susceptibility at 5.8 K, while Kida and Watanabe [37, 38] give $T_c = 7.5$ K. Moreover, ac magnetic susceptibility experiments show a sharp peak at 5.2 K [8]. Although there is no conclusive explanation for such a disagreement, it is worth mentioning that $NH_4MnF_4 \cdot H_2O$ has been reported to order at 7.22 K [46]. This compound has often been found as an undesired phase in the preparation of $(NH_4)MnF_5$ [8].

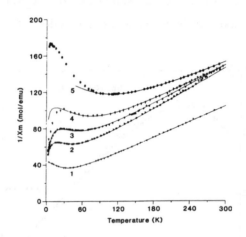

Fig. 8. Temperature dependence of the inverse magnetic susceptibility: (1) Li_2MnF_5; (2) Na_2MnF_5; (3) $SrMnF_5 \cdot H_2O$; (4) $BaMnF_5 \cdot H_2O$; (5) $Cs_2MnF_5 \cdot H_2O$. The full curves are the theoretical calculation from equation (1) (after [41]).

Magnetic ordering for $K_2MnF_5 \cdot H_2O$ and $Tl_2MnF_5 \cdot H_2O$ has been observed at 17.6 and 28 K from, respectively, ac susceptibility at zero external magnetic field and neutron diffraction experiments [8, 30]. Interestingly enough, magnetic ordering in $K_2MnF_5 \cdot H_2O$ was not observed in previous experiments carried out in a SQUID magnetometer, besides the fact that the measurements cover the same temperature range [39]. Similar disagreement occurs in the case of $Rb_2MnF_5 \cdot H_2O$. Mössbauer experiments on diluted $Rb_2Mn_{1-x}Fe_xF_5 \cdot H_2O$ ($x \leq 0.50$) permit to extrapolate the value of T_c for the undiluted compound to $T_c = 25$ K [40]. However, independent experiments carried out by Núñez et al. [39] and Andrés [8] indicate no magnetic ordering in this compound all the way down 1.7 K, the lowest temperature attained in the experiments. Moreover, the ac susceptibility measurements [8] show a sharp increase in the in–phase component accompanied by non-zero values and a sharp increase in the out-of-phase component as depicted in Fig. 9. The data permit to estimate a value for T_c of about 1.5 K. We quote this value in Table IV. The behavior observed in the proximities of T_c is indicative of the presence of net magnetic moments in the material, most likely caused by an small canting of the spins. The same canting phenomena has been observed in $(NH_4)_2MnF_5$ and $K_2MnF_5 \cdot H_2O$ [8]. It would be of great interest to unambiguously determine the ordering temperature for these compounds since this is an important parameter to calculate the interchain interaction, J'.

TABLE IV

Magnetic properties of Mn(III) pentafluromanganates

Compound	θ (K)	$T(\chi_{max})$ (K)	T_c (K)	$J/k^{(*)}$ (K)	J'/J (K) Oguchi	J'/J (K) Villain	T_c/θ
Li_2MnF_5	-66	30.5	<1.5	-6.0	$< 10^{-3}$	$< 1\times10^{-3}$	$< 2.3\times10^{-2}$
Na_2MnF_5	-91	43	?	-8.1	?	?	?
$(NH_4)_2MnF_5$	-153	60	5.2	-10.0	1.1×10^{-3}	3.6×10^{-3}	3.4×10^{-2}
$SrMnF_5 \cdot H_2O$	-134	53	?	-9.9	?	?	?
$BaMnF_5 \cdot H_2O$	-181	69	?	-12.2	?	?	?
$K_2MnF_5 \cdot H_2O$	-295	105	17.6	-18.2	7×10^{-3}	1.5×10^{-2}	6.0×10^{-2}
$Rb_2MnF_5 \cdot H_2O$	-445	117	≈ 1.5	-18.8	-	8.8×10^{-5}	3.4×10^{-3}
$Cs_2MnF_5 \cdot H_2O$	-360	115	?	-16.5	?	?	?
$Tl_2MnF_5 \cdot H_2O$	-470	115	28	-21.5	1.2×10^{-2}	2.3×10^{-2}	6.0×10^{-2}

(*) As determined from h.t.s. when available

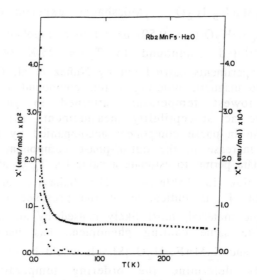

Fig. 9. Temperature dependence of the ac susceptibility of $Rb_2MnF_5 \cdot H_2O$: (x) in-phase component; (•) out-of-phase component.

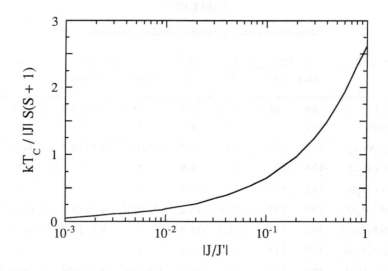

Fig. 10. T_c as a function of $|J'/J|$ as calculated from Oguchi's formula

The value of J' can be estimated from the values of J and T_c following two different approaches. Using a Green function method Oguchi has obtained a relation between T_c and the ratio |J'/J| for a tetragonal lattice [47]. In the corresponding Hamiltonian the sum extends only over interactions between in–chain nearest neighbors, J, and interactions between nearest neighbors belonging to nearest neighbor chains, J'. Oguchi gives numerical results for only four pairs of $(T_c, |J'/J|)$ values. In Fig. 10 we give the whole curve of T_c versus |J'/J| as calculated with Oguchi's formula.

An alternative approach has been developed by Villain and Loveluck [48, 49]. In a system of weakly coupled chains, the ordering temperature can be considered as the temperature at which the thermal energy equals the interaction energy between correlated chain segments:

$$k_B T_c/|J| = \xi(T_c) \ |J'/J| \ S(S+1)$$

where $\xi(T_c)$ is the correlation length within the chain. For the Heisenberg chain

$$\xi(T) = 2|J| \ S(S+1)/k_B T$$

so that

$$k_B T_c/|J| = S(S+1)(2|J'/J|)^{1/2} \tag{3}$$

Magnetic anisotropy, always present in a certain degree in Mn(III), tends to rise T_c, therefore, the above relation provides an upper bound for |J'/J|. The values of |J'/J| calculated from both methods are collected in Table IV, where the ratio T_c/θ is also given for comparative purposes.

3.2.2. $AMnF_4 \cdot H_2O$, $(A = NH_4, K, Rb, Tl)$

With the only exception of the NH_4 derivative, there has been much interest in the study of the magnetic properties of these compounds, in part because it has been possible to grow them as large single crystals suitable for magnetic experiments. Susceptibility measurements have been reported in both powder and single crystal samples [31, 34, 35, 39, 46, 50, 51]. The compounds exhibit very alike magnetic behavior as corresponds to the closed similarity of their respective crystal structures. The case of the NH_4 derivative is somewhat special since it has been indirectly synthesized after partial dehydratation of $NH_4MnF_4 \cdot 2H_2O$, its crystal structure not being as yet determined [8, 46]. However, the magnetic susceptibility of this compound is analogous to the rest of the series. The properties of the NH_4 derivative will be further discussed in §5.2.

At about 52, 42 and 25 K for, respectively, the K, Rb and Tl derivatives the susceptibility curve exhibits a broad maximum corresponding to the one-dimensional character of the compounds. In the case of the Tl derivative the data have been fitted to the Fisher formula, equation (1), while in the case of the K and Rb derivatives h.t.s. for a Heisenberg S = 2 linear chain [45] have been used, equation (2). The fitted values of the g factor are 2.05 for the K and Tl derivatives and 2.03 for the Rb one, indicative of full population of the quintet orbital ground state at these temperatures, as in the case of the pentafluoromanganates. The fitted values of the intrachain exchange interaction, J/k, are summarized in Table V together with other magnetic parameters.

TABLE V

Magnetic properties of linear chain Mn(III) tetrafluromanganates

Compound	θ (K)	$T(\chi_{max})$ (K)	T_c (K)	$J/k^{(*)}$ (K)	J'/J (K) Oguchi	Villain	T_c/θ
$KMnF_4 \cdot H_2O$	-	52	8.45	-6.5	1.1×10^{-2}	2.3×10^{-2}	-
$RbMnF_4 \cdot H_2O$	-58	42	7.3	-6.2	1.3×10^{-2}	1.9×10^{-2}	0.12
$TlMnF_4 \cdot H_2O$	-40	25	8	-3.9	3×10^{-2}	5.8×10^{-2}	0.2
$NH_4MnF_4 \cdot H_2O^{(\#)}$	-	-	7.8	-	-	-	-

(*) As determined from h.t.s. when available; (#) See §5 for the description of this compound

Ac susceptibility measurements on single crystals of the K and Rb derivatives oriented parallel to a direction (hereafter designed as *d* axis) placed in the crystallographic *ac* plane, show a rapid increase of the in-phase component and the appearance of an out-of-phase component at temperatures below 10 K [31, 50]. The data present a sharp peak at 8.45 K in the case of $KMnF_4 \cdot H_2O$ and at 7.3 K in the case of $RbMnF_4 \cdot H_2O$. Instead, the data observed parallel to the *b* crystallographic axis remain essentially constant in the whole temperature range below the broad maximum of the susceptibility. In addition, the data observed parallel to the third principal magnetic axis, perpendicular to the *d* and *b* directions, (hereafter designed as *e* axis) tends to zero as the temperature decreases. These results, exemplified in Fig. 11 for the case of $KMnF_4 \cdot H_2O$, are the fingerprint of weak ferromagnetic behavior of the compounds, the *e* axis being the easy axis of antiferromagnetic alignment, *b* the direction of the hard axis and *d* the direction along which the net magnetic moments align. The precise direction of the magnetic moments has been calculated from the magnetic structure as detailed in §3.3 [35].

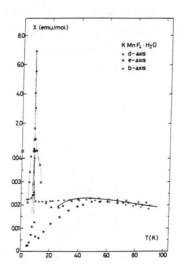

Fig. 11. Temperature dependence of zero-field magnetic susceptibility of $KMnF_4 \cdot H_2O$ along the three principal magnetic axes. For T < 10 K a dotted line is used as guide to the eye. Solid line is the theoretical calculation for a Heisenberg S = 2 linear chain. Observe the change in the scale of the susceptibility axis.

Magnetic ordering has also been observed in the case of a powder sample of $TlMnF_4 \cdot H_2O$ as a peak near 8 K in the susceptibility curve [34]. The occurrence of canting in the magnetic moments can not be excluded and either ac susceptibility measurements or magnetization experiments at low magnetic fields in single crystal samples should be desirable to elucidate this point.

Magnetization measurements on single crystals oriented with the direction of the weak ferromagnetic moment parallel to the external magnetic field have been reported as a function of magnetic field and temperature for the K and Rb derivatives [35]. Field dependent magnetization indicates that fields as small as 50 Oe are sufficient to saturate the weak ferromagnetic moments. The temperature dependence of the spontaneous magnetization, as represented in Fig 12, has been derived by extrapolating for each temperature in the field dependent magnetization curves the saturation of the ferromagnetic moment down to H = 0 Oe. The magnitude of the spontaneous (weak) ferromagnetic moment, $M_S(0)$, has been calculated by extrapolating the temperature dependence of the spontaneous magnetization down to T = 0 K. Values of $M_S(0) = 625$ emu.Oe/mol (0.11 μ_B) and 675 emu.Oe/mol (0.12 μ_B) have been calculated for, respectively, $KMnF_4 \cdot H_2O$ and $RbMnF_4 \cdot H_2O$. From the simple relation $\gamma = tg^{-1}[M_S(0)/Ng\mu_B S]$ the canting angle is calculated to be $\gamma = 1.5°$ and 1.7° for the K and Rb derivatives.

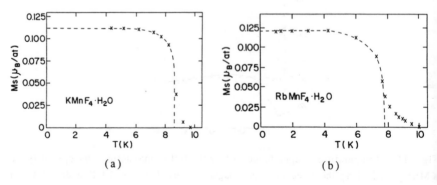

Fig. 12 Temperature dependence of the spontaneous magnetization of: (a) $KMnF_4 \cdot H_2O$; (b) $RbMnF_4 \cdot H_2O$

3.3. Magnetic structures

3.3.1. $A_2MnF_5.xH_2O$ (x = 0 for A = Li, Na, NH_4 and x = 1 for A = K, Rb, Cs, Tl) and $BMnF_5.H_2O$ (B = Sr , Ba)

Only the magnetic structure of $Tl_2MnF_5.H_2O$ has been described within these series of pentafluoromanganates(III). Neutron powder diffraction experiments indicate that this compound orders as a collinear antiferromagnet below $T_c = 28 \pm 1$ K [30]. The magnetic moments align along the chain direction (see Fig. 13) and the nuclear and magnetic unit cell parameters have the same values.

b
c
a

Fig. 13. Magnetic structure of $Tl_2MnF_5 \cdot H_2O$ at $T = 2$ K showing the antiferromagnetic arrangement of the magnetic moments in the unit cell (after [30]).

3.3.2. $AMnF_4.H_2O$ (A = K, Rb, Tl)

The magnetic structures of the members of the $AMnF_4.H_2O$ (A = K, Rb, Tl) family have been study by neutron powder diffraction. These tetrafluorides are isomorphic from the point of view of the nuclear and magnetic structures being the propagation vector of the magnetic structures $k=0$. The three compounds order as collinear antiferromagnets below 8.3 and 8.5, for the K and Rb derivatives [35], and 8.1 for the Tl derivative [34]. The two magnetic sublattices correspond to the two types of octahedra, $[MnF_4F_{2/2}]^{2-}$ and

$[MnF_4(H_2O)_2]^-$. The magnetic moments of each sublattice are located in the xz plane and the angle of the spin direction with respect to the chain axis is about 40° (see Fig. 14). However, a.c. susceptibility and magnetization measurements, as a function of the temperature, indicate that the K and Rb derivatives order as canted antiferromagnets with canting angles of 1.5° and 1.7°, as already mentioned in §3.2.2. In fact, the Shubnikov group C2'/c' allows the presence of a weak ferromagnetic component as the result of a canting of the spins of a sublattice with respect to the other. The small canting angle between the two magnetic sublattices observed in the macroscopic magnetic measurements is far below the resolution of the neutron powder diffraction experiments.

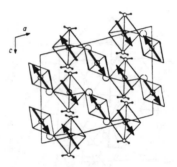

Fig. 14. Magnetic structure of $TlMnF_4 \cdot H_2O$ at T = 1.3 K showing the antiferromagnetic arrangement of the magnetic moments in the unit cell (after [34]).

3.4. Magneto-structural correlations

3.4.1. $A_2MnF_5.xH_2O$ (x = 0 for A = Li, Na, NH_4 and x = 1 for A = K, Rb, Cs, Tl) and $BMnF_5.H_2O$ (B = Sr , Ba)

The magnetic properties of this series of pentafluoromanganates(III) are largely influenced by their 1-d character. The one-dimensional magnetic behavior of these compounds is enhanced by shorter intrachain Mn–Mn distances between neighboring octahedra and intrachain Mn–F–Mn bonds. Table VI summarizes the main structural and magnetic parameters concerning the magneto–structural correlations within this series. In this Table, it is

TABLE VI

Main structural and magnetic parameters concerning the magneto-structural correlations in Mn(III) pentafluoromanganates. Distances are given in Å, angles in degrees and temperatures in K

Compound	Mn–F$_{bridge}$ (l)	Mn–F$_{eq}$ (m)	Mn–F$_{eq}$ (s)	ϕ	Mn–Mn intra	Mn–F–Mn	J/k intra	Mn–Mn inter	|J'/J|	T$_c$
Li$_2$MnF$_5$	2.123(1)[a]	1.852(1)[a]	1.841(1)[a]	28.0	3.7[a]	121.5(1)[a]	– 6.1[a]	5.0[a]	< 10^{-3}	< 1.5[b]
Na$_2$MnF$_5$	2.103(1)[c]	1.872(1)[c]	1.829(1)[c]	21.6	3.9[c]	132.5(1)[c]	– 8.2[a]	5.2[c]	?	?
	2.114(1)[c]	1.849(1)[c]	1.841(1)[c]	28.5						
(NH$_4$)$_2$MnF$_5$	2.091(5)[d]	1.842(4)[d]	1.838(9)[d]	29.3	4.0[d]	143.4(8)[d]	– 10.6[a]	6.2[d]	1.1x10^{-3}[e]	5.2[e]
SrMnF$_5$.H$_2$O	2.108(5)[f]	1.872(8)[f]	1.818(9)[f]	19.9	4.0[f]	139.8[f]	– 10.1[a]	5.1[f]	?	?
BaMnF$_5$.H$_2$O	2.127(1)[f]	1.869(2)[f]	1.838(3)[f]	24.4	4.1[f]	147.7(2)[f]	– 12.8[a]	5.4[f]	?	?
K$_2$MnF$_5$.H$_2$O	2.072(4)[g]	1.842(7)[g]	1.821(7)[g]	25.7	4.1[g]	163.3[g]	– 18.2[a]	6.0[g]	8x10^{-3}[e]	17.6[e]
Rb$_2$MnF$_5$.H$_2$O	2.089(1)[h]	1.860(2)[h]	1.835(2)[h]	24.9	4.2[h]	175.4(2)[h]	– 19.4[a]	6.2[h]	7.4x10^{-5}[e]	≈1.5[e]
Cs$_2$MnF$_5$.H$_2$O	2.127(1)[i]	1.868(9)[i]	1.835(10)[i]	24.1	4.3[i]	180.0[i]	– 17.8[a]	6.5[i]	?	?
Tl$_2$MnF$_5$.H$_2$O	2.085(1)[j]	1.845(9)[j]	1.818(7)[j]	24.7	4.2[j]	179.2(3)[j]	– 21.5[j]	6.3[j]	2.3x10^{-2}	28[j]

ϕ : indicates the ratio Q$_3$/Q$_2$ that is present in a given Jahn–Teller static distortion (see text). |J'/J| : ratio between inter and intra superexchange constants. There are two inequivalent Mn^{3+} ions in the unit cell of Na$_2$MnF$_5$. a) from [23]. b) from [36]. c) from [24]. d) from [25]. e) from [8]. f) from [26]. g) from [27]. h) from [28]. i) from [29]. j) from [30].

interesting to remark that the interchain constant $|J'|$ is much stronger for $(NH_4)_2MnF_5$ than for $Rb_2MnF_5.H_2O$ although the Mn-Mn distances involved are very similar. However, the interaction between chains in $(NH_4)_2MnF_5$ is enhanced by $N - H \cdots F$ bonds while no hydrogen bonds have been reported to link neighboring chains in $Rb_2MnF_5.H_2O$.

Mechanisms for the sign of the magnetic interaction between cation moments, via an anion intermediary, in a ionic crystal have been proposed and given semiquantitative justification [52-55]. These rules depend upon the number and configuration of the d electrons at the cations on both sides of the intermediary anion. Therefore, to apply properly these rules it is necessary to know the electron configuration of the cation. We have already explained (see §2) that the outer-electron configuration of Mn^{3+} in these fluoromanganates is $3d^4$ with an E_g orbital ground state which degeneracy is further removed because of the Jahn–Teller effect This means that the cubic symmetry is not stable and the crystal field at the $3d^4$ cation is less symmetrical, as experimentally observed on the $AMnF_5.xH_2O$ series (see Table VI).

Van Vleck (1939) showed that the normal vibration modes that split the E_g levels are those illustrated in Fig. 15. Positive Q_3 stabilizes the d_{z^2} orbital, negative Q_3 stabilizes the $d_{x^2-y^2}$ orbital and Q_2 stabilizes a mixture of the two. Moreover, Kanamori has obtained that the ratio Q_3/Q_2 that is present in a given static distortion is given by $\tan \phi = [(2/\sqrt{6})(2m - l - s)] / [\pm (2/\sqrt{2})(l - s)]$ where s, m and l are the short, medium and long cation–anion bond lengths of the distorted octahedra. If the static distortion is determined only by Q_3 (local tetragonal distortion) then $m = s$ and $\phi = 30°$, while $\phi = 0°$ $(2m = l + s)$ if it is determined only by Q_2 [56]. If $0° \leq \phi < 30°$, the distortion due to pure Jahn–Teller effect is orthorhombic. If the local symmetry is lower, other effects, either steric or electronic (e.g.: strong spin–orbit coupling), are superimposed.

The essential point for the magnetic coupling between $3d^4$ ions is that the electron ordering associated to the Q_3 mode gives completely empty orbitals directed along the s and m bonds and half filled orbitals directed along the l bonds, while the electron ordering associated to the Q_2 mode gives electron density not only along the l bonds but also, although smaller, along the m bonds [54]. Then, it is possible to deduce the sign of the isotropic exchange interaction along the Mn-F-Mn direction by applying the Goodenough–Kanamori rules [52-55].

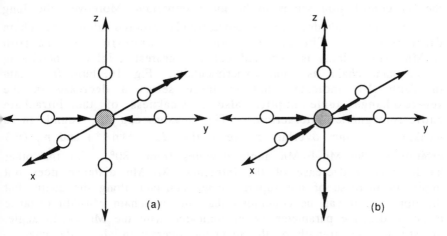

Fig. 15. The normal vibration modes: (a) Q_2 ($Q_2 > 0$) and (b) Q_3 ($Q_3 > 0$)

If a pure $Q_3 > 0$ mode is present in adjacent octahedra, which are antiferrodistortively ordered (...Mn–l–F–s–Mn...) and connected via an angle of 180°, then ferromagnetic interaction takes place via pσ–overlap of half–filled d_{z^2} (l–bond) with empty (s–bond) $d_{x^2-y^2}$ orbitals. Such an interaction weakens as the Mn–F–Mn angle decreases, due to the enhancement of the pπ-overlap between adjacent half–filled t_{2g} orbitals which gives an antiferromagnetic contribution. A ferrodistortive ordering (...Mn–l–F–l–Mn...) should give a strong (half–filled d_{z^2} – half–filled d_{z^2}) antiferromagnetic coupling along the z direction and weak (empty $d_{x^2-y^2}$ – empty $d_{x^2-y^2}$: ...Mn–s–F–s–Mn...) antiferromagnetic coupling along the orthogonal directions. On the other hand, a pure Q_2 mode (or hybrid Q_3/Q_2) on adjacent octahedra can give either ferro or antiferromagnetic coupling depending on the degree of orbital filling and the particular l, m, s spacial ordering.

As a result, from the particular structural arrangement of the distorted $[MnF_4F_{2/2}]^{2-}$ octahedra along the Mn(III) pentafluoromanganates it is possible to deduce the sign of the isotropic superexchange interaction along the chain direction. All the members of this series exhibit both Q_2 and Q_3 contributions, although the Q_3 contribution seems to be more important. Moreover, the long axis of the octahedra are ferrodistortively ordered along the chain direction. Therefore, antiferromagnetic interaction (...Mn–l–F–l–Mn...) is expected between nearest neighbors belonging to the same chain, as found experimentally. Fig. 16, built from data in Table VI, indicates that in these series a decrease in the superexchange angle implies also a weakness of the intrachain antiferromagnetic interaction. The reason can be found in a weakness in the overlap between the d_{z^2} (Mn^{3+}) and p_z (F^-) orbitals as the Mn–F–Mn angle deviates from 180°. On the other hand, since a decrease of the intrachain Mn-Mn distance does not imply an increase of the superexchange constant along the chain, but the opposite, it can be concluded that the intrachain Mn–Mn distance is not a decisive parameter, when compared with the Mn–F–Mn angle, to estimate the strength of the magnetic interaction along the chains.

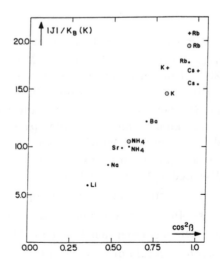

Fig. 16. Correlation between magnetic exchange interaction along the chains and Mn–F–Mn pathway angle. (•) data from [23], (+) data from [39] and (o) data from [8].

3.4.2. $AMnF_4.H_2O$ (A = K, Rb, Tl)

The similar magnetic behavior of these three isomorphic fluorides indicates that the nature of the alkaline cation does not play an important role in the establishment of the magnetic interactions. Superexchange paths along the chain consist of zigzag Mn–F–Mn bridges with very similar angles for the three compounds. O – H $\cdots$ F hydrogen bonds propagate superexchange interactions between Mn ions of neighboring chains, through Mn–F $\cdots$ O–Mn paths, thus permitting the establishment of a tridimensional magnetic ordering. The distances and angles concerning these intra and inter pathways are listed in Table VII clearly indicating the one dimensional character of these tetrafluoromanganates(III). The similarity in the superexchange pathways of the three compounds explain the similar values of critical temperatures (T_c) and interchain (J') and intrachain (J) exchange constants.

As mention in §3.3.2. the two magnetic sublattices correspond to the two types of octahedra, $[MnF_4F_{2/2}]^{2-}$ and $[MnF_4(H_2O)_2]^-$. The small canting angle between the two magnetic sublattices observed in the macroscopic magnetic measurements performed in $KMnF_4.H_2O$ and $RbMnF_4.H_2O$ (see §3.2.2.), can be understood as a result of the competition between the single–ion anisotropy of the Mn(III) in each sublattice. The isotropic exchange interaction does not impose any particular direction to the magnetic moments with respect to the crystal frame. This interaction tends just to keep the spins exactly in the same direction. The elongated axis of the two types of octahedra are not parallel (see Fig. 6 and 14) and one can expect that terms of the form $D_1(u_1S_1)^2$ and $D_2(u_2S_2)^2$ are present in the Hamiltonian, D_1 and D_2 being the strength of the single–ion anisotropy and u_1 and u_2 the unit vectors along the directions of anisotropy in each of the two types of octahedra. The competition between the isotropic exchange and the single–ion anisotropy is, therefore, a plausible mechanism that can explain the presence of a weak ferromagnetic component in these compounds. Of course antisymmetric (Dzyaloshinsky–Moriya) and anisotropic exchange terms are not excluded by symmetry and can also contribute to the actual magnetic structure of these compounds.

TABLE VII

Main structural and magnetic parameters concerning the magneto-structural correlations in linear chain Mn(III) pentafluoromanganates. Distances are given in Å, angles in degrees and temperatures in Kelvin

Compound	Mn-F$_{bridge}$ (l)	Mn-F$_{eq}$ (m)	Mn-F$_{eq}$ (s)	Mn-O	φ	Mn-Mn intra	Mn-Mn-F-Mn	Magnetic ordering	J/k intra	Mn-Mn inter	|J'/J| intra	T$_c$
KMnF$_4$.H$_2$O	2.092(13)[a]	1.883(37)[a]	1.789(35)[a]	2.156(23)[a]	20.8	3.76[a]	135.1(8)[a]	c-AF[a]	- 6.5[b]	5.69[a]	1.1x10^{-2}[c]	8.3[a]
	1.975(13)[a]		1.865(16)[a]									
RbMnF$_4$.H$_2$O	2.120(4)[a]	1.852(13)[a]	1.837(13)[a]	2.130(7)[a]	27.3	3.80[a]	137.7(2)[a]	c-AF[a]	- 6.2[d]	5.79[a]	1.3x10^{-2}[c]	8.5[a]
	1.956(4)[a]		1.828(4)[a]									
TlMnF$_4$.H$_2$O	2.171(4)[e]	1.945(12)[e]	1.719(11)[e]	2.210(5)[e]	0.0	3.79[e]	136.6(2)[e]	c-AF[e]	- 3.9[e]	5.75[e]	3x10^{-2}	8.1[e]
	1.911(4)[e]		1.787(6)[e]									

φ : indicates the ratio Q_3/Q_2 that is present in a given Jahn–Teller static distortion (see text). \|J'/J\| : ratio between inter and intra superexchange constants. c-AF : collinear antiferromagnet (from neutron diffraction). There are two inequivalent Mn^{3+} ions in the unit cell of the three compounds. a) from [35]; structural data at T = 1.5 K. b) from [31]. c) from [8]. d) from [50]. e) from [34]; structural data at T = 20 K.

4. Two-dimensional magnetic systems

The series of layered compounds $AMnF_4$, (A = Na, NH_4, K, Rb, Cs, Tl) are the only Mn(III) fluorides whose magnetic behavior exhibit two-dimensional properties. They are part of a much larger series of compounds of general formula AMF_4, where A is an alkaline ion and M a trivalent transition metal ion from the first raw or a trivalent metal ion from the group III of the Periodic Table. All them are structurally related to $TlAlF_4$ and all but the $CsMnF_4$ order as antiferromagnets. However, the structural and magnetic behavior of these AMF_4 compounds is by no means homogeneous, since the peculiarities of the Mn(III) ion introduce large differences with respect to the rest of the series. As it will be explained below in sections §4.1 to §4.3, the crystal structures of $AMnF_4$ derivatives are far more distorted than any other compound in the AMF_4 series, this feature strongly affecting to their magnetic properties. This makes this group of compounds another important series to investigate how magnetic and structural properties correlate. The magneto-structural correlations in this series will be described in the last part of this section.

4.1. Structural properties

The structural properties of this series at room temperature have been study by means of x-ray crystal diffraction in the case of $LiMnF_4$, $NaMnF_4$ and $TlMnF_4$ [57-59] and by neutron powder diffraction in the case of the K, Rb and Cs derivatives [60-62]. The crystal structures of the members of this family belong to the layered-perovskite structure [63], the spatial arrangement consisting of layers of $[MnF_2F_{4/2}]^-$ corner-sharing octahedra separated by the A atoms as represented in Fig. 17. All the members of the series exhibit a doubling of the a and b unit cell parameters when compared with the unit cell of the ideal tetragonal structure of $TlAlF_4$ or aristotype structure [63]. However, only the Tl derivative has been reported to double the c parameter [59]. The increase of the separation between layers, which corresponds to the c unit cell parameter, is directly related to the increase of the size of the alkaline ion, as shown in Table VIII.

258

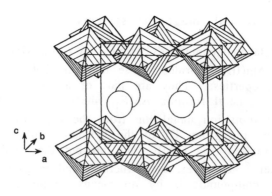

Fig. 17. View of the unit cell of RbMnF$_4$ showing the layered character of this compound. Open circles represent the Rb$^+$ ions while [MnF$_2$F$_{4/2}$]$^-$ units are represented by octahedra.

On the other hand, the [MnF$_2$F$_{4/2}$]$^-$ octahedra show a distortion induced by both steric and Jahn–Teller effects. The three different Mn–F distances are shown in Table X (see §4.4). As expected, the shortest Mn–F distance corresponds to the axial fluorine atoms since they are not shared with neighboring octahedra. The long axis of the octahedra are antiferrodistortively ordered within the layers, largely influencing the magnetic properties of these series of compounds (see Figs. 21 and 23 below in §4.3).

The [MnF$_2$F$_{4/2}$]$^-$ octahedra are also tilted by an angle which depends of the size of the alkaline ion. As an example, the Bond–Valence Sum (BVS) rule [64, 65] predicts for KMnF$_4$ and R b M n F$_4$ (in the absence of electronic and steric effects) : Mn–F = 1.92 Å, K–F = 2.76 Å and Rb–F = 2.93 Å. Therefore, the unit cell parameter of the aristotype structure corresponding to K M n F$_4$ should be $a_{Mn-F} = 3.84$ Å, as calculated from the Mn–F distance, but $a_{K-F} = 3.19$ Å, as calculated from the K-F distance. In a similar way, the unit cell parameter of the aristotype structure corresponding to RbMnF$_4$ should be $a_{Mn-F} = 3.84$ Å but $a_{Rb-F} = 3.38$ Å. The epitaxy between the A cation and the MnF$_4$ layers implies the existence of a certain stress in the structure,

TABLE VIII

Crystallographic data of layered Mn(III) tetrafluoromanganates

Compound	A radius (Å)	space group	a (Å)	b (Å)	c (Å)	β (°)	Z	Ref
$LiMnF_4$	0.92	$P2_1/a$	5.694(1)	4.629(1)	5.414(1)	113.24(2)	2	57
$NaMnF_4$	1.18	$P2_1/a$	5.748(2)	4.892(1)	5.736(2)	108.07(2)	2	58
		$P2_1/a$	5.760(2)[a]	4.892(1)[a]	5.755(2)[a]	108.62(1)[a]	2	58
		$P2_1/a$	5.755(1)[b]	4.889(1)[b]	5.755(1)[b]	108.67(1)[b]	2	58
$KMnF_4$	1.51	$P2_1/a$	7.7062(2)	7.6568(2)	5.7889(1)	90.434(2)	4	60
		$P2_1/a$	7.6830(1)[c]	7.6290(1)[c]	5.7444(1)[c]	90.402(2)[c]	4	60
$TlMnF_4$	1.50	$I2/a$	5.397(2)	5.441(2)	12.484(5)	90.19(3)	4	59
$RbMnF_4$	1.61	$P2_1/a$	7.8119(4)	7.7761(4)	6.0469(3)	90.443(4)	4	60
		$P2_1/a$	7.7865(3)[d]	7.7447(3)[d]	5.9968(2)[d]	90.434(3)[d]	4	60
$CsMnF_4$	1.74	$P4/nmm$	7.9440(6)	7.9440(6)	6.3376(9)	-	4	61
		$P4/n$	7.9148(2)[e]	7.9148(2)[e]	6.3069(2)[e]	-	4	62

a) T = 70 K. b) T = 4 K. c) T = 19.1 K. d) T = 9.6 K. e) T = 24.3 K.

proportional to $d_A = a_{Mn-F} - a_{A-F}$, which is partially relieved by tilting the octahedra. In this way, the atoms find new crystallographic positions minimizing d_A and satisfying, as better as possible, the BVS rule. Since $d_K > d_{Rb}$, the magnitude of the tilt angles should be more important for the K than for the Rb compounds, as found experimentally [60]. The tilting of the octahedra changes as a function of the temperature giving place to different structural phase transitions which are characterictis of the layered–perovskite structure [63].

Finally, the increase of the superexchange Mn–F–Mn angle along the series is directly related to the increase of the size of the A ion (see Table X in §4.4) and plays an important role in the nature of the magnetic interactions within the $AMnF_4$ family. No structural phase transitions have been detected in these compounds between room temperature and 1.5 K except in $LiMnF_4$ for which no low temperature data have been reported [58-62].

4.2. Magnetic properties

Magnetic experiments on these compounds have been so far carried out on powder samples only [11, 58]. Their magnetic characteristics are summarized in Table IX. The most remarkable feature is the ferromagnetic ordering of the $CsMnF_4$ below 21 ± 2 K. The χ' and χ'' components of the ac susceptibility at zero external field are represented in Fig. 18.

TABLE IX

Magnetic properties of $AMnF_4$ series

A	$T_c(K)$	$\theta(K)$	Comments	Ref.
Na	13	-19	$M_S(0)=0.16\mu_B$	[58]
NH_4	10	-12		[11]
K	6	-45		[11]
Tl	4,2	-7	J/k=-0.45	[59]
Rb	3,7(*)	-14		[11]
Cs	21	+27	Ferromagnetic	[11]

(*) T_c from neutron diffraction experiments [60]

The reciprocal magnetic susceptibility of these compounds follows the Curie-Weiss law at high temperatures. However, below about 80 K, 60 K and 150 K for, respectively, the Rb, NH_4 and K derivatives, the experimental susceptibility values tend to smoothly increase over the C-W law predictions, while the anomaly is not observed in the cases of the Cs and Na derivatives [11].

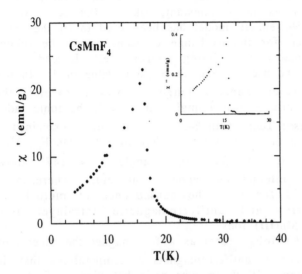

Fig. 18 Temperature dependence of the χ' component of the ac magnetic susceptibility of $CsMnF_4$ as measured at zero external magnetic field. The χ'' component is represented in the insert. ([66])

The origin of the increase of the susceptibility is unclear. The presence of short range ferromagnetic interactions should certainly tend to rise the susceptibility values over the C-W law predictions. Accordingly, ferromagnetic $CsMnF_4$ should present the strongest deviations, while antiferromagnetic $KMnF_4$ and $NaMnF_4$ should, at the most, present deviations in the opposite sense, the susceptibility values tending to be lower than the C-W law predictions. As we will show in §4.3, both K and Na derivatives lack nearest neighbor ferromagnetic interactions. However, reality is very different since experimental values indicate that for the Cs derivative the susceptibility does not deviate from the C-W law all the way down to very near T_c, whereas the K derivative is the member of the series which susceptibility shows the largest deviations (increased values) from the C-W law.

In the case of $TlMnF_4$, experimental susceptibility shows a broad maximum characteristic of two dimensional antiferromagnets [59]. The data have been fitted to the quadratic layer Heisenberg antiferromagnetic model using high temperature series expansions as given by equation (2) [45]. The fit yields a rather low value for the intralayer exchange constant, $J/k = -0.45$ K. Since a two-dimensional Heisenberg model has been used to fit the data, the low value obtained for the exchange constant cannot be related to the low three-dimensional ordering temperature. Instead, an structural argument can be used to explain the low value of J. In the $AMnF_4$ series intralayer superexchange pathways extend throughout $Mn-F_{eq}-Mn$ bridges which angle varies with the ionic radius of the alkaline ion (see Table X in §4.4.). As it will be explained in §4.4 in more detail, in the case of the Tl derivative the $Mn-F_{eq}-Mn$ angle is $146.5°$, very close to the critical angle $\alpha_c \approx 147°$ above which ferromagnetic interactions become dominant. Therefore, the low value obtained for J is the consequence of mutual cancelation between ferro- and antiferromagnetic intralayer interactions between the Mn(III) ions.

Whilst $CsMnF_4$ orders as a ferromagnet the other members of the series order as antiferromagnets at temperatures that depend of the alkaline ion. Weak ferromagnetic behavior at T_c has also been observed in all these compounds with the exception of the Tl and Rb derivatives. Since a vibrating sample magnetometer was used to make the measurements, the absence of weak ferromagnetic phenomena in $TlMnF_4$ and $RbMnF_4$ does not necessarily mean that no canting in the magnetic moments should be expected for saturation of the weak ferromagnetic moments may occur in the presence of rather weak magnetic fields [35]. In fact, this would be an exception in the trend shown by Mn(III) fluorides. In order to clarify this point, ac susceptibility measurements at zero external magnetic field would be desirable.

Thermodynamic arguments predict the absence of ordering in the 2-dimensional Heisenberg and XY models. However, a minute amount of uniaxial anisotropy or interlayer coupling in the system will permit the appearance of magnetic ordering. This is a qualitative difference between 1-d and 2-d Heisenberg models, since in the former only interchange interactions can be responsible for the ordering of the magnetic system. It is therefore of interest to determine which effect triggers the transition to long range order in the case of these layered Mn(III) fluorides. According to the universality law [67] the critical behavior of a 2-d Heisenberg system will change over to 2-d Ising type and then to 3-d when T_c is approached closed enough if the anisotropy, H_A, is larger than interlayer interactions, $g\mu_B H_A > J'$. The occurrence of these crossovers should be reflected in the values of the critical exponents. This approach has been followed in the case of $KMnF_4$ and $RbMnF_4$ analysing the form of the staggered magnetization, determined from neutron diffraction experiments, in the proximities of T_c [60]. Although the analysis made clear that both compounds belong to the same universality class, the presence of critical scattering due to the presence of short range interactions at the onset of magnetic ordering did not permit to stablish unambiguously the origin of the magnetic ordering of the systems. The value of $\beta = 0.26$ obtained for the critical exponent is too high for the Ising model ($\beta = 1/8$) and too low for the Heisenberg one ($\beta = 0.36$). More specific experiments, such as the determination of the magnetic phase diagram, should be desirable to clarify the ordering mechanism of these compounds..

The temperature dependence of the saturation magnetization of the weak ferromagnetic component has been determined in the case of $NaMnF_4$ [58]. Extrapolation of this curve down to T = 0 K gives $M_S(0) = 0.16 \, \mu_B$ which permits to estimate a canting of the magnetic moments of about 5°. This value is somewhat lower than the calculated from the magnetic structure (see §4.3)

4.3. Magnetic structures

With the exception of the Li derivative, the magnetic structures of these family of compounds have been determined from neutron powder diffraction experiments. The setting of the three dimensional magnetic ordering is exemplified for the case of $KMnF_4$ in Fig. 19, where the enhancement observed in some of the Bragg reflections corresponds to the 3-d magnetic ordering of the sample. Following the evolution of the intensity of the magnetic Bragg reflections with temperature (Fig. 20) it is possible to determine the critical temperature at which the magnetic ordering takes place.

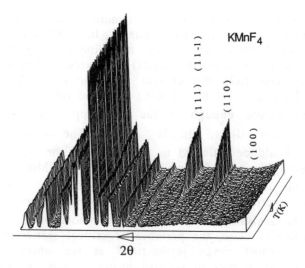

Fig. 19. 3-d plot of the neutron powder diffraction patterns collected between 1.5 K and 9.4 K in KMnF$_4$. The enhancement of some of the Bragg reflections corresponds to the magnetic ordering of the sample.

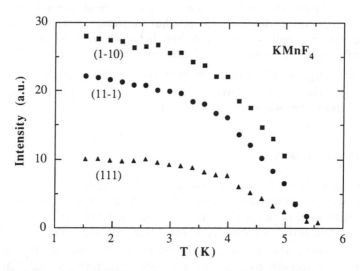

Fig. 20. Integrated intensities of a series of Bragg reflections, with large magnetic contribution, as a function of the temperature for KMnF$_4$. From these data a value of T$_c$ = 5.2±.2 can be deduced.

Then, $NaMnF_4$ and $KMnF_4$ order as non–collinear antiferromagnets below, respectively, 13.0 and 5.2 K [58, 60]. The angle between the two spins directions is 13° for the Na derivative and 17° for the K derivative (Fig. 21). The propagation vector of the magnetic structures is $k = (0, 0, 1/2)$ for $NaMnF_4$ and $k = 0$ for the rest of the members of the series.

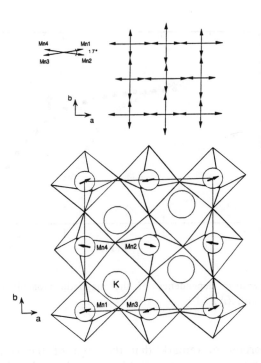

Fig. 21. [001] view of the unit cell of $KMnF_4$ showing the antiferrodistortive ordering of the octahedra and the orientation of the magnetic moments.

On the other hand, $RbMnF_4$ and $TlMnF_4$ order as collinear antiferromagnets below 3.7 and 4.2 K, respectively [59, 60]. It is worth remarking that neutron powder diffraction experiments may not be sensitive enough to small canting angles between magnetic moments and therefore other experiments, as a.c. susceptibility measurements at zero external magnetic field, should be required to verify the collinear antiferromagnetic ordering below T_c in the Rb

266

and Tl derivatives. The temperature dependence of the magnitude of the sublattice magnetic moments is represented in Fig. 22 for $KMnF_4$ and $RbMnF_4$. It has been calculated by means of magnetic structure refinement by using the Rietveld method applied to multiphase patterns [68].

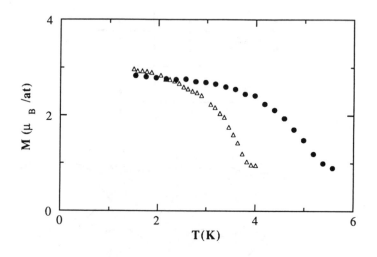

Fig. 22. Temperature dependence of the magnetic moment of each sublattice for : (•) $KMnF_4$ and (Δ)

It is important to remark that the sign of the isotropic exchange interaction between nearest neighbors within the layer is always antiferromagnetic in $TlMnF_4$, $NaMnF_4$ and $KMnF_4$. However, in the case of $RbMnF_4$, the exchange interaction may be either ferro- or antiferromagnetic, depending of which direction, either a or b axis respectively, is considered (see Fig. 23) [60]. The interest of $RbMnF_4$ comes also from the fact that there are two active irreducible representations in the magnetic structure of this compound. The mixture of irreducible representations is an unusual feature exhibited by around 10% of the total number of magnetic structures for which a symmetry analysis has been performed [69].

Finally, $CsMnF_4$ is probably the most remarkable member of the series since it has been described to order as a ferromagnet [61].

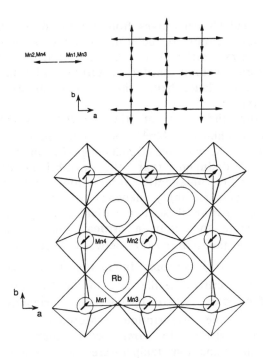

Fig. 23. [001] view of the unit cell of RbMnF$_4$ showing the antiferrodistortive ordering of the octahedra and the orientation of the magnetic moments.

4.4. Magneto-structural correlations

As already mentioned (see §3.4), the sign of the isotropic superexchange interaction is largely influenced by the magnitude of the ϕ parameter and the superexchange angle. The ϕ parameter together with the Mn–F–Mn angles and l, m and s bonds are shown in Table X for each member of the series. The four compounds exhibit both Q_2 and Q_3 contributions, although the Q_3 contribution seems to be more important for the Na and Tl derivatives. However, it is worth to remark that the crystal structures of these two compounds have been refined from x-ray diffraction data while those of the K, Rb and Cs derivatives from neutron data. The atomic position of the fluor atoms is determined more accurately with neutron than with x-ray diffraction. On the other hand, all the members of the series exhibit antiferrodistortive ordering of l and m bonds in the basal plane, being the s–bond always slightly canted from the c axis.

Therefore, from the sign of the isotropic superexchange constants between neighboring Mn^{3+} ions in the same layer, which are presented in Table X, and the Goodenough–Kanamori rules (see §3.4), it can be deduced that the establishment of an antiferromagnetic or ferromagnetic interaction between nearest neighbors in the layers strongly depends of the value of the Mn–F–Mn angle. Comparing these two variables along the series in Table X it is possible to obtain a critical superexchange angle ($\alpha_c \approx 147°$) at which the *crossover* between ferro ($\alpha > \alpha_c$) and antiferromagnetic ($\alpha < \alpha_c$) isotropic interaction takes place in this series. This is nicely verified in the case of the Rb compound, for which $\alpha_x > \alpha_c$ and $\alpha_y < \alpha_c$, where the observed magnetic structure follows closely the above prescription. Moreover, the isotropic interaction can be written, to first order approximation, as $J(\alpha) \propto (\alpha_c - \alpha)$. The transition temperatures also scale with the 'strength' of the magnetic interaction which increases with the departure from α_c.

Another point to be consider is the relationship between canting angle and magnetic anisotropy in this family of compounds. The most common spin–spin interactions found in insulating systems are of superexchange and dipolar types. The effects of the magnetic dipole–dipole interaction are generally negligible except at very low temperatures. Isotropic superexchange interaction does not impose any particular direction of the magnetic moments with respect to the crystal frame. This interaction tends just to keep the spins exactly parallel or antiparallel depending on its sign. Except in cases of topological frustration, or competition between nearest and next–nearest neighbor interactions, the pure isotropic exchange interaction gives collinear magnetic structures. Therefore, anisotropic terms are required in order to explain non–collinear structures.

Single ion anisotropy at inequivalent lattice sites as well as antisymmetric exchange interaction, which is anisotropic in character, are two possible causes of spin canting [70, 71]. An important source of magnetic anisotropy in octahedral Mn^{3+}

TABLE X

Main structural and magnetic parameters concerning the magneto–structural correlations in layered Mn(III) pentafluoromanganates. Distances are given in Å, angles in degrees and temperatures in Kelvin

Compound	Mn-F$_{ax}$ (s)	Mn-F$_{eq}$ (m)	Mn-F$_{eq}$ (l)	φ	Mn-F-Mn	T$_c$	J$_{13}$	J$_{14}$	Magnetic ordering	Canting angle	Ref
LiMnF$_4$	1.817(2)	1.868(2)	2.136(2)	21.6	132.6(1)	There are no magnetic data for LiMnF$_4$					57
NaMnF$_4$[a]	1.818(5)	1.860(6)	2.179(5)	23.9	138.4(3)	13.0±5	-	-	nc-AF	13	58
KMnF$_4$[b]	1.808(2)	1.923(2)	2.102(1)	7.2	146.4(1)[e]	5.2±2	-	-	nc-AF	17	60
	1.806(2)	1.910(1)	2.134(2)	11.8	140.1(1)[f]						
TlMnF$_4$	1.78(1)	1.86(1)	2.15(1)	18.1	146.5(7)	4.2(5)	-	-	c-AF	0	59
RbMnF$_4$[c]	1.802(3)	1.960(4)	2.108(3)	1.0	150.3(1)[e]	3.7±2	+	-	c-AF	0	60
	1.824(3)	1.920(3)	2.094(4)	9.6	145.5(1)[f]						
CsMnF$_4$[d]	1.816(2)	1.924(3)	2.095(2)	7.4	159.9(1)	18.9±5	+	+	c-F	0	61,62

φ : indicates the ratio Q$_3$/Q$_2$ that is present in a given Jahn–Teller static distortion. J$_{13}$: Sign of the isotropic exchange interaction between nearest neighbors (+ or - for ferro- or antiferromagnetic, respectively) along the a axis (see figures 21 and 23) considering that the overall spin arrangement is mainly due to isotropic exchange between nearest neighbors. J$_{14}$: As J$_{13}$ but along the b axis. a) structural data at 70 K. b) structural data at 19.1 K. c) structural data at 9.6 K. d) structural data from [62] at 24.3 K . There are two inequivalent Mn^{3+} ions in the unit cell of KMnF$_4$ and RbMnF$_4$. e) along the a axis. f) along the b axis (see figures 21 and 23). nc-AF : non–collinear antiferromagnet. c-AF: collinear antiferromagnet. c-F : collinear ferromagnet.

compounds is the distortion of the octahedra due to the Jahn–Teller effect [35]. When the orientation of adjacent octahedra is very different (as happens in the $AMnF_4.H_2O$ series), single ion anisotropy can compete with isotropic exchange to give non–collinear structures. In the case of $AMnF_4$, it seems that this interaction does not play an important role. More probably the antisymmetric Dzyaloshinsky–Moriya term of the general exchange hamiltonian, $\mathbf{D}_{ij}(\mathbf{S}_i \times \mathbf{S}_j)$, is the agent of the spin canting observed in the Na and K compounds. The reason for this can be found in the collinear structures observed for Rb, Tl and Cs, where the departure from tetragonal symmetry is very small and, therefore, $\mathbf{D} \approx 0$. Moreover, the anisotropic (symmetric) exchange has a strength proportional to $(\Delta g/g)^2 J$, which is weaker than $D \approx (\Delta g/g)J$, where g is the gyromagnetic ratio and Δg its departure from the free electron value ($g = 2$).

5. Three-dimensional magnetic systems

Only two compounds, $CsMnF_4 \cdot 2H_2O$ and $NH_4MnF_4 \cdot 2H_2O$, whose magnetic behavior lacks lower dimensional properties have so far been reported [46, 51]. In addition, we should quote $Cu_3Mn_2F_{12} \cdot 12H_2O$ which orders as a 3-d ferrimagnet at 3.8 K [72], however, since this compound does not belong to any series of structurally related Mn(III) fluorides we will not include the discussion of its magnetic properties in this review. In fact, three dimensional magnetic behavior is rather unusual in Cr(II), Mn(III) and Cu(II) compounds, since the Jahn-Teller distortion tends to favour lower dimensionality magnetic properties.

It is interesting to observe that the dimensionality of the magnetic lattice in the $AMnF_4 \cdot xH_2O$ (x = 0, 1, 2) is strongly correlated to the value of x. In consequence a careful study of the dehydratation processes of these compounds may lead to new linear chain compounds such as $NH_4MnF_4 \cdot H_2O$ and $CsMnF_4 \cdot H_2O$ whose direct synthesis have not been reported yet. This will be discussed in the second part of this section.

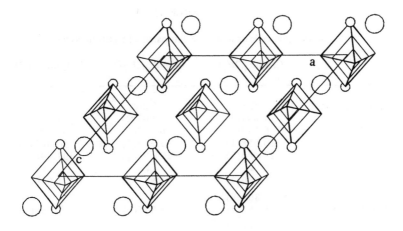

Fig. 24. Crystal structure of $NH_4MnF_4 \cdot 2H_2O$ showing the isolated $[MnF_4(H_2O)_2]^-$ octahedra where the circles represent the oxygen atoms. Large circles stand for NH_4^+ ions.

5.1. Structural properties

Contrary to the 1-d character of the $AMnF_5 \cdot xH_2O$ and $AMnF_4 \cdot H_2O$ series and the 2-d character of the $AMnF_4$ family, the crystal structure of $AMnF_4 \cdot 2H_2O$ (A = Cs, NH_4) contains separate $[MnF_4(H_2O)_2]^-$ anions, linked by a tridimensional network of O–H $\cdots$ F hydrogen bonds (Fig. 24). The Jahn-Teller distortion in these two isomorphic compounds is of the same type as in the Mn^{3+} fluorides above described with long trans Mn–O bonds and shorter Mn–F bonds (Table XI). It is necessary to remark that although x–ray single crystal diffraction experiments carried out in these two compounds indicate that both crystallize in the monoclinic space group C2/c [8, 73]. However, Dubler *et al.* [74] claims that the correct space group to describe the spatial arrangement of the atoms in $CsMnF_4 \cdot 2H_2O$ is the monoclinic noncentrosymmetric C2. Further careful diffraction experiments are required to analyze this discrepancy.

TABLE XI

Crystallographic data of three dimensional Mn(III) fluorides

	$CsMnF_4.2H_2O$	$CsMnF_4.2H_2O$	$(NH_4)MnF_4.2H_2O$
space group	C 2	C2/c	C2/c
a (Å)	11.891(2)	11.907(2)	11.986(1)
b (Å)	6.589(1)	6.597(1)	6.2760(6)
c (Å)	10.558(1)	9.316(2)	10.382(1)
β (°)	131.46(1)	121.77(1)	131.846(4)
Z	4	4	4
$Mn-F_{eq}(l)$	1.870(7)	1.847(2)	1.849(2)
	1.801(8)		
$Mn-F_{eq}(s)$	1.852(7)	1.846(2)	1.833(1)
	1.817(8)		
Mn-O	2.268(6)	2.211(2)	2.208(2)
	2.146(6)		
Ref	a	b	c

a) from [74], (note that C2 is a non-centrosymmetric space group); b) From [73]; c) from [8].

5.2. Magnetic properties

In the absence of an external dc field, the temperature dependence of the ac magnetic susceptibility of these compounds shows a smooth increment in the susceptibility down to 4 K. Below this temperature measurements carried out on single crystal samples indicate that in a given orientation of the crystal with respect to the direction of the alternating field, the susceptibility curve abruptly increases giving rise to a very sharp peak at 1.55 K and 1.75 K for, respectively, the Cs and NH_4 derivatives. The peak in the in-phase component of the susceptibility is accompaigned by an also very sharp peak in the out-of-phase component at slightly lower temperature. The susceptibility curves corresponding to a single crystal (T ≤ 4.2 K) and a powder (T ≥ 4.2 K) sample of the Cs derivative are depicted in Fig. 25. The magnetic behavior at T_c is characteristic of weak ferromagnetism in closed similarity with the rest of the compounds included in this review.

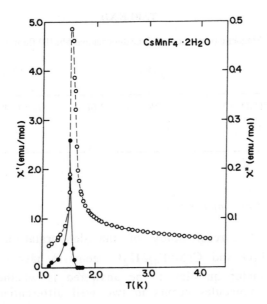

Fig. 25. Temperature dependence of the ac magnetic susceptibility of $CsMnF_4 \cdot 2H_2O$ at H = 0. The measurements corresponding to temperatures below 4.2 K belong to a single crystal oriented in the direction where the susceptibility peaks. For temperatures above 4.2 K the measurements correspond to a powder sample. The continuous line in the paramagnetic region is the fit to the sc Heisenberg antiferromagnetic model with S = 2 (see text).

As detailed in §5.1, the crystal structure of these isomorphic compounds consists in isolated $[MnF_4(H_2O)_2]^-$ octahedra interconnected by a network of hydrogen bonds extending throughout the unit cell. Each Mn ion possess six nearest neighbors. Magnetic interactions between Mn ion neighbors are transmitted by a superexchange mechanism through double bridges Mn-O-F···F-Mn. Pathways of this type have been already described in related systems [1]. In the paramagnetic regime the susceptibility curves fit well the sc Heisenberg antiferromagnetic model with S = 2 as depicted in Fig. 25 for the case of the $CsMnF_4 \cdot 2H_2O$ compound. The resulting fitted parameters calculated using h.t.s. expansions [75] are summarized in Table XII.

TABLE XII

Magnetic parameters of three dimensional Mn(III) fluorides

Compound	T_c (K)	g	J/k_B (K)	Ref.
$NH_4MnF_4 \cdot 2H_2O$	1.75	2.00	-0.08	[76]
$CsMnF_4 \cdot 2H_2O$	1.55	2.00	-0.06	[51]

5.3. Dehydratation processes

A detailed and careful study of the dehydratation processes of $NH_4MnF_4 \cdot 2H_2O$ and $CsMnF_4 \cdot 2H_2O$ seems to have not been as yet reported. A major question to be answered is whether the loss of the two water molecules occurs in two well differentiated steps, thus permitting the appearance of the $AMnF_4 \cdot H_2O$ phases, or in just one. This question has been partially answered in the case of the $NH_4MnF_4 \cdot 2H_2O$ compound [46]. In the process of studying the magnetic properties of $NH_4MnF_4 \cdot 2H_2O$ single crystals below T_c, the magnetic susceptibility of one particular crystal, that had been left in a desiccator for over two months, exhibited a temperature dependence depicted in Fig. 26(b). Instead of one peak, as expected from freshly prepared crystals, the susceptibility curve shows two, at 1.75 K and 7.22 K. The behavior of the peak at higher temperature resembles that of the $KMnF_4 \cdot H_2O$ and $RbMnF_4 \cdot H_2O$ compounds whose susceptibility curves are depicted in Fig. 26(a) and (c), respectively, to facilitate the comparison.

Thermal analysis has confirmed that in the temperature interval between 60 and 105 C this compound losses one molecule of water, the second one being lost at higher temperatures in a more continues process that includes decomposition [8]. These results indicate that $NH_4MnF_4 \cdot H_2O$ does, indeed, exists and its magnetic behavior at the ordering temperature should be close to that of the K and Rb analogous derivatives.

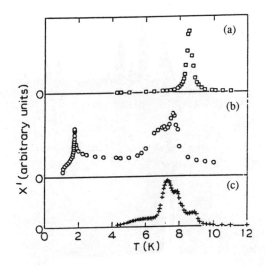

Fig. 26. In-phase susceptibility *versus* temperature at zero external field of (a) $KMnF_4 \cdot H_2O$; (b) $NH_4MnF_4 \cdot 2H_2O$; (c) $RbMnF_4 \cdot H_2O$. The curves are given in arbitrary units that differ in the three cases in order to facilitate the comparison.

According with neutron diffraction experiments, the dehydratation processes in the case of the $CsMnF_4 \cdot 2H_2O$ seems to be rather different than in the NH_4 analogous derivative [77]. The high incoherent background from hydrogenous samples is usually considered very inconvenient in neutron diffraction experiments when powder samples are used, since it severely decreases the quality of the pattern (peak–to–background ratio). This inconvenience can sometimes be turned into an advantage in real–time neutron diffraction experiments because the incoherent scattering provides a straightforward measure of the proton content of the material under investigation as an external parameter (e.g., the temperature) is varied. This affords the possibility of investigating simultaneously the composition (proton content) and structural characteristics of a sample as a function of the temperature.

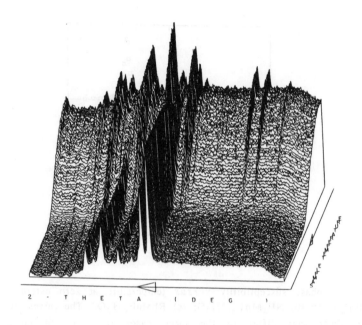

Fig. 27. Neutron powder thermodiffractogram of the dehydratation process of $CsMnF_4 \cdot 2H_2O$

Neutron thermodiffraction experiments on a powder sample of $CsMnF_4 \cdot 2H_2O$ are shown in Fig. 27. Diffraction patterns at the lowest temperatures correspond to the $CsMnF_4 \cdot 2H_2O$ phase while those at the highest temperature correspond to the $CsMnF_4$ one. It is notorious the decrease of the background as the sample losses water in the heating process. As water is removed from the sample the intensities of Bragg peaks corresponding to the low temperature phase smoothly decrease while those corresponding to the high temperature phase increase. Contrary to the case of the NH_4 analogous derivative, there seems to be no evidence of the existence of an intermediate phase in the dehydratation process of $CsMnF_4 \cdot 2H_2O$.

6. Concluding remarks and future prospects

Mn(III) fluorides have proved to be a very reach source of exciting new series of magnetic systems exhibiting a rather large variety of interesting magnetic properties. Apart canting of the magnetic moments, which seems to occur in all these compounds, there are many good examples of lower dimensional systems, both linear chains and layers.

The structural simplicity of these compound series makes them quasi-ideal to investigate detailed correlations between their structural and magnetic properties.

Although much has been recently investigated on the properties of these compounds, this chapter in the Magnetochemistry of Mn(III) fluorides is far to be closed. Many problems still remain obscure waiting for new experiments and/or new approaches to have them solved. To list some few examples, we may start by the lack of precise knowledge of the magnetic behavior of the single Mn^{3+} ion. Detailed EPR experiments would be very valuable to determine the zero field splitting of these compounds accurately, therefore providing a good clue to estimate their single ion anisotropy. In addition, although several electronic spectroscopy experiments have been made in the past, there is a clear need of systematic spectroscopic studies in a wider wavelength range using more modern techniques.

Many linear chain systems have been characterized, some of them showing rather weak interchain interactions. In particular, $Rb_2MnF_5 \cdot H_2O$ is a good representative of the 1-d Heisenberg model with S = 2 and $Cs_2MnF_5 \cdot H_2O$ could be even better, since no magnetic ordering has been observed yet in this compound. The interest in a good, quasi-ideal, system of this model is that it might provide the right material to explore the existence of the Haldane gap in a S = 2 system, something that has not been experimentally verified yet.

The series of layered $AMnF_4$ compounds require further detailed magnetic investigation on single crystal samples. In particular, $CsMnF_4$ is a model example of an S = 2 layer ferromagnet which may be extremely useful to investigate crossover phenomena associated with the magnetic ordering of the system. Moreover, the likely existence of structural transitions above room temperature in the $AMnF_4$ needs to be studied. Comparison of such phase transitions with those extensively reported ones occurring in AMF_4 (M other than Mn) compounds might give better insight to the influence of the Jahn-Teller effect in these lattices.

Finally, a detailed study of the dehydratation processes of $NH_4MnF_4\cdot2H_2O$ and $CsMnF_4\cdot2H_2O$ is still waiting to be carried out. There are several reasons supporting the interest of such study. Thus, it could provide a synthetic route for the preparation of $NH_4MnF_4\cdot H_2O$ and $CsMnF_4\cdot H_2O$, whose direct synthesis have not been reported so far. It would also permit to investigate the dehydratation mechanisms and the structural evolution of the compounds. In addition, a more complete characterization of the magnetic properties of $NH_4MnF_4\cdot H_2O$ would be desirable.

Acknowledgements. The authors gratefully acknowledge Drs. M. Andrés and R. Cases for their permission to report unpublished data in this article. The authors' research has been supported by a succession of grants from the Comision Interministerial de Ciencia y Tecnología, the most recent being MAT91-681.

References

1. R.L. Carlin and F. Palacio, *Coord. Chem. Rev.*, **65** (1985) 141.
2. R.L. Carlin, C.B. Lowe, and F. Palacio, *Anales de Química B*, **87** (1991) 5.
3. R.L. Carlin and R. Burriel, *Gazz. Chim. Italiana*, **121** (1991) 171.
4. R.L. Carlin, *this vol.*
5. R. Dingle, *Inorg.Chem.*, **4** (1965) 1287.
6. C. Rudowicz, in *Physics of magnetic garnets,* Proceedings of the International School of Physics "Enrico Fermi", Course LXX (North Holland, 1978) p. 467.
7. C. Rudowicz, *Physica B*, **155** (1989) 336.
8. M. Andrés, *Thesis*, Univ. of Zaragoza, (1989).
9. M. Andrés, R. Cases, A. Labarta, and F. Palacio, (Results to be published).
10. M.N. Bhattacharjee, M.K. Chaudhuri, H.S. Dasgupta, and A. Kathipri, *Polyhedron*, **4** (1985) 621.
11. P. Köhler, W. Massa, D. Reinen, B. Hofmann, and R. Hoppe, *Z. Anorg. Allg. Chem.*, **446** (1978) 131.
12. D. Oelkrug, *Angew. Chem. Intern. Ed. Engl.*, **5** (1966) 744.

13. T.S. Davis, J.P. Fackler, and M.J. Weeks, *Inorg. Chem.*, 7 (1968) 1994.
14. M.N. Bhattacharjee and M.K. Chaudhuri, *Indian J. Chem.*, 23 A (1984) 424.
15. T.S. Piper and R.L. Carlin, *Inorg. Chem.*, 2 (1963) 260.
16. R.J.H. Clark, *J. Chem. Soc.*, (1964) 417.
17. J.P. Fackler, T.S. Davis, and I.D. Chawla, *Inorg. Chem.*, 4 (1965) 130.
18. R. Dingle, *Acta Chem. Scand.*, 20 (1966) 33.
19. A. Abragam and B. Bleaney, *Electron Paramagnetic Resonance of Tansition Metal Ions*. (Clarendon Press, Oxford, 1970).
20. C.K. Jörgensen, *Absortion Spectra and Chemical Bonding*. (Pergamon Press, Oxford, 1962).
21. J. Owen and J.H.M. Thornley, *Rep. Prog. Phys.*, 29 (1966) 675.
22. A.D. Liehr and C.J. Ballhausen, *Ann. Phys.*, 3 (1958) 304.
23. J. Pebler, W. Massa, H. Lass, and B. Ziegler, *J. Solid State Chem.*, 71 (1987) 87.
24. W. Massa, *Acta Cryst. C*, 42 (1986) 644.
25. D.R. Sears and J.L. Hoard, *J. Chem. Phys.*, 50 (1969) 1066.
26. W. Massa and V. Burk, *Z. anorg. allg. Chem.*, 516 (1984) 119.
27. A.J. Edwards, *J. Chem. Soc. A*, (1971) 2653.
28. P. Bukovec and V. Kaucic, *Acta Cryst.*, B34 (1978) 3339.
29. V. Kaucic and P. Bukovec, *Acta Cryst. B*, 34 (1978) 3337.
30. P. Núñez, A. Tressaud, J. Darriet, P. Hagenmuller, G. Hahn, G. Frenzen, W. Massa, D. Babel, A. Boireau, and J.L. Soubeyroux, *Inorg. Chem.*, 31 (1992) 770.
31. F. Palacio, M. Andrés, C. Esteban-Calderón, and M. Martínez-Ripoll, *J. Solid State Chem.*, 76 (1988) 33.
32. W. Massa, G. Baum, and S. Drueeke, *Acta Cryst. C*, 44 (1988) 167.
33. V. Kaucic and P. Bukovec, *J. Chem. Soc. Dalton*, (1979) 1512.
34. P. Núñez, A. Tressaud, F. Hahn, W. Massa, D. Babel, A. Boireau, and J.L. Soubeyroux, *phys. stat. sol. (a)*, 127 (1991) 505.
35. F. Palacio, M. Andrés, J. Rodríguez–Carvajal, and J. Pannetier, *J. Phys.: Condens. Matter*, 3 (1991) 2379.
36. S. Emori, M. Inoue, M. Kishita, and M. Kubo, *Inorg. Chem.*, 8 (1969) 1385.
37. J. Kida, *J. Phys. Soc. Japan*, 30 (1971) 290.
38. J. Kida and T. Watanabe, *J. Phys. Soc. Japan*, 34 (1973) 952.
39. P. Núñez, J. Darriet, P. Bukovec, A. Tressaud, and P. Hagenmuller, *Mat. Res. Bull.*, 22 (1987) 661.
40. J. Pebler, *Inorg. Chem.*, 28 (1989) 1038.

280

41. J. Pebler, W. Massa, H. Lass, and B. Ziegler, *J. Solid State Chem.*, **71** (1987) 87.
42. M.E. Fisher, *Amer. J. Phys.*, **32** (1964) 343.
43. G.R. Wagner and S.A. Friedberg, *Phys. Lett.*, **9** (1964) 11.
44. T. Smith and S.A. Friedberg, *Phys. Rev.*, **176** (1968) 660.
45. G.S. Rushbrooke and P.J. Wood, *Molec. Phys.*, **1** (1958) 257.
46. F. Palacio, M. Andrés, D. Noort, and A.J. van Duyneveldt, *J. de Phisique*, **C8** (1988) 819.
47. T. Oguchi, *Phys. Rev.*, **133** (1964) 1098.
48. J. Villain and J. Loveluck, *J. Phys. Lett.*, **38** (1977) L77.
49. G. Mennenga, L.J. de Jongh, W.J. Huiskamp, and J. Reedijk, *J. Magn. Magn. Mat.*, **44** (1984) 89.
50. F. Palacio, M. Andrés, R. Horne, and A.J. van Duyneveldt, *J. Magn. Magn. Mat.*, **54-57** (1986) 1487.
51. F. Palacio and M. Andrés, in *Organic and Inorganic Low-dimensional Crystalline Materials*, Ed. P. Delhaes and M. Drillon, NATO ASI series B, 168 (Plenun Press, 1987) p. 425.
52. J.B. Goodenough, *Phys. Rev.*, **100** (1955) 564.
53. J.B. Goodenough, *J. Phys. Chem. Solids*, **6** (1958) 287.
54. J.B. Goodenough, *Magnetism and the Chemical Bond.* (John Wiley-Interscience, New York, 1963).
55. J. Kanamori, *J. Phys. Chem. Solids.*, **10** (1959) 87.
56. J. Kanamori, *Suppl. J. Appl. Phys.*, **31** (1960) 14.
57. K.H. Wandner and R. Hoppe, *Z. anorg. allg. Chem.*, **546** (1987) 113.
58. M. Molinier, W. Massa, S. Khairoun, A. Tressaud, and J.L. Soubeyroux, *Z. Naturforsch., B,* **46** (1991) 1669.
59. P. Núñez, A. Tressaud, J. Grannec, P. Hagenmuller, W. Massa, D. Babel, A. Boireau, and J.L. Soubeyroux, *Z. anorg. allg. Chem.*, **609** (1992) 71.
60. M.C. Morón, F. Palacio, and J. Rodríguez–Carvajal, *J. Phys.: Condens. Matter*, (1993) (in press).
61. W. Massa and M. Steiner, *J. Solid State Chem.*, **32** (1980) 137.
62. J. Rodríguez–Carvajal, F. Palacio, and M.C. Morón (to be published).
63. K.S. Aleksandrov, B.V. Beznosikov, and S.V. Misyul, *Ferroelectrics*, **73** (1987) 201.
64. I.D. Brown and D. Altermat, *Acta Cryst. B*, **41** (1985) 244.
65. N.E. Brese and M. O'Keeffe, *Acta Cryst. A*, **47** (1991) 192.
66. F. Palacio, *unpublished results.*
67. L.J. de Johng and A.R. Miedema, *Experiments on Simple Magnetic Model Systems.* (Taylor and Francis, 1974).

68. Rodríguez–Carvajal. *FULLPROF.* in *Satellite Meeting on Powder Diffraction (XVth Conf. Int. Union of Crystallography.* 1990. Toulouse (Francia).

69. Y.A. Izyumov, V.E. Naish, and S.B. Petrov, *J. Magn. Magn. Mater.*, **13** (1979) 275.

70. T. Moriya, *Phys. Rev.*, **117** (1960) 635.

71. T. Moriya, *Phys. Rev.*, **120** (1960) 91.

72. P. Núñez, J. Darriet, A. Tressaud, P. Hagenmuller, W. Massa, S. Kummer, and D. Babel, *J. Solid State Chem.*, **77** (1988) 240.

73. P. Bukovec and V. Kaucic, *J. Chem. Soc. Dalton*, (1977) 945.

74. E. Dubler, L. Linowsky, J.P. Matthiew, and H.R. Oswald, *Helv. Chim. Acta*, **60** (1977) 1589.

75. G.S. Rushbrooke and P.J. Wood, *Molec. Phys.*, **6** (1963) 409.

76. F. Palacio, M. Andrés, F. Lahoz, C.E. Westphal, and V. Barbeta, *results to be published*.

77. F. Palacio, M.C. Morón, and J. Rodríguez-Carvajal, in *Spanish Research using Neutron Scattering Techniques: 1986-1991*, Ed. J.C. Gómez Sal, *et al.*, (Servicio de Publicaciones Universidad de Cantabria, Santander, 1991) p. 60.

MAGNETIC PROPERTIES OF METALLOCENIUM-BASED ELECTRON-TRANSFER SALTS

Joel S. Miller
Science & Engineering Laboratories
Du Pont
Experimental Station-E328
Wilmington, DE 19880-0328 U. S. A.

Arthur J. Epstein
Department of Physics and Department of Chemistry
The Ohio State University
Columbus, OH 43210-1106 U. S. A.

Molecular and/or organic-based ferromagnetic compounds were postulated in the 1960's, but it has been only during the past few years that several examples have been synthesized and characterized.[1-5] As ferromagnetism is a cooperative property of the bulk and not a of a property molecule, this discovery parallels the discovery of molecular and/or organic-based superconductors and extends the studies of cooperative phenomena in molecular/organic materials. The broad range of phenomena in the molecular/organic solid state combined with the anticipated modification of the physical properties via conventional synthetic organic chemistry as well as the ease of fabrication enjoyed by soluble materials may ultimately lead to utility in future generations of electronic and/or photonic devices.

The first molecular-based material determined to have a ferromagnetic ground state was decamethylferrocenium tetracyanoethylenide, $[Fe^{III}(C_5Me_5)_2]^{.+}[TCNE]^{.-}$.[6,7] The ferromagnetic ground state was characterized by magnetization, M, and neutron diffraction studies.[8a] The critical (or Curie) temperature, T_c, of 4.8 was determined from magnetization[7] as well as heat capacity measurements.[8b] With this observation the systematic study of the structure-function relationship of metallocenium electron-transfer salts of polycyano anions evolved. Since ferromagnetism is a bulk property, the materials primary, secondary, and tertiary structures governs the bulk magnetic properties. Just as for proteins, an understanding of the primary, secondary, and tertiary (3-D) structures is crucial to understanding and modulating the magnetic behavior of a material.

In this review we summarize the structure-magnetic property relationship for the electron-transfer salts based on metallocenium cations and polycyano anions. The common idealized magnetic behaviors expected in materials[2,5] as well as a more comprehensive discussion on the several models for ferromagnetic coupling in molecular/polymeric materials can be found in reviews.[2-5,9,10]

1. $[Fe^{III}(C_5Me_5)_2]^{\cdot+}[TCNE]^{\cdot-}$ as a Model Ferromagnetic System

Experimental evidence for ferromagnetic ground state behavior in a molecular compound has been most extensively studied for the electron-transfer salt of the decamethylferrocene, $Fe^{II}(C_5Me_5)_2$, donor (D) with the tetracyanoethylene, TCNE, acceptor (A).[1-8] This salt has the alternating $\cdots D^{\cdot+}A^{\cdot-}D^{\cdot+}A^{\cdot-}D^{\cdot+}A^{\cdot-}\cdots$ crystal, Figure 1, and electronic structures necessary for the configuration mixing mechanism for ferro- and antiferromagnetic exchange.[2-5,9,10g,11,12]

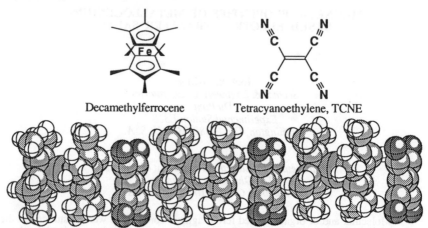

Decamethylferrocene   Tetracyanoethylene, TCNE

Figure 1. Alternating donor/acceptor, $\cdots[D]^{\cdot+}[A]^{\cdot-}\cdots$, linear chain structure of $[Fe^{III}(C_5Me_5)_2]^{\cdot+}[A]^{\cdot-}$ [A = TCNQ, TCNE, DDQ, $C_4(CN)_6$ etc. (see Tables 2 and 3)] and $[Fe^{II}(C_5H_5)_2][TCNE]$, and $[Fe^{III}(C_5Me_5)_2]^{\cdot+}[C_3(CN)_5]^-$. The above structure shows a chains for A = TCNE.

The high-temperature susceptibility of single crystals of $[Fe^{III}(C_5Me_5)_2]^{\cdot+}[TCNE]^{\cdot-}$ aligned parallel to the C_5 molecular axis fits the Curie-Weiss expression with $\theta = +30$ K indicating dominant ferromagnetic interactions.[7] The susceptibility and saturation magnetization calculated as the sum of the contributions from $[Fe^{III}(C_5Me_5)_2]^{\cdot+}$ parallel to the C_5 molecular axis and $[TCNE]^{\cdot-}$ is 6.46 memu/mol at 290 K and 16,300 emuG/mol, respectively. This is in excellent agreement with the observed values of 6.67 memu/mol and 16,700 emuG/mol for single crystals aligned parallel to the chain axis.[7] A spontaneous magnetization is observed for polycrystalline samples below 4.8 K in the Earth's magnetic field.[7] The latter is 36% greater than iron metal on a per iron basis. The critical temperature, T_c, is 4.8 K and hysteresis loops characteristic of ferromagnetic materials are observed.[7] A large coercive field of 1 kOe is recorded at 2 K.[7] The physical properties are summarized in Table 1.

The single crystal susceptibility can be compared with different physical models to aid the understanding of the microscopic spin interactions. For samples oriented parallel to the applied magnetic field, H, the susceptibility above 16 K fits an 1-D Heisenberg model with ferromagnetic exchange of 19 cm^{-1} (27 K).[7] Variation of the low field magnetic susceptibility with temperature for an unusually broad temperature range above T_c [$\chi \propto (T$

Table 1

Summary of Properties of the Molecular/Organic Bulk Ferromagnet [Fe(C$_5$Me$_5$)$_2$]·$^+$[TCNE]·$^-$

Formula (MW):	C$_{26}$H$_{30}$FeN$_4$ (454.4 daltons)
Structure:	1-D ···D·$^+$A·$^-$D·$^+$A·$^-$D·$^+$A·$^-$··· Chains
Solubility:	Conventional Organic Solvents (*e.g.*, THF, CH$_2$Cl$_2$, MeCN)
Critical/Curie Temperature:	4.8 K
Curie-Weiss θ Constant ∥ (⊥) to 1-D chains:	+30 (+10) K
Spontaneous Magnetization:	Yes - in zero applied field
Magnetic Susceptibility ∥ (⊥)to 1-D chains:	0.00667 (0.00180) emu/mol (Observed, 290 K)
Magnetic Susceptibility ∥ (⊥)to 1-D chains:	0.00640 (0.00177) emu/mol (Calculated, 290 K)
Saturation Magnetization ∥ (⊥) to 1-D chains:	16,300 (6,000) emuG/mol (Calculated16,700 emuG/mol)
Intrachain Exchange Interaction ∥ (⊥) to 1-D chains):	27.4 K (19 cm^{-1}) [8.1 K (5.6 cm^{-1})]
Hysteresis Curves:	Yes (1000 Oe Coercive Field; cf. 1 Oe for iron metal)
α Critical constant:	0.09
β Critical constant:	~0.5
γ Critical constant ∥ (⊥) to 1-D chains:	1.22 (1.19)
δ Critical constant:	4.4
Ferromagnetic Ordering:	Yes - Neutron Diffraction Studies on polycrystalline-d$_{30}$ samples
^{57}Fe Mossbauer Zeeman Splitting:	Yes - in zero applied Field (Large Internal Field: 424,000 G (4.2K)

- $T_c)^{-\gamma}$], magnetization with temperature below T_c [$M \propto (T_c - T)^{-\beta}$], and the magnetization with magnetic field at T_c ($M \propto H^{1/\delta}$) enabled the estimation of the β, γ and δ critical exponents. The values of 1.2, ~0.5 and 4.4 respectively were determined for the magnetic field parallel to the chain axis. These values are consistent with a 3-D behavior. Thus, above 16 K 1-D nearest neighbor spin interactions are sufficient to understand the magnetic coupling, however, near T_c 3-D spin interactions are dominant.[7]

The [57]Fe Mössbauer spectra of the TCNE electron-transfer salt of Fe(C_5Me_5)$_2$ give insight into the development of local internal magnetic fields. Atypical six line Zeeman split spectra are observed in zero applied magnetic field at low temperature. The radical anions provide an internal dipolar field. For example, a Zeeman split spectrum with an internal field of 424 kG is observed for the [TCNE]·⁻ salt at 4.2 K. The internal fields are substantially greater than the usual expectation of 110 kG/spin/Fe.[6]

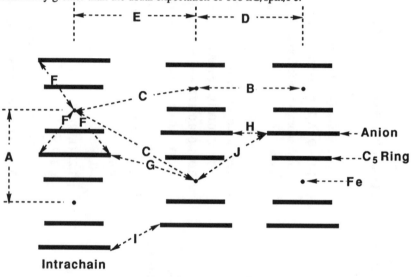

Figure 2   Schematic illustration of the 1-D ···D·⁺A·⁻D·⁺A·⁻D·⁺A·⁻··· and in-registry and out-of-registry interchain interactions.

2. Structure-Magnetic Property Relationship

In order to understand the structural features necessary to stabilize bulk ferromagnetic behavior for [Fe^{III}(C_5Me_5)$_2$]·⁺[TCNE]·⁻ a systematic study of the structure magnetic properties of metallocene based electron-transfer salts with various cyanocarbon anions was undertaken. The metallocenes studied include those with substitution of the Me groups on the cyclopentadienide ring with H, and substitution of Fe with Ru and Os. Numerous materials have been prepared and studied by crystallography as well as magnetic susceptibility. A summary of the observed unit cell parameters as well as general structural motif for 1:1 salts are reported in Table 2. Although not isomorphous, several of the 1:1

Table 2
Summary of the Unit Cell Parameters and General Structural Motif for 1:1 [Metallocene][Acceptor] Complexes

Donor	Acceptor	Structure	D:A Space Group	Volume, Å³	a, Å	b, Å	c, Å	α,°	β,°	γ,°	Z	Sᵃᵗ	T, C
[FeCp*₂]·⁺	[C(CN)₃]⁻	Not 1-D	1:1 Pma2	4496	22.357	21.331	9.420	90.00	90.00	90.00	8	S	-100
[Cr(C₆H₆)₂]·⁺	[TCNE]⁻	1-D .DAAD.	1:1 P2₁/m	1528	10.347	12.423	12.763	90.00	111.33	90.00	4	S	23
[Cr(C₆Me₃H₃)₂]·⁺	[TCNE]⁻	1-D .DAAD.	1:1 P1̄	1266	10.475	11.427	8.419	64.20	78.59	81.56	2	S	23
[CrCp*₂]·:⁺	[TCNE]⁻	1-D .DADA.	1:1 Cmc2₁	2491	10.6	16.1	14.6	90.00	90.00	90.00	4	P	23
[CrCp*₂]·:⁺	[TCNE]⁻	1-D .DADA.	1:1 P2₁/n	2465	10.953	14.170	15.900	90.00	92.63	90.00	4	S	-70
[Fe(C₅Et₅)₂]·⁺	[TCNE]⁻	Not Solved	1:1	3387	11.769	11.769	24.45	90.00	90.00	90.00		S	
[Fe(C₅H₄)₂C₃H₆]·⁺	[TCNE]⁻	1-D DD AA	1:1 P2₁/c	1117	6.833	18.385	12.214	90.00	90.56	90.00	4	S	
[Fe(C₅Me₄H)₂]·⁺	[TCNE]⁻	1-D .DADA.	1:1 Cmca	2170	14.990	11.580	12.503	90.00	90.00	90.00	4	S	23
[FeCp₂]·⁺	[TCNE]⁻	1-D .DADA.	1:1 P1̄	373	7.863	7.852	6.767	113.45	96.33	76.96	1	S	
[FeCp₂]·⁺	[TCNE]⁻	1-D .DADA.	1:1 P1̄	370	7.770	7.870	6.780	113.60	96.70	77.00	1	S	
[FeCpCp*]·⁺	[TCNE]⁻	1-D .DADA.	1:1 P2₁/m	1063	8.349	13.338	10.253	90.00	111.47	90.00	2	S	-65
[FeCpCp*]·⁺	[TCNE]⁻·ᵇ	DAAD dimer	1:1 P2₁/n	2159	8.826	17.603	13.900	90.00	90.64	90.00	4	S	-70
[MnCp*₂]·⁺	[TCNE]⁻	1-D .DADA.	1:1 Cmc2₁	2492	10.6	16.1	14.6	90.00	90.00	90.00	4	P	
[NiCp*₂]·⁺	[TCNE]⁻	1-D .DADA.	1:1 Cmc2₁	2488	10.530	16.130	14.640	90.00	90.00	90.00	4	P	
[FeCp*₂]·⁺	[TCNE]⁻	1-D .DADA.	1:1 Cmc2₁	2493	10.621	16.113	14.558	90.00	90.000	90.00	4	S	
[FeCp*₂]·⁺	[TCNE]⁻	1-D .DADA.	1:1 C2/c	5443	16.250	10.415	32.851	90.00	101.76	90.00	8	S	
[FeCp*₂]·⁺	[TCNQ]⁻·ᵇ	1-D .DADA.	1:1 P1	652	8.639	9.577	10.028	63.75	70.21	62.78	1	S	
[CoCp*₂]·⁺	[C₃(CN)₅]⁻	1-D .DADA.	1:1 C2/c	2534	14.060	14.234	12.873	90.00	100.27	90.00	4	S	
[CrCp*₂]·:⁺	[C₃(CN)₅]⁻	1-D .DADA.	1:1 C2/c	2610	14.269	14.210	13.157	90.00	101.91	90.00		S	
[FeCp*₂]·⁺	[C₃(CN)₅]⁻	1-D .DADA.	1:1 C2/c	2501	13.950	14.160	12.870	90.00	100.35	90.00	4	S	-100
[CrCp*₂]·:⁺	[C₄(CN)₆]⁻	1-D .DADA.	1:1 P2₁/n	1394	11.185	8.694	14.343	90.00	91.17	90.00	2	S	23
[FeCp*₂]·⁺	[C₄(CN)₆]⁻	1-D .DADA.	1:1 P2₁/n	1340	10.783	8.719	14.266	90.00	91.22	90.00	2	S	-106
[FeCp*₂]·⁺	[C₅(CN)₅]⁻	Not 1-D	1:1 Cmc2₁	2702	14.704	13.090	14.037	90.00	90.00	90.00	4	S	23

[FeCp*2].+	[C5(CN)4Cl]-	1D .DAD.	1:1	Imm2	1343	14.484	10.556	8.781	90.00	90.00	90.00	2	S	25
[FeCp*2].+	[C6(CN)6]-	1-D DD AA	1:1	Pī	1442	7.532	14.776	14.981	115.90	95.76	100.74	2	S	
[CoCp*2].+	[C6(CN)6]-	1-D DD AA	1:1	Pī	1413	7.421	14.763	14.696	115.90	96.02	101.22	2	S	
[FeCp*2].+	[C6(CN)6]-	1-D DD AA	1:1	C2/c	2801	15.207	26.500	7.287	90.00	107.46	90.00	4	S	-108
[Cr(C6Me3H3)2].+	[TCNQ]-	1-D .DDAA...	1:1	C2/c	5019	14.014	16.357	22.964	90.00	107.44	90.00	8	S	
[Cr(C6Me3H3)2].+	[TCNQ]-	1-D .DADA.	1:1	P21/c	1266	9.588	16.390	8.419	90.00	106.84	90.00	2	S	
[Cr(C6H5Me)2].+	[TCNQ]-	1-D DD AA	1:1	P21/n	2200	7.000	15.450	20.500	90.00	97.00	90.00	4	S	
[CrCp*2].:+	[TCNQ]-	DAAD dimer	1:1	P21/c	2812	9.690	12.320	23.610	90.00	95.54	90.00	4	S	
[CrCp*2].:+	[TCNQ]-	1-D .DADA.	1:1	P21/n	2815	10.972	30.941	8.511	90.00	102.96	90.00	4	S	
[Fe(C5Me4H)2].+	[TCNQ]-	1-D .DADA.	1:1	Pī	652	8.636	9.574	10.025	63.77	70.22	162.79	1	S	23
[Fe(C5Et5)2].+	[TCNQ]-	1-D .DADA.	1:1	Pī	961	9.416	9.966	11.710	93.83	109.29	109.05	1	S	-70
[MnCp*2].+	[TCNQ]-	1-D .DADA.	1:1	P21/n	2792	10.829	31.014	8.544	90.00	103.39	90.00	4	S	
[FeCp*2].+	[TCNQ]-	1-D .DADA.	1:1	Pī	702	8.635	9.384	10.635	116.76	112.58	72.49	1	S	-106
[FeCp*2].+	[TCNQ]-	DAAD dimer	1:1	P21/c	2785	9.708	12.211	23.585	90.00	95.012	90.00	4	S	23
[CrCp*2].:+	[TCNQF4]-	DAAD dimer	1:1	P21/c	2922	9.825	12.458	23.963	90.00	94.810	90.00	4	S	23
[CoCp*2].+	[TCNQF4]-	DAAD dimer	1:1	P21/c	2875	9.806	12.310	23.879	90.00	94.010	90.00	4	S	23
[Fe(C5Et5)2].+	[TCNQF4]-	DAAD dimer	1:1	P2	3841	11.688	11.958	27.572	90.00	94.683	90.00	4	S	
[FeCp2].+	[TCNQF4]-	1-D DD AA	2:3	P21/n	1995	6.886	13.774	21.085	90.00	93.870	90.00	4	S	
[FeCp*2].+	[TCNQF4]-	DAAD dimer	1:1	P21/c	2882	9.790	12.350	23.880	90.00	94.520	90.00	4	S	
[CoCp*2].+	[DDQCl2]-	1-D .DADA.	1:1		2603	17.060	14.520	10.520	90.00	90.000	90.00	4	P	
[Fe(C5Me4H)2].+	[DDQCl2]-	Not Solved	1:1	Pī	2264	7.056	12.745	25.823	100.87	95.47	92.88	4	S	
[FeCp*2].+	[DDQCl2]-	1-D .DADA.	1:1	Pbna	2719	17.050	14.890	10.710	90.00	90.000	90.00	4	P	23
[FeCp*2].+	[DDQCl2]-	1-D .DADA.	1:1	Pbna	2620	17.027	14.497	10.616	90.00	90.000	90.00	4	S	23
[FeCp*2].+	[DDQCl2]-	1-D .DADA.	1:1	Pbna	2574	16.954	14.427	10.524	90.00	90.00	90.00	4	S	-70
[MnCp*2].+	[DDQCl2]-	1-D .DADA.	1:1	Pbna	2631	17.000	14.480	10.690	90.00	90.000	90.00	4	P	23
[FeCp*2].+	[DDQBr2]-	1-D .DADA.	1:1	Pbna	2583	17.066	14.381	10.526	90.00	90.000	90.00	4	S	-70

288

Cation	Anion	Structure	Ratio	Space Group	V	a	b	c	α	β	γ	Z	S/P	T
[MnCp*2]·+	[DDQBr2]·-	1-D .DADA.	1:1	Pbna	2652	17.146	14.438	10.714	90.00	90.000	90.00	4	P	23
[FeCp*2]·+	[DDQI2]·-	1-D .DADA.	1:1	Pbna	2665	17.257	14.560	10.606	90.00	90.000	90.00	4	S	-70
[MnCp*2]·+	[DDQI2]·-	1-D .DADA.	1:1	Pbna	2724	17.302	14.621	10.768	90.00	90.000	90.00	4	P	23
[FeCp*2]·+	[Cyanil]·-	1-D DDD AA	1:1	P2$_1$/n	2595	7.401	25.000	14.079	90.00	94.77	90.00	4	S	-70
[FeCp*2]·+	[CyanilH]	1-D .DADA.	1:1	C2/c	2679	18.494	9.935	14.671	90.00	96.46	90.00	4	S	-70
[CoCp*2]+	[CyanilH]	1-D .DDAA.	1:1	C2/c	2700	18.374	10.043	14.731	90.00	96.60	90.00	4	S	23
[CrCp*2]·:+	[CyanilH]	1-D .DDAA...	1:1	C2/c	2738	19.02	9.85	14.71	90.00	96.33	90.00	4	S	23
[FeCp*2]·-	[Ni(bds)2]·-	1-D .DAA.	1:1	Pī	1723	11.270	16.282	9.606	100.66	106.03	81.75	2	S	
[FeCp*2]·+	[Ni(dmit)2]·-	1-D .DDAA.	1:1	Pī	1575	11.347	14.958	10.020	97.68	94.36	109.52	2	S	-120
[FeCp*2]·+	[Ni(mnt)2]·-	DAAD dimer	1:1	P2$_1$/c	3040	9.959	12.338	25.086	90.00	99.540	90.00	4	S	
[FeCp*2]·-	[Pt(mnt)2]·-	1-D .DADA.	1:1	Pī	2316	12.106	14.152	14.374	108.94	96.370	90.51	3	S	
[FeCp*2]·+	[Pt(mnt)2]·-	1-D .DAA.	1:1	C2/m	4473	16.802	21.095	12.942	90.00	94.520	90.00	6	S	20
[FeCp*2]·+	[Ni(S2C4F6)2]·-	1-D .DADA.	1:1	C2/c	2354	14.417	12.659	18.454	90.00	95.17	90.00	4	S	-70
[FeCp*2]·+	[Pt(S2C4F6)2]·-	1-D .DADA.	1:1	Pī	846	8.490	10.278	10.936	106.79	103.95	101.98	1	S	-70
[CrCp*2]·+	[Mo(S2C4F6)3]·-	1-D .DADA.	1:1	C2/m	4083	22.25	12.896	14.24	90.00	91.38	90.00	4	S	23
[FeCp*2]·+	[Mo(S2C4F6)3]·-	1-D .DADA.	1:1	C2/m	4171	22.551	12.997	14.236	90.00	91.23	90.00	4	S	25
[FeCp*2]·+	[C4(CF3)4O]·-	1D .DAD.	1:1	C2/c	2887	11.751	16.105	16.018	90.00	107.76	90.00	4	S	-100
[FeCp2]·+	[FeBr4]·:--	Not 1-D	1:1	Pnma	1549	9.030	14.175	12.102	90.00	90.000	90.00		S	
[FeCp*2]·+	[FeBr4]·:--	Not 1-D	1:1	Pī	1290	9.348	7.936	17.591	93.94	85.890	96.76	2	S	20
[FeCp2]·+	[FeCl4]·:--	Not 1-D	1:1	Pna2$_1$	1450	13.837	11.966	8.762	90.00	90.000	90.00	4	S	
[FeCp*2]·+	[FeCl4]·:--	Not 1-D	1:1	Pī	4769	18.538	20.384	15.712	110.17	108.42	60.43	8	S	
[FeCp*2]·+	[TCNQMe2]·-	Not Solved			1011	3.870	9.150	28.510	90.00	90.000	90.00	-	S	
[FeCp*2]·+	[TCNQ(OPh)]·-	Not Solved	1:1		948	9.271	9.609	13.55	64.79	64.56	68.58	-	S	
[FeCp*2]·+	[TCNQBr2]·-	Not Solved	1:1		2906	7.480	14.870	26.680	88.18	82.180	81.04	-	S	
[FeCp*2]·+	[TCNQI2]·-	1-D DDADA	1:1	P2$_1$/n	3143	11.131	32.929	8.728	90.00	105.06	90.00	4	S	-100

a S = Single Crystal P = Powder Diffraction b ·MeCN

Table 3

Summary of the In-registry[a] and Out-of-Registry[b] Inter- and Intrachain Interactions for 1-D Metallocene Donor/Acceptor Complexes

Donor	Anion	Intrachain M···M, Å, In-[a] (A)	Out-of-[b] (B)	Interchain Separation, In- (C)	In- (D)	Out-of- (E)	Intrachain Fe···N (F)	Interchain Fe···N In- (J)	Out-of- (G)	Interchain N···N In- (H)	Out-of- (I)	Misc
MnCp*2	TCNQ	10.829	8.544	9.526	8.312	8.436	5.805	6.429	5.229	4.455	5.054	N·H 2.804
				10.591		10.083	6.595	6.924	5.562		5.510	N·H 2.845
				10.656			6.980	7.459	5.623		5.510	N·H 3.315
				12.496			7.066	8.500	6.053		6.765	N·H 3.345
							7.203		6.141		7.311	
							7.229		7.690			
							8.023					
							8.790					
FeCp*2	TCNQ	10.549	8.628	9.384	8.517	8.447	6.102	6.591	5.249	3.922	5.478	N·H 2.527
				10.635		9.362	6.508	7.371	5.451	5.730	6.212	N·H 2.779
				10.670			6.729	7.627	5.829		6.974	N·H 3.427
				10.824			6.806	8.414			7.750	N·H 3.512
							7.347					
							7.518					
							8.228					
							8.543					
FeCp*2 Orthorhombic	TCNE	10.621	8.689	9.618								
				9.649								
FeCp2	TCNE	9.78	6.77	7.85	6.58	6.15	4.81		4.23	3.50		
				7.86		7.36	5.33		4.40			
				8.07			5.89		4.67			
							6.01		4.88			
									5.19			

FeCp*2 TCNE Monoclinic	10.415	8.722	9.473 10.028	8.732	8.232	5.628 5.666 5.969 5.773 6.143 6.266 6.471 6.357	8.378 8.476 10.212 10.384	5.670 5.707	4.721 6.578	5.256 6.566 6.676	
FeCp*2 TCNQI2	11.131	8.728	9.778 10.994	8.428	8.716	6.659 6.963 6.958 7.096 7.211 7.473 7.718 8.872	6.638 7.711	5.111 5.152 5.880	5.236 5.253 7.233	5.099 6.151 6.201 7.981	I·I 5.208 N·I 3.115
Fe(C5HMe4)2 TCNE	11.580	9.760	8.521 8.521 9.471 9.471	9.760	6.251 7.495	5.534 5.534 7.357 7.357	8.445 9.739	5.660 5.660	4.173	5.967 6.432 6.432 6.564 6.564 7.795 7.795	
Fe(C5HMe4)2 TCNQ	10.360	8.636	9.520 9.574 10.025 10.793	8.600	8.310 8.670	6.919 6.937 7.093 7.094	6.474 6.474 7.532 7.532	4.786 5.102 5.522	3.895	4.040 5.009 6.718	

FeCp*2 RT   DDQCl2

10.616 8.692 9.723 8.692 8.146
                      9.723   8.514
                     10.033
                     10.033

O 5.546	Cl 7.883	N 5.369	Cl-Cl 4.164	Cl-O 5.824
Cl 6.007	Cl 8.168	N 5.501	Cl-O 3.634	Cl-O 6.085
N 6.317	O 8.173	N 5.501	Cl-O 3.634	N-N 7.006
O 6.357	O 8.744	O 5.945	O-O 5.116	O-O 6.819
Cl 6.376	Cl 10.398	Cl 5.969		Cl-N 5.358
N 6.815	Cl 10.616	Cl 5.969		Cl-N 6.314
	N 10.961			Cl-N 6.714
	N 11.255			Cl-N 6.858
				N-O 6.869
				N-O 7.960

FeCp*2 -70 C   DDQCl2

10.524 8.647 9.668 8.647 8.111
                      9.668   8.477
                      9.977
                      9.977

O 5.578	Cl 7.811	N 5.330	Cl-O 3.610	N-N 6.971
Cl 5.949	Cl 8.120	N 5.451	Cl-O 3.610	Cl-O 5.757
N 6.280	O 8.097	N 5.451	Cl-Cl 4.139	Cl-O 6.040
O 6.337	O 8.701	O 5.923	O-O 5.090	O-O 6.776
Cl 6.350	Cl 10.331	Cl 5.940		Cl-N 5.264
N 6.800	Cl 10.567	Cl 5.940		Cl-N 6.281
	N 10.907			Cl-N 6.664
	N 11.215			Cl-N 6.803
				N-O 7.468
				N-O 8.391

FeCp*2 -70 C   DDQBr2

10.526 8.631 9.671 8.631 8.114
                      9.671   8.533
                     10.026
                     10.026

O 5.512	Br 7.699	N 5.290	O-Br 3.465	N-Br 5.251
Br 6.031	Br 7.999	N 5.463	O-Br 3.465	N-Br 6.221
O 6.305	O 8.097	N 5.463	Br-Br 3.93	O-Br 5.752
N 6.326	O 8.701	O 5.902	O-O 5.155	O-Br 6.004
N 6.814	Br 10.622	Br 5.939		O-O 6.745
	N 10.511			N-Br 6.711
	N 10.743			N-Br 6.837
				N-O 7.591
				N-O 8.289

FeCp*2 -70 C	DDQl2	10.614	8.690	9.817 9.817 10.132 10.132	8.690	8.259 8.631	O 5.522 I 6.141 O 6.374 N 6.357 I 6.534 N 6.923	I 7.665 I 7.983 O 8.210 O 8.806 I 10.582 N 10.814 I 10.982 N 11.319	N 5.370 N 5.516 N 5.516 I 5.980 I 5.980 O 6.022	I·I 3.837 I·O 3.409 I·O 3.409 O·O 5.281	N·I 5.202 I·O 5.840 N·N 7.010 I·O 6.061 N·I 6.297 O·O 6.810 N·I 6.792 N·I 6.966 N·O 6.752 N·O 8.297
FeCp*2	C4(CN)6	10.783	8.719	9.865 10.030	8.719	8.358	5.536 6.482 6.540 6.581 7.320 7.602	6.791 6.791 7.917 7.917	5.153 5.856	3.311 3.934 4.031	4.498 6.685 6.697 7.109 7.402
FeCp*2	[C3(CN)5]-	10.305	8.598	9.939 9.567	10.271	8.903	5.859 6.036 6.087 6.240 6.284	6.937	5.328 5.649 5.763 7.498	3.609 3.780	5.163 6.349
FeCp*2	C5(CF3)4O	11.751	8.664	9.618 9.968	8.312	8.052 Fe-O 4.957 8.985 Fe-O 8.178					
Fe(C5Et5)2	TCNQ	13.224	9.416	9.966 11.257 11.710 12.367	8.842	9.273 9.376	7.839 7.871 8.374 8.413	5.344 5.678 5.769	5.158	7.287 11.710	

FeCpCp* TCNE 10.253 8.133 7.677 7.770 4.1
8.977 7.777
9.651

Cr(C6Me3H3)2 TCNQ 10.772 8.025 7.172 6.126 5.250 3.951
8.419 8.870 6.784 5.266 4.336
9.213 8.946 7.514 5.329 4.470
9.588 8.025 6.755 5.839
7.870 7.457

FeCp*2 Ni(S2C4F6)2 12.097 12.695 9.389 12.695 12.709 Fe:S 5.863 F3C·CF3 4.199
13.051 9.605 Fe:S 5.924 F3C·CF3 4.668
Fe:S 5.933
Fe:S 6.862
Fe:Ni 6.049
Fe:Ni 10.589
Fe:Ni 12.345
Ni:Ni 9.605
Ni:Ni 11.185
Ni:Ni 12.210

FeCp*2 Pt(S2C4F6)2 10.936 8.490 10.278 8.240 9.839 Fe:S 5.598
11.895 10.786 Fe:S 5.707
12.304 Fe:S 6.116
12.661 Fe:S 6.212
Fe:Pt 5.468
Fe:Pt 8.922
Fe:Pt 10.152
Fe:Pt 10.795
Fe:Pt 11.895
Pt:Pt 8.490
Pt:Pt 10.936

FeCp*$_2$		10.385	7.509	Fe-S 5.635
		10.470	7.509	Fe-S 5.989
	Mo(S$_2$C$_4$F$_6$)$_3$ 14.238			Fe-S 6.773
				Fe-S 7.159
				Fe-Mo 7.110
				Fe-Mo 7.128
				Fe-Mo 7.509
				Mo-Mo 10.210
				Mo-Mo 10.320
				Mo-Mo 14.236

a In- is In-registry b Out-of- is Out-of-registry

salts possess the alternating $\cdots D \cdot^+ A \cdot^- D \cdot^+ A \cdot^- D \cdot^+ A \cdot^- \cdots$ crystal structure, Figure 1, that is observed for $[Fe^{III}(C_5Me_5)_2] \cdot^+ [TCNE] \cdot^-$ and $[Fe^{III}(C_5Me_5)_2] \cdot^+ [TCNQ] \cdot^-$. These materials comprise the aforementioned alternating D/A chains with parallel chains being both in-registry and out-of-registry as depicted in Figure 2. The distances designated in Figure 2 are reported in Table 3 for representative materials studied by single crystal X-ray analysis.

Since the prime requirement to form ferromagnetically coupled chains is to have stable radicals, electron transfer must occur for closed shell donors and acceptors to be candidates for magnetic materials. The one-electron solution reversible reduction potential, E^o, provides a means to gauge whether or not electron transfer might occur for a solid.[13] For example, ferrocene is more difficult to oxidize (by 0.5 V) than decamethylferrocene and does not reduce TCNE in either solution or the solid state. Nevertheless, the diamagnetic ferrocene analog of $[Fe^{III}(C_5Me_5)_2] \cdot^+ [TCNE] \cdot^-$, *i. e.*, $[Fe^{II}(C_5H_5)_2][TCNE]$, forms[14,15] and possesses the identical structural motif,[16] Figure 1. Perhaps a temperature or pressure induced 'neutral-ionic' transition[17] might be sufficient to lead to the stabilization of ferromagnetic behavior; however, above 2 K at ambient pressure only Fe^{II} is observed via Mössbauer spectroscopy and no discontinuity is observed in the susceptibility data.[15]

The Co^{III} analog, $[Co^{III}(C_5Me_5)_2]^+ [TCNE] \cdot^-$, with $S = 0$ $[Co^{III}(C_5Me_5)_2]^+$ has been prepared and exhibits essentially the Curie susceptibility anticipated for $S = 1/2$ $[TCNE] \cdot^-$ ($\theta = -1.0$ K).[6] Likewise, $[Fe^{III}(C_5Me_5)_2] \cdot^+ [C_3(CN)_5]^-$ with $S = 0$ $[C_3(CN)_5]^-$ exhibits essentially the Curie susceptibility anticipated for $S = 1/2$ $[Fe^{III}(C_5Me_5)_2] \cdot^+$ ($\theta = -1.2$ K).[6] Hence the $\cdots D \cdot^+ A \cdot^- D \cdot^+ A \cdot^- \cdots$ structure type, not $\cdots D^+ A \cdot^- D^+ A \cdot^- \cdots$ or $\cdots D \cdot^+ A^- D \cdot^+ A^- \cdots$ (*i. e.*, both $S \geq 1/2$ D's and $S \geq 1/2$ A's are present) is necessary, but insufficient, for stabilizing cooperative highly magnetic behavior.

Attempts to prepare $[M^{III}(C_5Me_5)_2] \cdot^+$ (M = Ru, Os) salts of $[TCNE] \cdot^-$ have yet to lead to suitable compounds for comparison with the highly magnetic Fe^{III} phase.[18] Formation of $[Ru^{III}(C_5Me_5)_2] \cdot^+$ is complicated by disproportionation to $Ru^{II}(C_5Me_5)_2$ and $[Ru^{IV}(C_5Me_5)(C_5Me_4CH_2)]^+$.[19] The Os^{III} analog has lead to the preparation of a low susceptibility salt with TCNE; however, crystals suitable for single crystal X-ray studies[18] have not as yet been isolated and limiting progress in this area.

Replacement of Fe^{III} in $[Fe^{III}(C_5Me_5)_2] \cdot^+ [TCNE] \cdot^-$ with Ni^{III} ($S = 1/2$), Mn^{III} ($S = 1$), or Cr^{III} ($S = 3/2$) leads to compounds exhibiting cooperative magnetic properties. $[Ni^{III}(C_5Me_5)_2] \cdot^+ [TCNE] \cdot^-$ is antiferromagnetic with $\theta = -11.5$ K. $[M^{III}(C_5Me_5)_2] \cdot^+ [TCNE] \cdot^-$ (M = Mn[20] and Cr[21]) are ferromagnetic. $[M^{III}(C_5Me_5)_2] \cdot^+ [TCNQ] \cdot^-$ (M = Mn[22] and Cr[23]) are also ferromagnetic, whereas the 1-D kinetic phase of $[Fe^{III}(C_5Me_5)_2] \cdot^+ [TCNQ] \cdot^-$ is metamagnetic with a critical field of 1600 Oe.[24] The key magnetic properties are summarized in Table 4. The T_cs for these of materials increase as $Mn^{III} > Fe^{III} > Cr^{III}$. This is in contrast to the expectation of $T_c \propto S(S + 1)$,[26] *i. e.*, $Cr^{III} > Mn^{III} > Fe^{III}$.

Table 4

Curie-Weiss θ's and Critical Temperatures, T_c, of Some Metallocene-Based Materials

Salt with •••D⁺A⁻D⁺A⁻••• Structure	θ,[a] K	T_c,[b] K	H_c, Oe (K)	ref
$[Mn^{III}(C_5Me_5)_2]^{\cdot+}[TCNE]^{\cdot-}$	+22.6	8.8	1200 (2 K)	20
$[Mn^{III}(C_5Me_5)_2]^{\cdot+}[TCNQ]^{\cdot-}$	+10.5	6.5 [c]	3600 (3 K)	22
$[Fe^{III}(C_5Me_5)_2]^{\cdot+}[TCNE]^{\cdot-}$	+16.9 [d]	4.8	1000 (2 K)	6
$[Cr^{III}(C_5Me_5)_2]^{\cdot\cdot+}[TCNE]^{\cdot-}$	+22.2	3.65 [o]	q	21
$[Cr^{III}(C_5Me_5)_2]^{\cdot\cdot+}[TCNQ]^{\cdot-}$	+12.8	3.5 [f]	e	23
$[Mn^{III}(C_5Me_5)_2]^{\cdot+}\{Pd[S_2C_2(CF_3)_2]_2\}^{\cdot-}$	+3.7	2.8		25
$[Fe^{III}(C_5Me_5)_2]^{\cdot+}[TCNQ]^{\cdot-}$	+11.6	2.55 [g]		h
$[Mn^{III}(C_5Me_5)_2]^{\cdot+}\{Ni[S_2C_2(CF_3)_2]_2\}^{\cdot-}$	+2.6	2.4		25
$[Mn^{III}(C_5Me_5)_2]^{\cdot+}\{Pt[S_2C_2(CF_3)_2]_2\}^{\cdot-}$	+1.9	2.3		25
$[Fe^{III}(C_5Me_5)_2]^{\cdot+}[C_4(CN)_6]^{\cdot-}$	+35			i
$[Fe^{III}(C_5Me_5)_2]^{\cdot+}\{Pt[S_2C_2(CF_3)_2]_2\}^{\cdot-}$	+27			h
$[Mn^{III}(C_5Me_5)_2]^{\cdot+}[DDQCl_2]^{\cdot-}$	+26.8			30,31
$[Fe^{III}(C_5Me_5)_2]^{\cdot+}[C_5(CF_3)_4O]^{\cdot-}$	+15.1			j
$[Fe^{III}(C_5Me_5)_2]^{\cdot+}\{Ni[S_2C_2(CF_3)_2]_2\}^{\cdot-}$	+15			33
$[Fe^{III}(C_5Me_5)_2]^{\cdot+}[DDQI_2]^{\cdot-}$	+12			h
$[Fe^{III}(C_5Me_5)_2]^{\cdot+}[DDQCl_2]^{\cdot-}$	+10.3	8.5 [p]		h, 29
$[Fe^{III}(C_5Me_5)_2]^{\cdot+}[DDQBr_2]^{\cdot-}$	+10			h
$[Fe^{III}(C_5Me_5)_2]^{\cdot+}[TCNQI_2]^{\cdot-}$	+9.5			k
$[Fe^{III}(C_5Me_5)_2]^{\cdot+}\{Mo[S_2C_2(CF_3)_3]_3\}^{\cdot-}$	+8.4			32
$[Fe(C_5Et_5)_2]^{\cdot+}[TCNE]^{\cdot-}$	+7.5			l
$[Fe(C_5Et_5)_2]^{\cdot+}[TCNQ]^{\cdot-}$	+6.1			l
$[Fe(C_5H_5)(C_5Me_5)]^{\cdot+}[TCNE]^{\cdot-}$	+3.3			m
$[Fe(C_5Me_4H)_2]^{\cdot+}[TCNQ]^{\cdot-}$	+0.8			n
$[Co^{III}(C_5Me_5)_2]^{+}[TCNE]^{\cdot-}$	-1			6
$[Fe^{III}(C_5Me_5)_2]^{+}[C_3(CN)_5]^{-}$	-1.2			6
$[Ni^{III}(C_5Me_5)_2]^{\cdot+}[TCNE]^{\cdot-}$	-11.5			h

[a] For polycrystalline samples [b] T_c determined from a linear extrapolation of the steepest slope of the M(T) data to the temperature at which M = 0. [c] 15 G, T_c from the maximum slope of dM/dT is reported as 3.1 K. [d] For crystals aligned parallel and perpendicular to the applied magnetic field: $\theta_{||}$ = +30 K and $\theta_\perp$ = +10 K.[7] [e] not observed. [f] 50 G, T_c from the maximum slope of dM/dT is reported as 6.2 K. [g] metamagnetic with a 1600 G critical field. [h] J. S. Miller and R. S. McLean, unpublished observations. [i] J. S. Miller, J. H. Zhang, and W. M. Reiff, *J. Am. Chem. Soc. 109*, (1987) 4584. [j] M. D. Burk, J. C. Calabrese, and J. S. Miller, in preparation. [k] J. S. Miller, D. A. Dixon, J. C. Calabrese, R. L. Harlow, J. H. Zhang, W. M. Reiff, S. Chittippeddi, M. A. Selover, and A. J. Epstein, *J. Am. Chem. Soc. 112*, (1990) 5496. [l] K-M. Chi, J. C. Calabrese, W. M. Reiff, and J. S. Miller, *Organometallics 10*, (1991) 688. [m] J. S. Miller, J. C. Calabrese, W. M. Reiff, D. T. Glatzhofer, in preparation. [n] J. S. Miller, D. T. Glatzhofer, D. M. O'Hare, W. M. Reiff, A. Chackraborty, and A. J. Epstein, *Inorg. Chem. 27*, (1989) 2930. [o] 0.15 G. [p] metamagnetic with a ~900 G critical field which is temperature dependent. [q] <0.15 mOe.

To test of the critical importance of the one-dimensionality, spinless $S = 0$ $[Co^{III}(C_5Me_5)_2]^+$ cations were randomly substituted for the cation in $[Fe^{III}(C_5Me_5)_2]^{.+}$-$[TCNE]^{.-}$ structure. This leads to the formation of finite magnetic chain segments imbedded randomly into the linear chains.[27] This results in the dramatic reduction of T_c with increasing $[Co(C_5Me_5)_2]^+$ content (Table 5) and is in excellent agreement with theoretical concepts.[28] With 14.5% replacement of Co^{III} for Fe^{III}, the T_c dropped from 4.8 to 0.75 K. The extreme sensitivity of the 3-D ordering temperature T_c stands as a cautionary note in attempts to observe a high T_c for solids comprised of high spin oligomers. For such systems the ratio of the intra- to interoligomer exchange may be very large. However, given the finite length of oligomers, the results of the aforementioned doping experiments suggest a significant suppression of 3-D ordering.

Table 5

Critical Temperatures, T_c, for $[Fe^{III}(C_5Me_5)_2]^{.+}$ Replaced with $[Co^{III}(C_5Me_5)_2]^+$ in $[Fe^{III}(C_5Me_5)_2]^{.+}[TCNE]^{.-}$ [27]

Compound	T_c,[a] K	T_c,[b] K
$[Fe(C_5Me_5)_2]^{.+}[TCNE]^{.-}$	4.8	4.9
$[Fe(C_5Me_5)_2]_{0.955}^{.+}[Co(C_5Me_5)_2]_{0.045}^+[TCNE]^{.-}$	4.4	4.4
$[Fe(C_5Me_5)_2]_{0.923}^{.+}[Co(C_5Me_5)_2]_{0.077}^+[TCNE]^{.-}$	3.8	
$[Fe(C_5Me_5)_2]_{0.915}^{.+}[Co(C_5Me_5)_2]_{0.085}^+[TCNE]^{.-}$	2.75	
$[Fe(C_5Me_5)_2]_{0.855}^{.+}[Co(C_5Me_5)_2]_{0.145}^+[TCNE]^v$		0.75

[a] Faraday Method   [b] ac (100 Hz)

The effect of a subtle alteration in the structure on the magnetic properties was also probed with the study of the isomorphous 2,3-dihalo-5,6-dicyanoquinone (DDQX$_2$; X = Cl, Br, I) electron-transfer salts of decamethylferrocene. The key inter- and intramolecular separations are summarized in Table 3. At -70 C there are slight differences in the solid state structure, however, the differences in the observed θ values 10 to 12 K, are insufficient to make a meaningful comparison. Likewise the magnetic properties of the isomorphous $[Mn(C_5Me_5)_2]^{.+}[DDQX_2]^{.-}$ (X = Cl, Br, I) are similar to each other (vide infra), but not to be properties of the $[Fe(C_5Me_5)_2]^{.+}[DDQX_2]^{.-}$ system.

Although bulk magnetic behavior is not observed, some 1:1 bis(dithiolato)metallate salts of decamethylferrocenium studied to date exhibit magnetic coupling as evidenced by Weiss θ constant which range from 0 to +27 K[33] (Table 6). These materials provide insight into a structure-function relationship. Of the compounds studied only $[Fe(C_5Me_5)_2]-\{M[S_2C_2(CF_3)_2]_2\}$ (M = Ni and Pt) has an 1-D chain structure. They also have the greatest θs, largest effective moments, and the most pronounced field dependencies of the susceptibility. The Pt analog with θ = +27 K possesses 1-D $\cdots D^{.+}A^{.-}$ $D^{.+}A^{.-}D^{.+}A^{.-}\cdots$ chains whereas the Ni analog with only a θ of +15 K possesses zig-zag 1-D $\cdots D^{.+}A^{.-}D^{.+}A^{.-}D^{.+}A^{.-}\cdots$ chains and longer M$\cdots$M separations (11.19 Å vs. 10.94 Å for the Pt analog). Thus, the enhanced magnetic coupling arises from the stronger intrachain coupling.

Table 6

Summary of the Curie-Weiss θ's and μ_{eff} for $[Fe^{III}(C_5Me_5)_2]^{.+}\{M[S_2C_2R_2]_n^{.-}\}$ [32,33]

Anion	Structural Arrangement	Spin Repeat Unit	Susceptibility	
			θ, K	μ_{eff}, μ_B
$\{Ni[S_2C_2(CN)_2]_2\}^{.-}$	$D^{\bullet+}[A]_2^{2-}D^{\bullet+}$ Dimer [a]	$D^{\bullet+}$	0	2.83
α-$\{Pt[S_2C_2(CN)_2]_2\}^{.-}$	$\bullet\bullet D^{\bullet+}A^{\bullet-}D^{\bullet+}\bullet\bullet$ Sheets	$D^{\bullet+} + 1/3\ A^{\bullet-}$	+6.6	3.05
	1-D $\bullet\bullet D^{\bullet+}[A]_2^{2-}\bullet\bullet$ Chains [a]			
β-$\{Pt[S_2C_2(CN)_2]_2\}^{.-}$	1-D $\bullet\bullet D^{\bullet+}A^{\bullet-}\bullet\bullet$ Chains	$D^{\bullet+} + 1/3\ A^{\bullet-}$	+9.8	3.10
	$D^{\bullet+}[A]_2^{2-}D^{\bullet+}$ Dimers [a]			
$\{Ni[S_2C_2(CF_3)_2]_2\}^{.-}$	1-D $\bullet\bullet D^{\bullet+}A^{\bullet-}\bullet\bullet$ [b]	$D^{\bullet+} + A^{\bullet-}$	+15	3.73
$\{Pt[S_2C_2(CF_3)_2]_2\}^{.-}$	1-D $\bullet\bullet D^{\bullet+}A^{\bullet-}\bullet\bullet$ Chains	$D^{\bullet+} + A^{\bullet-}$	+27	3.76
$\{Mo[S_2C_2(CF_3)_2]_3\}^{.-}$	1-D $\bullet\bullet D^{\bullet+}A^{\bullet-}\bullet\bullet$ Chains	$D^{\bullet+} + A^{\bullet-}$	+8.4	3.85

[a] $[A]_2^2$ = isolated $S = 0$ $\{M[S_2C_2(CN)_2]_2\}_2^{2-}$ dimer   [b] Zig-Zag Chains

In the opposite extreme $[Fe(C_5Me_5)_2]\{Ni[S_2C_2(CN)_2]_2\}$ possesses isolated $D^{.+}A_2^{2-}D^{.+}$ dimers and has a zero θ, the lowest μ_{eff}, and has no field dependence of the susceptibility. This is consistent with one spin per repeat unit. Intermediate between the 1-D chain and dimerized chains structures are the α- and β-phases of $[Fe(C_5Me_5)_2]\{Pt[S_2C_2(CN)_2]_2\}$ which have 1-D $\cdots D^{.+}A^.D^{.+}A^.D^{.+}A^{.-}\cdots$ strands in one direction and $\cdots DAAD\cdots$ units in another direction. For these materials θ, μ_{eff}, and M(H) are intermediate in value. This is consistent with the presence of one-third of the anions having a singlet ground state. Thus a correlation exists between the presence of 1-D $\cdots D^{.+}A^.D^{.+}A^.D^{.+}A^{.-}\cdots$ chains and the presence and magnitude of ferromagnetic coupling as evidenced by a θ value.

In an attempt to evaluate the effect of increasing the intrachain separation a tris(dithiolato)metallate salt of decamethylferrocenium, namely $[Fe(C_5Me_5)_2]^{.+}$-$\{Mo[S_2C_2(CF_3)_2]_3\}^{.-}$, was prepared.[32] The salt only possesses parallel out-of-registry 1-D $\cdots D^{.+}A^.D^{.+}A^.D^{.+}A^{.-}\cdots$ chains with intrachain Mo$\cdots$Mo separations of 14.24 Å. The θ value is reduced to 8.4 K which probably reflects the enhanced shielding of the spin and reducing the spin-spin interactions by the bulky -CF_3 groups. 3-D ordering is not observed to the lowest temperature studied, i. e., 2 K, as expected for the weak ferromagnetic inter- and interchain interactions anticipated. Consequently, $[Fe(C_5Me_5)_2]^{.+}\{Mo[S_2C_2(CN)_2]_3\}^{.-}$ is sought to ascertain, if it possesses a salt-like structure, if it has $\theta > 27$ K. These salt-like structures are to-date the only structural motif that does not have parallel chains in-registry and thus the antiferromagnetic $A^{.-}/A^{.-}$ interactions are minimized.

In contrast to the aforementioned bis(perfluoromethyldithiolato)metallate salts of decamethylferrocenium, the bis(perfluoromethyldithiolato)metallate salts of decamethyl-manganocenium exhibit 3-D cooperative magnetic order. For example, $[Mn^{III}(C_5Me_5)_2]^{.+}$-$\{M'[S_2C_2(CF_3)_2]_2\}$ (M' = Ni, Pd, Pt)[25] exhibit metamagnetic behavior with $T_c = 2.5 \pm 0.3$ K and θ of $+2.8 \pm 0.8$ K. The T_c is comparable to that of the metamagnetic behavior reported for $[Fe(C_5Me_5)_2]^{.+}[TCNQ]^{.-}$.[24] Since $[M(C_5Me_5)_2]^{.+}\{Ni[S_2C_2(CF_3)_2]_2\}$, but not $[M(C_5Me_5)_2]^{\cdot+}\{Pt[S_2C_2(CF_3)_2]_2\}$ (M = Fe, Mn), are isomorphous, Table 2, a greater ferromagnetic coupling would be expected as $T_c \propto S(S+1)$.[26] Perhaps $[Fe(C_5Me_5)_2]$-$\{Ni[S_2C_2(CF_3)_2]_2\}$ is also metamagnetic with a T_c substantially below 2 K and thus yet to be observed. Its substantially greater θ could arises in part from a greater anisotropy in g-values. Nonetheless, the actual reason is unclear at the present time.

With the goal of preparing additional molecular-based ferromagnets, the ferromagnetically coupled $[Fe(C_5Me_5)_2]^{\cdot+}[DDQCl_2]^{\cdot-}$ electron-transfer salt was characterized (θ = +10 K).[29] As T_c is proportional to the spin magnitude as $S(S+1)$, $[Mn(C_5Me_5)_2]^{\cdot+}[DDQCl_2]^{\cdot-}$ was prepared anticipating that T_c might occur at temperatures accessible in our laboratories.[30] The magnetic susceptibility of $[Mn(C_5Me_5)_2]^{\cdot+}$-$[DDQCl_2]^{\cdot-}$ can be fit by the Curie-Weiss expression with a +26.8 K θ-value suggesting primary ferromagnetic exchange interactions. Hysteretic magnetic field dependent behavior, albeit complex, was observed below ~7 K. The magnetic field dependence of the magnetization for a zero-field cooled sample previously aligned by 19.5 kG magnetic field is observed for increasing and decreasing magnetic fields. Above ~3.8 K the magnetization exceeds the expectation calculated from the Brilluoin function for fully aligned S = 1 and S = 1/2 spins with a dramatically different behavior below this temperature. Thus, a complex magnetic phase diagram is present at low temperature. Assuming complete alignment of the crystals with magnetic field parallel to the C_5 molecular axis, a sample-history dependent saturation magnetization, M_s, is observed with values up to 24,200 emuG/mol. The data are consistent with strong ferromagnetic coupling between adjacent radicals within each chain and a net weak antiferromagnetic coupling between the chains.[31] This leads to metamagnetic behavior. Thus, when an applied magnetic field is sufficiently large, it becomes energetically favorable for the spins in all the chains to align ferromagnetically. Below ~4 K there is an anomalous behavior with large hysteresis and remnant magnetization.[31] At ~4 K the magnetization abruptly drops by more than an order of magnitude depending on the applied field to a value lower than calculated from the Brilluoin function. At high temperature there is a field dependent crossover from low magnetization to a high magnetization state. This is suggestive of the presence of perhaps both metamagnetic and possible lattice distortion (spin-Peierls-like) transitions. However, since spin-Peierls transitions occur only in antiferromagnetic states, complex magnetic behaviors must be operative for the material. Similar complex magnetic behavior was observed for $[Mn(C_5Me_5)_2]^{\cdot+}[DDQX_2]^{\cdot-}$ (X = Br, I) at lower temperatures. Simple metamagnetic behavior has been reported for $[Fe(C_5Me_5)_2]^{\cdot+}[TCNQ]^{\cdot-}$ and $[Mn(C_5Me_5)_2]^{\cdot+}$-$[\{M[S_2C_2(CF_3)]_2\}]^{\cdot-}$ (M = Ni, Pd, Pt) (*vide supra*).

These results support the necessity of 1-D $\cdots D^{\cdot+}A^{\cdot-}D^{\cdot+}A^{\cdot-}\cdots$ chain structure for achieving ferromagnetic coupling and ultimately bulk ferromagnetic behavior as observed for $[Fe(C_5Me_5)_2]^{\cdot+}[TCNE]^{\cdot-}$.

3. Conclusion

The quest for s/p-orbital based ferromagnets remain the focus of intense interest worldwide. The magnetic data on $[Fe(C_5Me_5)_2]^{\cdot+}[TCNE]^{\cdot-}$ demonstrates that ferromagnetism is achievable in organic-based molecular systems. Replacement of the doublet organic acceptor with a diamagnetic acceptor demonstrates that the organic species is crucial for achieving bulk ferromagnetism. This system contains low spin Fe^{III} not high spin Fe^{II} or Fe^{III} or iron metal. The ferrocenes possess chemical (*i. e.*, reactivity similar to aromatic organic compounds, *e. g.*, benzene) and physical (*e. g.*, solubility in conventional polar organic solvents) properties akin to organic compounds not inorganic network solids. A summary of representative linear chain metallocenium containing studied to date with their observed Curie-Weiss θ values and where appropriate ordering temperatures are presented in Table 4.

Accidental or intrinsic orbital degeneracies, albeit rare for organic molecules, are needed for stabilization of ferromagnetic coupling by the extended McConnell mechanism. Thus, stable D_{2d} or $C_{\geq 3}$ symmetry $S \geq 1/2$ radicals with a degenerate POMO are required. It is a challenge to the synthetic chemist to prepare radicals that have nondegenerate POMO's and do not undergo a Jahn-Teller distortion which eliminates the desired electronic configuration. In addition to preparing the desired radicals appropriate secondary and tertiary solid state structures must be achieved. Finally, suitably large single crystals enabling the study of their anisotropic magnetic properties are necessary for a quantitative studies.

4. Acknowledgment

The authors gratefully acknowledge partial support by the Department of Energy Division of Materials Science (Grant No. DE-FG02-86ER45271.A000). We deeply thank our co-workers over the years for the important contributions they have made enabling the success of the work reported herein.

5. References

1. J. S. Miller, A. J. Epstein, and W. M. Reiff, *Mol. Cryst., Liq. Cryst.* **120** (1986) 27.
2. J. S. Miller, A. J. Epstein, and W. M. Reiff, *Chem. Rev.* **88** (1988) 201.
3. J. S. Miller, A. J. Epstein, and W. M. Reiff, W. M. *Acc. Chem. Res.* **23** (1988) 114.
4. J. S. Miller, A. J. Epstein, and W. M. Reiff, *Science* **240** (1988) 40.
5. J. S. Miller and A. J. Epstein, *Angew. Chem. internat. ed.* in press. A. J. Epstein, and J. S. Miller, *Mol. Cryst., Liq. Cryst.* in press. J. S. Miller and A. J. Epstein, in press.
6. J. S. Miller, J. C. Calabrese, H. Rommelmann, S. Chittipeddi, J. H. Zhang, W. M. Reiff, and A. J. Epstein, *J. Am. Chem. Soc.* **109** (1987) 769.
7. S. Chittipeddi, K. R. Cromack, J. S. Miller, and A. J. Epstein, *Phys. Rev. Lett.* **58** (1987) 2695.
8. (a) S. Chittappeddi, M. A. Selover, A. J. Epstein, D. M. O'Hare, J. Manriquez, and J. S. Miller, *Synthetic Metals*, **27**, (1989) B417. (b) A. Chackraborty, A. J. Epstein, W. N. Lawless, and J. S. Miller, *Phys. Rev. B*, **40** (1989) 11422.
9. Proceedings of the Conference on *Ferromagnetic and High Spin Molecular Based Materials*, J. S. Miller and D. A. Dougherty, eds., *Mol. Cryst., Liq. Cryst.* **176** (1988). Proceedings on the Conference on *Molecular Magnetic Materials*, O. Kahn, D. Gatteschi, J. S. Miller, and F. Palacio, eds. *NATO ARW Molecular Magnetic Materials*, **E198** (1988).
10 (a) A. L. Buchachenko, *Russ. Chem. Rev.* **59** (1990) 307; *Usp. Khim.* **59** (1990) 529. (b) O. Kahn, *Structure and Bonding*, **68** (1968) 89. O. Kahn, and Y. Journaux, in press. (c) A. Caneschi, D. Gatteschi, R. Sessoli, and P. Rey *Acc. Chem. Res.* **22** (1989) 392. (d) L. Dulog, *Nachr. Chem. Tech. Lab.* **38** (1990) 448. (e) H. Ishida, *Encyl. Poly. Sci. & Engg. (Supp. Vol.)* **S446** (1989). (f) T. Sugawara, *Yuki Gos. Kag.* **47** (1989) 306. (g) J. S. Miller and A. J. Epstein, *New Aspects of Organic Chemistry*, Z. Yoshida, T. Shiba, and Y. Ohsiro, eds, VCH Publishers, New York, NY (1989), p 237.
11. H. M. McConnell, *Proc. R. A. Welch Found. Chem. Res.* **11** (1967) 144.

302

12. J. S. Miller and A. J. Epstein, *J. Am. Chem. Soc.* **109** (1987) 3850.
13. Electron transfer may occur for a solid even for D/As which from solution redox potentials have $\Delta E^o < 0$. For example, the $\Delta E^o < 0$ for $Fe^{II}(C_5H_5)_2$ and TCNQ is $\Delta E^o = -240$ mV, however, $[Fe^{III}(C_5H_5)_2]^+[TCNQ]_2^-$ can be isolated. M. D. Ward, *Electro. Anal. Chem.* **16** (1988) 181. R. C. Wheland and J. L. Gillson, *J. Am. Chem. Soc.* **98** (1976) 3916. R. C. Wheland, *J. Am. Chem. Soc.* **98** (1976) 3926.
14. (a) J. L. Robbins, N. Edelstein, B. Spencer, and J. C. Smart, *J. Am. Chem. Soc.* **104** (1982) 1882. (b) O. W. Webster, W. Mahler, and R. E. Benson, *J. Am. Chem. Soc.* **84** (1962) 3678. (c) M. Rosenblum, R. W. Fish, and C. Bennett, *J. Am. Chem. Soc.* **86** (1964) 5166. (d) R. L. Brandon, J. H. Osipcki, and A. Ottenberg, *J. Org. Chem.* **31** (1966) 1214.
15. J. S. Miller, J. H. Zhang, and W. M. Reiff, in preparation.
16. E. Adman, M. Rosenblum, S. Sullivan, and T. N. Margulis, *J. Am. Chem. Soc.* **89** (1967) 4540. B. Foxman, private communication. B. W. Sullivan and B. Foxman, *Organometallics* **2** (1983) 187.
17. P. Batail, S. J. LaPlaca, J. J. Mayerle, J. B. Torrance, *J. Am. Chem. Soc.* **103** (1981) 951.
18. D. M. O'Hare and J. S. Miller, *Organometallics* **7** (1988) 1335.
19. U. Kolle, J. Grub, *J. Organomet. Chem.* **289** (1985) 133.
20. G. T. Yee, J. M. Manriquez, D. A. Dixon, R. S. McLean, D. M. Groski, R. B. Flippen, K. S. Narayan, A. J. Epstein, and J. S. Miller, *Adv. Mater.* **3** (1991) 309.
21. F. Zuo, S. Zane, P. Zhou, A. J. Epstein, R. S. McLean, J. S. Miller, *J. Appl. Phys.* in press. P. Zhou, A. J. Epstein, F. Zuo, S. Zane, R. S. McLean, J. S. Miller, in preparation.
22. W. E. Broderick, J. A. Thompson, E. P. Day, and B. M. Hoffman, *Science* **249** (1990) 410.
23. W. E. Broderick and B. M. Hoffman, *J. Am. Chem. Soc.* **113** (1991) 6334.
24. G. A. Candela, L. Swartzendruber, J. S. Miller, and M. J. Rice, *J. Am. Chem. Soc.* **101** (1979) 2755.
25. W. E. Broderick, J. A. Thompson, and B. M. Hoffman, *Inorg. Chem.* **30** (1991) 2960.
26. D. A. Dixon, A. Suna, J. S. Miller, and A. J. Epstein, in *NATO ARW Molecular Magnetic Materials*, O. Kahn, D. Gatteschi, J. S. Miller, and F. Palacio, eds. **E198** (1991) 171.
27. K. S. Narayan, K. M. Kai, A. J. Epstein, and J. S. Miller, *J. Appl. Phys.* **69** (1991) 5953. K. S. Narayan, G. Morin, J. S. Miller, and A. J. Epstein, *Phys. Rev. B* in press.
28. R. B. Stinchcombe, in *Phase Transitions and Critical Phenomena*, C. Domb and J. L. Lebowitz, eds. (Academic Press, London, UK, 1983) **7**, 152.
29. E. Gerbert, A. H. Reis, J. S. Miller, H. Rommelmann, and A. J. Epstein, *J. Am. Chem. Soc.* **104** (1982) 4403. J. S. Miller, P. J. Krusic, D. A. Dixon, W. M. Reiff, J. H. Zhang, E. C. Anderson, and A. J. Epstein, *J. Am. Chem. Soc.* **108** (1986) 4459.
30. J. S. Miller, R. S. McLean, C. Vazquez, G. T. Yee, K. S. Narayan, and A. J. Epstein, *J. Mat. Chem.* **1** (1991) 479.
31. K. S. Narayan, O. Heres, A. J. Epstein, and J. S. Miller, *J. Magn. Magn. Mat.* **110** (1992) L6.
32. W. B. Heuer, D. O'Hare, M. L. H. Green, and J. S. Miller, *Chem. Mat.* **2** (1990) 764.
33. J. S. Miller, J. C. Calabrese, and A. J. Epstein, *Inorg. Chem.* **27** (1989) 4230.

MAGNETIC PROPERTIES OF ORGANIC DI-, OLIGO- AND POLYRADICALS

HIIZU IWAMURA

Department of Chemistry, The University of Tokyo,
7-3-1 Hongo, Bunkyo-ku, Tokyo 113, Japan

Electronic Structure of Organic Di-, Oligo- and Polyradicals

Studies of strongly magnetic organic materials are still at their embryonic stage in that establishment of the knowledge of how to align electron spins in parallel in short range is still at issue. The large majority of organic molecules have closed-shell electronic structures and therefore most organic compounds are diamagnetic. There are some organic molecules that have open shell structures where not all of the electrons are paired. These molecules, called free radicals, have one or more unpaired electrons and are often accepted as reactive intermediates. Unlike typical organic compounds, an assembly of these free radicals is paramagnetic. The interaction between two doublet centers becomes important in diradicals [1] and radical pairs [2]. The Coulombic repulsion between electrons lifts the zeroth-order degeneracy and gives rise to singlet and triplet states of different total energy for these chemical entities. It is the singlet state which is usually more stable, as described by the Heitler-London spin exchange between spins of opposite sign. Chichibabin's hydrocarbon 1 with a singlet ground state is a classical example [3]. When the interacting spins are in orthogonal orbitals, however, the triplet states may be favored. The presence of the half-filled orthogonal orbitals is dictated by symmetry and can be achieved for organic molecular systems in two ways: by geometrical and topological symmetry [4]. One-centered diradicals, such as carbenes and and nitrenes, are examples of the former. In π-conjugated diradicals, the intervening π-electrons are polarized and, therefore, if the periodicity of the spin polarization is in phase, the parallel alignment of the two spins can become favored. Schlenk's hydrocarbon 2 has a triplet ground state and is the first representative of such non-Kekulé hydrocarbons [5]. On the basis of these considerations, a number of di-, tri-, tetra-, and polyradicals with high-spin states have been designed, prepared and characterized. These

1   **2**

high-spin states have been designed, prepared and characterized. These molecules are of importance in designing organic molecular magnetic materials [4].

It was as early as in the mid-thirties when magnetic susceptibility and effective magnetic moment data of organic diradicals were determined and discussed for the first time [6]. The diamagnetic susceptibility (χ = -260 $\times 10^{-6}$ cm^{3}mol^{-1}) of **1**, for example, was found to be greater than that (χ = -300$\times 10^{-6}$ cm^3 mol^{-1}) of the precursor dichloride. However, the increment was interpreted in terms not of the paramagnetic contribution but correctly of the changes in the diamagnetic Pascal constants from a benzene ring to a quinoid structure. The molar paramagnetic susceptibility of **2** was found to be +(280~300)$\times 10^{-6}$ cm^{3}mol^{-1} at 74 K [6].

Triplet and higher-spin paramagnetic species are often studied by means of EPR spectroscopy. The EPR fine structure due to the dipolar interaction of the electron spins can yield important pieces of information about the structure and energy of the high-spin species. Since bulk magnetic properties of these compounds are of profound interest from the point of view of creating strongly magnetic organic materials [4], a series of magnetization/magnetic susceptibility measurements on di-, oligo- and polyradicals have been carried out. These classical techniques are found to afford a number of important results that are difficult to obtain otherwise, or, complementary to EPR data, and will be discussed in this chapter.

Molecular Paramagnetism

The magnetization I of a paramagnetic substance that contains, in a unit volume, weight or mole, N magnetic atoms, ions or molecules each of which has a magnetic moment M is given by Eq.1, where $B_J(\alpha)$ is the Brillouin function and α is defined by Eq.2 [7].

$$I = NgJ\mu_B B_J(\alpha) \qquad (1)$$

$$\alpha = gJ\mu_B H/kT \qquad (2)$$

Plots of I vs. H/T are dependent only on gJ as shown in Figure 1. When $\alpha \gg 1$, $B_J(\alpha) = 1$ and when $\alpha \ll 1$, $B_J(\alpha)$ may be expanded to a series of odd powers of α, and if we take only the first term, then Eq. 3 results.

$$\chi = I/H = Ng^2J(J + 1)\mu_B^2/3kT = C/T \qquad (3)$$

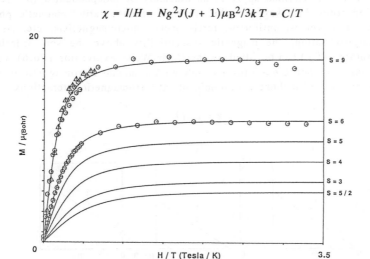

Figure 1. Typical M/μ_B vs. H/T curves for paramagnetic species represented by Brillouin functions. The exerimental data for hexa- and nonacarbenes (vide infra, **10**: $m = S = 6$ and $m = S = 9$, respectively) are also included.

The paramagnetic susceptibility χ per cm^3, g or mole is inversely proportional to the absolute temperature T. This relation is called a Curie law, and the proportionality constant C is a Curie constant that should be 0.37 and 1.0 K emu/mol for $J = 1/2$ and 1, respectively. For a species in degenerate singlet and triplet states, C should be 0.75.

By comparison with the classical mechanical expression for the magnetic moment, the effective magnetic moment μ_{eff} is defined by Eq. 4:

$$\mu_{eff}/\mu_B = 3\chi kT/N\mu_B^2 = g\sqrt{J(J + 1)} \qquad (4)$$

Plots of inverse χ and μ_{eff} vs. temperature are given figuratively in Figure 2; lines with a slope of $1/C$ and 0 are obtained, respectively. The heights of the horizontal μ_{eff} J obtained by use of Eq. 4:

$$\mu_{eff}^2 = \sum n_J \mu_{eff\,J}^2 / \sum n_J \qquad (5)$$

where n_J is the multiplicity due to J. A degenerated singlet and triplet state is expected to have a μ_{eff}/μ_B value of 2.45 and should clearly be distinguished from 2.83 of ground triplet species.

The above relations apply to an assembly of independent spins. Any deviation from them suggests the presence of cooperative magnetic phenomena, i.e., ferro-, antiferro-, ferri-, meta-, mictomagnetism, etc. As a first approximation, the magnetic susceptibility above the spin-ordering temperature ($>T_C$) can be fitted by the Curie-Weiss expression (Eq. 6) with the Weiss temperature $\theta > 0$ for a sample with dominant ferromagnetic interaction, and $\theta < 0$ for those with dominant antiferromagnetic interactions.

$$\chi = C / (T - \theta) \tag{6}$$

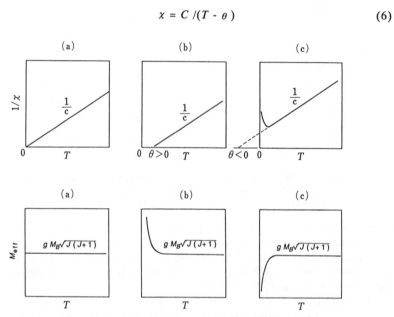

Figure 2. Figurative plots of inverse χ and μ_{eff} vs. T for paramagnetic species without and with magnetic interaction among the isolated spins.

The ferro- and antiferromagnetic interactions between the paramagnetic spins are represented by negative and positive intercepts, respectively, of the inverse χ vs. temperature plots extrapolated to 0 K. As cooperative effects set in at the lower temperatures, the μ_{eff} value will start to deviate from the horizontal line: upward for the ferro- and downward for the antiferromagnetic interaction (Figure 2).

More rigorous expression for the long-range order can be obtained by one-, two-, or three-dimensional Heisenberg models, e.g., Eq.7 for one-dimensional systems.

$$\chi_g = \frac{Ng^2\mu_B^2}{4kT} \left(\frac{1+5.798X+16.903X^2+29.377X^3+29.833X^4+14.037X^5}{1+2.798X+7.0087X^2+8.654X^3+4.574X^4}\right)\frac{2}{3} \quad (7)$$

where $X = J/2kT$

For organic compounds that do not contain heavy atoms, the spin-orbit coupling is negligible $(L = 0)$ and therefore $J = S + L = S$ where S is the spin quantum number. In the following discussion S will be used in place of J in Eq.1~5 and 7.

Instrumentation and Sensitivity

In principle, any instrument capable of determining the magnetic force experienced by paramagnetic substances can be used. In practice, however, the capability of measurements at cryogenic temperature is strongly required for several reasons. Firstly, all the left-hand side measurables of Eqs.1~3 and 6~7 increase with lowering temperature indicating that the signal-to-noise ratio will be greatly enhanced at lower temperature.

Secondly, paramagnetic species of interest are very often reactive intermediates and have to be studied at low temperature. Finally, it is usually the case that the cooperative effects leading to spin ordering are typically weak relative to kT of ambient temperature and detected only at cryogenic temperature.

A quartz light guide has been installed on an Oxford Faraday balance so that a sample in a quartz sample basket suspended in the superconducting solenoid (≤ 7 T) can be irradiated with UV-vis-light of selected wavelength from outside (Figure 3) [8]. Thus, reactive intermediates generated photochemically at cryogenic temperature, e.g., 2.0 K, can be studied *in situ*.

In the Faraday method, the increase in the weight of a typical organic free radicals, e.g., bis(*p*-methoxyphenyl) nitroxide ($S = 1/2$ and 5 mg), at the field gradient of 0.05 T/cm, and external field strength of 1.0 T is 3.92 mg at 10 K.

A SQUID susceptometer/magnetometer determines the magnetic moment of a sample as a change in the superconducting current of a Josephson junction and is more sensitive (sensitivity 10^{-8} emu) than magnetic balances.

The high magnetic field strength is desirable for sensitivity (see Eq.3). Remember, however, that these equations are based on the assumption that

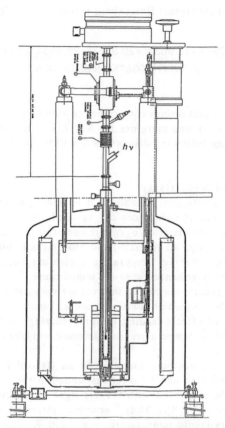

Figure 3. An example of a Faraday balance installed with a quartz light guide.

$\alpha \ll 1$, namely, lower H and higher T. It is often desirable to confirm the linearity of the magnetization on the field strength; for a triplet species, for example, the field strength not higher than 1 T is recommended for the measurement at 5 K. The use of low field is also desirable to minimize the effect of Zeeman splitting of the spin sublevels.

Magnetic Properties of Organic Di-, Oligo- and Polyradicals

Diradicals

A number of diradicals have been studied by magnetometry/susceptometry. When topology is appropriate [4], namely, a second radical center is placed in phase with the spin polarization of the intervening π-electrons due to the first radical center and vice versa, the π-conjugated diradicals may have triplet ground states. A m-phenylene unit is such an example in that the doublet and triplet spins as well as polarons attached to its two ends have a strong tendency to align in parallel. A dinitroxide 3 [9], a diverdazyl [10], a bis(nitronyl nitroxide) [11], and chloro- and hydrocarbons [12,13], analogous to 2 have been demonstrated to have triplet ground states by their Curie constants of 1.0 K emu/mol and μ_{eff} of 2.83 μ_B. When the topology is inappropriate for aligning the spins in parallel, a singlet state may become the ground state. Since it is very often the case that the first excited triplet state is thermally accessible, the energy gap between the two states can often be determined (for example, $2J/k$ = 490 and 19 K in the case of a bis(galvinoxyl)[14] and a dinitroxide 4 [9c], respectively). A set of examples for intermediate cases will be discussed in some detail.

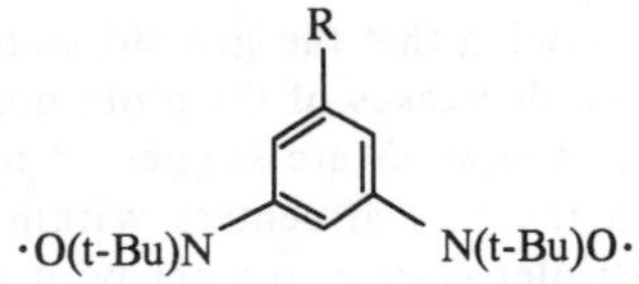

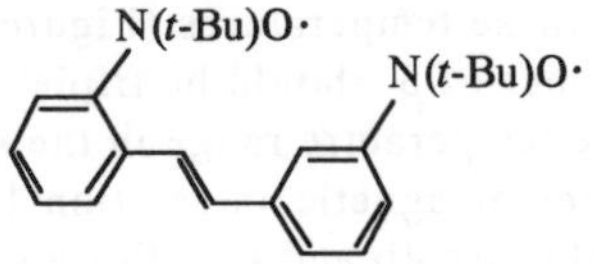

3 : R = H

3a : R = t-Bu

4

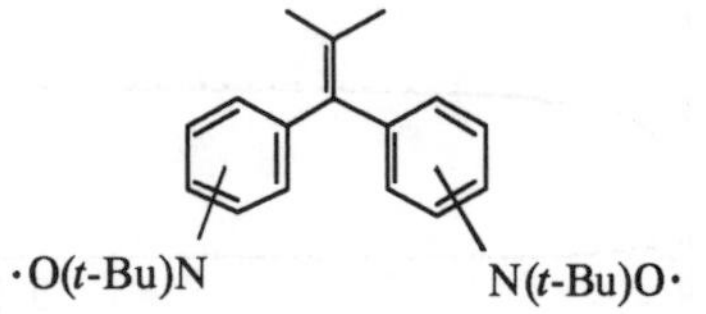

5

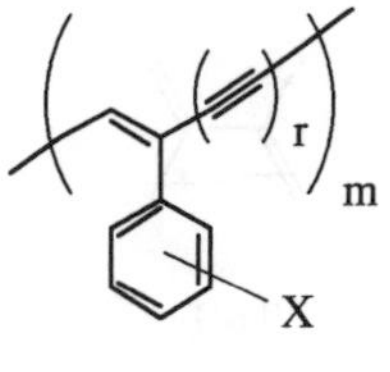

6 : r = 0

7 : r = 1

Determination of the Energy Gap between the Ground and Excited States of Diradicals

Delineation of the intramolecular exchange coupling between the radical centers in persistent dinitroxides 5 [15] is a subject of considerable interest since structures 5 constitute lower vinylogues of the dimer units of potentially very high-spin poly(phenylacetylenes) 6 [16] and poly(phenyldiacetylenes) 7 [17]. The knowledge of the sign and size of the coupling in the dimer units are instrumental for locating the appropriate sites of the radical centers X to be introduced on the phenyl rings of the polymers. The p,p, m,p' and m,m' isomers of 5 might also be considered to be related in connectivity to trimethylenemethane TMM, tetramethyleneethane TME, and pentamethylenepropane PMP, respectively. The ground spin states of the latter two are still subjects of some controversy in theoretical and physical organic chemistry [18].

In Figure 4 are reproduced μ_{eff}/μ_B vs. temperature plots for polycrystalline samples of isomeric 5 obtained on a SQUID susceptometer in the temperature range 5~300 K. The μ_{eff} values approaching 2.45 μ_B at room temperature are strongly indicative of near degeneracy of singlet and triplet states in all the three isomers at ambient temperature. A sharp decrease in the μ_{eff} values at the lowest temperature represents the antiferromagnetic exchange coupling between diradical molecules and taken into account by a Weiss field approximation. The presence of a maximum at 16 K in the plot for the p,p'-5 corresponds to an increased population of a triplet relative to a singlet at these temperatures (Figure 4a), showing that the ground state of the p,p' isomer of 5 should be triplet. Gradual decreases of the plots during intermediate temperature range in the m,p'- and m,m'-5 are suggested to be due to antiferromagnetic interaction between the radical centers within the molecules (Figure 4b and 4c). For a singlet-triplet system, the behavior of χ with respect to T can be described according to a modified Bleaney-Bowers equation (Eq. 8)[19]:

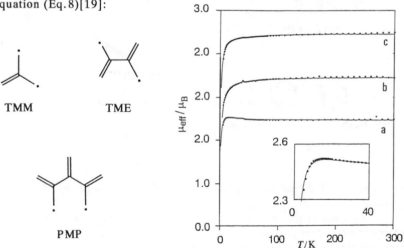

Figure 4. Plots of μ_{eff}/μ_B vs. T for a) p,p'-, b) m,p'- and c) m,m'-5

$$\chi = P\,\frac{2Ng^2\mu_B^2}{k(T - \theta)[3 + \exp(-2J/kT)]} \tag{8}$$

where N is the number of the spins per unit weight, g is the isotropic g factor, and θ is a Weiss temperature. The plots of μ_{eff}/μ_B vs. temperature were analyzed with a theoretical curve and the parameters obtained were refined by a least-squares method to give results summarized in Table 1. Whereas the dinitroxides 5 were analytically pure, better fit of the experimental data were obtained when an empirical factor P was introduced in Eq. 8 for correcting slight reduction in the effective magnetic moment of the samples used for the SQUID measurements due to the presence of non-diradical impurities; P's were 0.97, 0.88, and 0.94 for the p,p'-, m,p'- and m,m'-5, respectively. Thus, the open-shell centers on each phenyl ring in 1,1-diphenylethylenes 5 couple ferromagnetically in the p,p' isomer and antiferromagnetically in m,p' and m,m' isomers.

The coupling in polyradicals 6 and 7 should be ferromagnetic only when radical centers X, e.g., N($tert$-Bu)O·, are all at the para positions. Even in these cases, however, the interaction among the neighboring X's should not be very strong relative to kT at ambient temperature. There are two more messages from this work. Antiferromagnetic interaction found in the m,p'- and m,m'-topology suggests that the ground states of TME and PMP may be singlet with small magnitude of the singlet-triplet energy gap. Since the electron spins are more localized in the nitroxide moieties and the π-systems are enlarged in 5, it is slightly dangerous to put too much weight on this conclusion based on topological analogy. However, measurements of absolute μ_{eff} values and their temperature dependence over a wide temperature range should be a useful method for distinguishing between a triplet ground state and nearly degenerate singlet/triplet state in general and solving the controversy in TME and PMP in particular.

Table 1. Energy gap between the singlet and triplet states in isomeric dinitroxides 5.

5	ΔE_{S-T}/cm-1	θ/K	GS from VB	GS from MO
p,p'	10.6	-2.0	S = 1	non-disjoint
m,p'	-3.4	-2.0	S = 0	disjoint
m,m'	-1.8	-2.1	S = 1	doubly disjoint

Triradicals

A tris(verdazyl) and tris(nitronyl nitroxide) have been studied with inconclusive results [20]. When the topology of the intervening π-bonds is appropriate, and di- and triradicals have triplet and quartet ground state, respectively, it is more the rule rather than the exception that large energy gaps do not allow the population of the lower spin states thermally. A persistent 1,3,5-benzenetris(methyl radical) in a quartet ground state may be a typical example [13a]. When the energy gap of such a molecule is on the order 10^2 cm^{-1}, the value has be determined as follows.

Determination of the Energy Gap between the Ground and Excited States in Triradicals

The temperature dependence of the magnetic susceptibility and the effective magnetic moment of trinitroxide **8** in the range 2~300 K are shown in Figure 5 [9b]. As the temperature was increased, the moment of **8** increased from 1.05 μ_B at 2 K, reached a maximum 3.53 μ_B at ca. 140 K, and gradually decreased to 3.43 μ_B at 300 K. The latter decrease is due to the thermal population of the spins from a quartet state to excited doublet ones. The suppression at lower temperature is due to intermolecular antiferromagnetic coupling, which is confirmed by the fact that the lowest moment was smaller than 1.73 μ_B, a characteristic value of isolated doublet species.

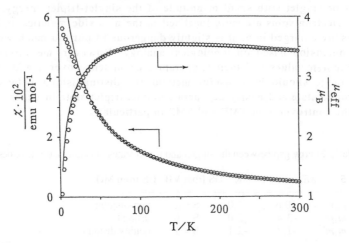

Figure 5. The temperature dependence of the magnetic susceptibility χ and effective magnetic moment μ_{eff} of trinitroxide **8** in the range 2 - 300 K.

The magnetic interaction in a linear triradical system can be described by the spin Hamiltonian:

$$\mathcal{H} = -2J\,(S_1\,S_2 + S_2\,S_3) \tag{9}$$

which assumes an isotropic exchange interaction and neglects intramolecular coupling (J') between the two terminal radicals. For poly(m-phenylenecarbenes) like **9**, the J' value is estimated to be two orders of magnitude smaller than J [21].

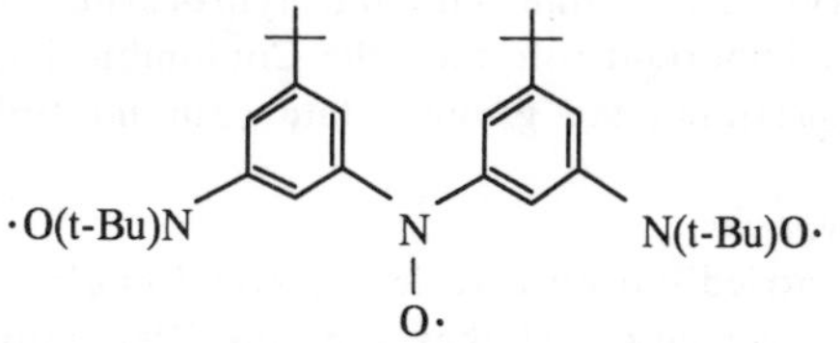

8 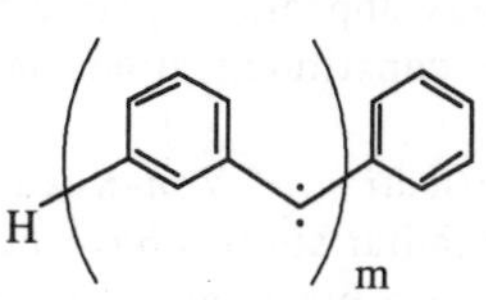   **9**

The eigenvalues of Hamiltonian Eq.9 are depicted in Figure 6. The molar susceptibility is given by Eq.10, where all symbols have their usual meaning. The temperature correction factor $T/(T - \theta)$ is introduced by means of the Weiss molecular field approximation. Best fit parameters are $\theta = -19 \pm 2$ K and $J/k = 240 \pm 20$ K. The energy gap between the quartet and the lower doublet state was estimated to be 240 K $= 167$ cm^{-1} $= 480$ cal/mol.

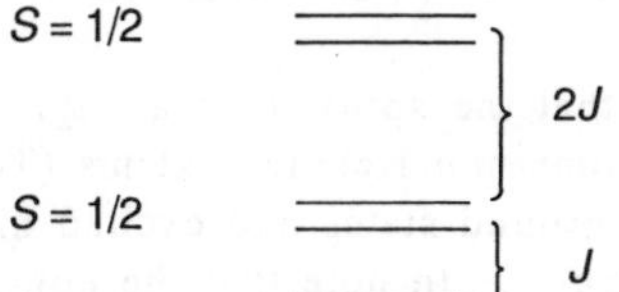

Figure 6. The eigenvalues for Eq. 9 describing **8**.

$$\chi = \frac{Ng^2\mu_B^2\,[10 + \exp(-J/kT) + \exp(-3J/kT)]}{4k(T - \theta)\,[2 + \exp(-J/kT) + \exp(-3J/kT)]} \tag{10}$$

The intramolecular exchange coupling is strongly ferromagnetic. Since the 2pπ-atomic orbital of more electronegative heteroatoms is lower in energy, the degeneracy of nonbonding molecular orbitals of hydrocarbon polyradicals will be lifted by introduction of a heteroatom radical center. The results obtained here suggest that the perturbation due to the introduction of heteroatoms is smaller in most cases and the topological consideration originally developed for alternant hydrocarbons can be extended to heteroatom-perturbed alternant systems. This statement is supported by the reports that m-phenylenedinitrene [22] and (m-nitrenophenyl)methylene [23] have indeed quintet ground state. Oxy analogues of TMM appear to constitute exceptional cases while phenoxyls are in line with the hydrocarbons. The topology appears to play a more important role than the Coulombic integral of the constituent atoms in determining the ground-state spin multiplicity [24].

Classification of Non-Kekulé Systems

Similar studies have been carried out on a series of non-Kekulé diradicals, dicarbenes and dinitrenes. We note that they are classified into two group: the one in which the intervening π-systems serve as a strongly ferromagnetic couplers and the other as a weakly antiferromagnetic couplers. They are summarized in Tables 2 and 3. In the former, it is usually the case that the ferromagnetic coupling is so strong that the ground state are highest-spin states and lower-spin excited state cannot be populated thermally. In aromatic nitroxide radicals, the electron spins are more localized and the spin polarization on the phenyl rings in very much reduced. By taking advantage of this attenuated interaction in di- and trinitroxides and the wider temperature range (5~300 K) applicable to the persistent dinitroxides, it was possible to obtain the absolute J values even for some ferromagnetic couplers (vide supra).

It is interesting to note that the spins have a slight tendency to align antiparallel to each other in some non-Kekulé systems (Table III), although the energy gaps between the ground states and excited high-spin states are not very large. It is also instructive to note that the non-Kekulé systems in Table II and III are similar in connectivity to TMM on the one hand and TME and PMP on the other, respectively.

Using valence bond theory and a Pariser-Parr-Pople Hamiltonian, Ovchinnikov has derived a method of predicting the ground states of alternant hydrocarbons by counting starred and unstarred atoms [25]. The results are summarized in Eq.11:

$$S = (n^* - n)/2 \qquad (11)$$

where S is the spin quantum number of the ground state, and n^* and n are,

Table 2. Ferromagnetic coupling units and the magnitude of the coupling

J/cm⁻¹	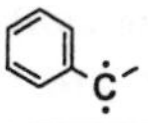 (↑) ⬡–Ċ–	(↑) –Ṅ·	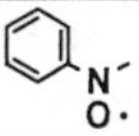 ⬡–N⁻–O·	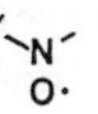 –N⁻–O·
[m-xylylene]	≫ 100	≫100	⎫ ≫ 100	
[diphenylmethylene]		≫10	167	5.3
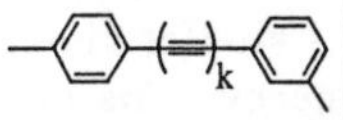 –⬡–(=)ₖ–⬡–	≫ 100	≫100 k= 0, 1, 2		
[stilbene-type]		≫100	≫100	
[stilbene-type]	≫ 100	≫100		

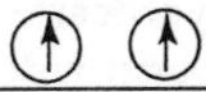

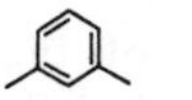

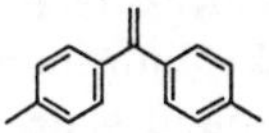

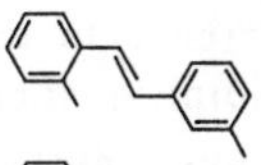

(left column label: **F** *)

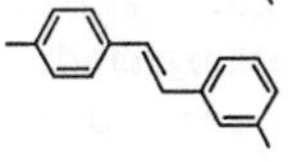

Table 3. Antiferromagnetic coupling units and the magnitude of the coupling

–J/cm⁻¹	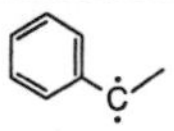 (↑) ⬡–Ċ– (↓)	–Ṅ·	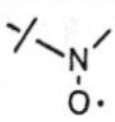 –N⁻–O·
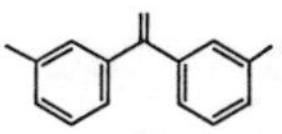 [benzophenone-type]		4.4	0.9
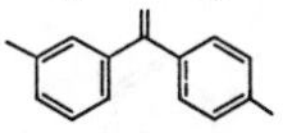 [benzophenone-type]		7.0	1.7
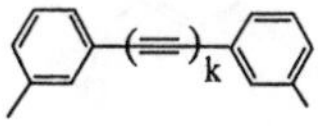 ⬡–(=)ₖ–⬡	(20, 45 (k=0))	12 (k=1) 6.0 (k=2)	
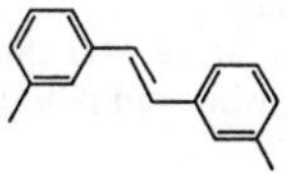 [stilbene-type]	10		6.5

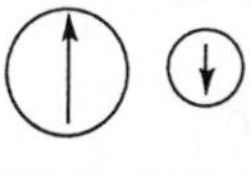

(left column label: **AF** *)

316

respectively, the numbers of starred and unstarred carbons. According to this rule, triplet ground states $S = 1$ are predicted for both p,p'- and m,m'-5 that have $n^* = 9$ and $n = 7$ (Figure 7). m,p'-5 has $n^* = n = 8$ and therefore a singlet ground state $S = 0$ is predicted. The validity of the extension of this rule to alternant non-hydrocarbons in general and nitrenes and nitroxides in particular has been amply demonstrated [21,23,24]. Note that each nitroxide group should be counted once as a pseudo atom [9b,21]. The high-spin p,p' and low-spin m,p' isomers are consistent with this theory but the observed low-spin m,m' isomer is contradictory.

Borden and Davidson noted the localizability of the Hückel non-bonding MO's in governing the ground states of non-Kekulé hydrocarbons and proposed a theory predicting that, if the Hückel non-bonding MO's cannot be localized to different sets of atoms, the ground state is expected to be high-spin [18,26]. If localization is possible, a singlet should be the ground state, or at least very close to the triplet in energy. The systems are called non-disjoint and disjoint, respectively. In reference to this theory, p,p' and m,p' isomers of 5 are non-disjoint and disjoint, respectively, as shown in Figure 7. The experimental results are in good agreement both with Ovchinnikov's rule and the MO theory. On the other hand, the result that the m,m' isomer of 5 has a singlet ground state is contradictory to Eq. 11. Since m,m' diradicals are doubly disjoint in a sense that the carbons with substantial positive spin density are separated by three carbons instead of two (Figure 7), they are predicted to have singlet ground states with the smallest magnitude of the energy gap of all. The experimental results are in line with the perturbative MO theory.

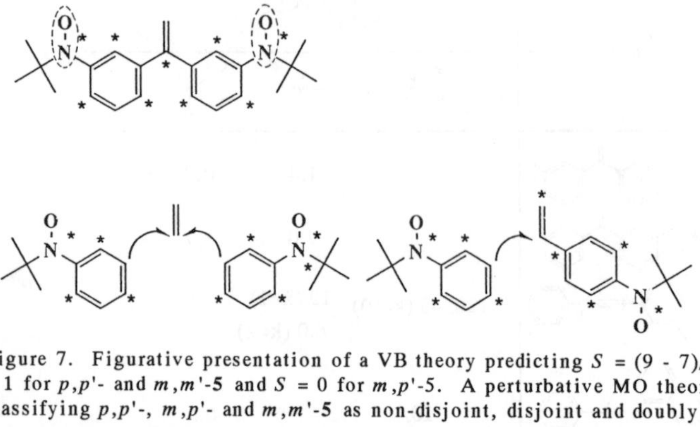

Figure 7. Figurative presentation of a VB theory predicting $S = (9 - 7)/2$ = 1 for p,p'- and m,m'-5 and $S = 0$ for m,p'-5. A perturbative MO theory classifying p,p'-, m,p'- and m,m'-5 as non-disjoint, disjoint and doubly disjoint, respectively, is also given.

Tetraradicals and Dinitrenes

A perchlorinated persistent tetraradical with a pair of triplet due to the ferro- and antiferromagnetic *m*-phenylene and *m*,*m*'-stibenediyl couplers has been reported [27]. A hydrocarbon polyarylmethyl quintet tetraradical is less well characterized by effective magnetic moment determined by means of Evans's method [28]. Reactive dinitrenes have been generated and studied in situ by susceptometry/magnetometry [29,30]. Dougherty and co-workers studied a poly(*m*-phenyleneoctatetraene) derivative doped with AsF_5 on a SHE SQUID magnetometer to find that the generated polarons couple ferromagnetically to form a segment in which $S > 2$ [31].

Tetracarbenes and Higher Analogues. Determination of the Ground-state Spin Multiplicities

Poly(*m*-phenylenecarbenes) **9** are geminate of the two concepts for constructing high spin organic molecules; triplet one-center diradical units are joined through a topologically robust ferromagnetic *m*-phenylene cou-pler. In 1984, the magnetization data of elusive reactive intermediate **9** (*m* = 4) were obtained for the first time in crystals and frozen solid solutions [32]. *I*/*I*s's correspond to theoretical Brillouin functions:

$$I/I_S = B_J(\alpha) \tag{12}$$

Therefore, by fitting a Brillouin function with the observed *I*/*I*s vs.*H*/*T* plot (Figure 1) the ground state spin quantum number was obtained, *J* = *S* = 4. Thus, it was firmly established that **9**(*m* = 4) is in a nonet ground state, a spin multiplicity which is even greater than those of typical 3d transition metal ions (*S* = 5/2) and 4f lanthanoids (*S* = 7/2) (Figure 1).

The spin alignment in **9** is after all one-dimensional. In order to realize the long-range order at finite temperature, it is imperative to increase the dimension of **9**. This could in principle be achieved either by imposing proper magnetic interaction between the molecular chains of **9** as in crystals or by increasing the dimension of the high-spin molecules themselves. Studies of prototypes of the first approach by controlling the stacking ori-entation of diphenylcarbene derivatives have been reported [33]. Pseudo-two-dimensional structure **10** has been introduced as a second possibility, although a two-dimensional honeycomb-like π-system appears to be ideal. The magnetization curves of typical paramagnetic samples given in Figure 1 contain data points for hexacarbene **10**(*m* = 6) at 2.1, 4.8 and 10.0 K, and nonacarbene **10**(*m* = 9) at 4.2 K. A least-squares analysis gave *S* = 6.0 and *S* = 9.0, respectively, the highest spins ever reported for a purely organic molecule [8].

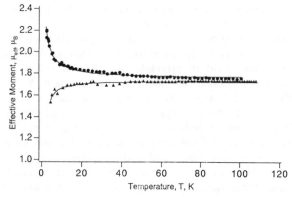

10

11

12

13

Long-range Spin Order in Crystals of Organic Radicals

As pointed out in the introduction and readily seen from the energy level diagram of the approach of two hydrogen atoms to form a hydrogen molecule, the exchange interaction between adjacent free radical molecules is usually antiferromagnetic. Crystals or solid solutions of organic free rad-

Figure 8. Temperature dependence of μ_{eff} for a crystalline sample of triacetylenic nitroxide 11 (•) in reference to that of similar diacetylenic nitroxide (▲).

icals exhibit such properties. There are very limited numbers of exceptional examples available in which such interaction between free radical molecules become ferromagnetic. A typical example of triacetylenic nitr-oxide 11 is shown in Figure 8. The magnetic susceptibility can be ex-pressed by means of a one-dimensional Heisenberg model (Eq. 7). It is not yet clear in this example if the ferromagnetic state is realized at lower tem-perature [34]. The γ-phase crystals of a 2-(p-nitrophenyl)-1-oxido-3-oxy-4,4,5,5-tetra-methhyl-2-imidazoline 12 showed similar magnetic susceptibility/magneti-zation behaviors and did undergo transition into a ferromagnetic state at 0.6 K [35].

Merits of Magnetic Susceptometry/Magnetometry over EPR Spectroscopy

EPR spectroscopy is more often used to detect triplet and higher spin species and characterize their electronic state. However, it is not without limitations. Firstly, fine structures are averaged out due to exchange nar-rowing for neat radical samples. High-spin species have to be diluted in solid solutions or in mixed crystals. It is therefore difficult to study interradical interactions and long-range orders.

For a second, the principal axis of the zero-field tensor must be ori-ented with respect to the external magnetic field for analyzing the zfs parameters. Otherwise, complex powder patterns are obtained.

A third limitation is the EPR spectral intensity. In principle, EPR in-tensities are related to the transition probability due to the spin sublevels. In practice, however, the efficiency of a microwave cavity is an experi-mental parameter and the measurement of the absolute intensities is in reference to that of standard samples. It is more the rule rather than exception that the transition probability and, therefore, the signal intensity is temperature-dependent. Since the Boltzmann factor $\exp(-\Delta W_m/kT)$ re-duces to $-\Delta W_m/kT$ under experimental conditions, where ΔW_m is the energy gap between the spin sublevels pertaining to transition m, the signal in-tensity (In_m) is linearly proportional to the inverse of the absolute tem-perature. This relation is called the Curie law in magnetic resonance and usually holds true for any transition m of any spin multiplicity, unless there are other states populated in equilibrium with the state of interest. In the latter case, the population of the magnetic state under consideration becomes subject to another Boltzmann distribution. Let us take for example, the equilibration of a triplet with a singlet state. Since the latter is EPR-silent, the signal intensities of the triplet are observables. Since In_m is inversely proportional to temperature, and the population of the triplet may be re-

320

presented by a Boltzmann factor, In_m is represented by a product of the two factors as in Eq. 13.

$$In_m = c \ \frac{\exp(-\Delta E/kT)}{T\,[1+3\exp(-\Delta E/kT)\,]} \tag{13}$$

where c is a proportionality constant and ΔE is the potential energy difference between the singlet and triplet states. Eq. 13 is an EPR version of Bleaney-Bowers equation (cf. Eq. 8), and some theoretical curves for representative ΔE-values are drawn in Figure 9. It is clearly seen that the plots are unique for large positive ΔE-values; it is rather difficult to differentiate between triplet ground states with and without a low-lying excited singlet state [4]. Even when a strictly linear In vs. $1/T$ relation is obtained, $\Delta E = 0$ (degenerate singlet/triplet states) and large negative ΔE value (triplet ground state) cannot be distinguished in principle. Since the signals due to the high-spin species can get more readily saturated at cryogenic temperatures, it is even more difficult in practice to confirm the linearity of EPR In with respect to microwave power and modulation.

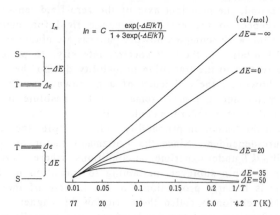

Figure 9. Theoretical curves representing the temperature dependences of the signal intensities due to a triplet species that has a singlet manifold in equilibrium. The ΔE values correspond to the energy gaps of the two states.

On the other hand, there can be more than one paramagnetic species present and yet their electronic states may be studied separately in EPR studies, as long as the fine structures are separated. In the magnetization/magnetic susceptibility method, however, since the magnetic properties as a whole are measured, high homogeneity of the paramagnetic samples to be

examined is required. *m*-Connected poly(arylmethyl) heptaradical and deca-
radical are good examples; while these compounds are well characterized by
EPR spectroscopy, magnetization studies on a SQUID susceptometer are
only partly successful [36].

Studies of Diradicals in Neat Crystals

Since neat crystalline samples of diradicals can be studied, any difference in the conformation of a diradicals in neat crystals or in solid solutions
can be studied by the magnetic susceptibility method [37].

In a frozen toluene matrix, 2,4-dimethoxy-*m*-phenylenbis(*N*-*tert*-butyl-
nitroxide) (**13**) showed X-band EPR fine structures characteristic of triplet
dinitroxides: $D/hc = 0.0179$ cm^{-1} and $E/hc = 0.0008$ cm^{-1}. A $\Delta m_s = 2$ transition was observed at 1692 mT. Deviation of the temperature dependence of
the signal intensities from a Curie law was not appreciable. The μ_{eff} obtained on a Faraday balance showed limiting values of 2.45 μ_B at higher
temperature, a value consistent with degenerate singlet and triplet states. As
the temperature was decreased, the μ_{eff} value for a sample in PVC film (2
mol%) remained constant in the range 45~300 K and decreased slightly at
the lower temperature (Figure 10). The results are interpreted in terms of
nearly degenerate triplet and singlet state: the χ value were simulated by a
Bleaney-Bowers equation (Eq.8) in which $2J = -7.0$ k_BK = -4.9 cm^{-1}; the
observed triplet lies slightly above the ground singlet state.

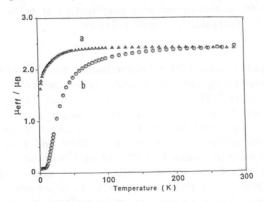

Figure 10. Temperature dependence of μ_{eff} values of **13** in (a) PVC films
and (b) in neat crystals.

Similar plots of μ_{eff} vs. temperature for a microcrystalline sample of **13**
are dominated by an antiferromagnetic interaction; as the temperature was
lowered, μ_{eff} started to decrease appreciably at ca. 50 K, and then decreased

more sharply (Figure 10). The data were interpreted in terms of a singlet/triplet model in which $2J = -73.8$ k_BK $= 51.3$ cm^{-1}; the ground state is decidedly a singlet.

It is often the case that μ_{eff} values of free radicals in neat crystals decrease at lower temperatures because of antiferromagnetic interradical exchange coupling. Diamagnetic dimers are sometimes formed in the extreme. An X-ray crystal structure analysis has been performed on a monoclinic single crystal of 13 to show that the *tert*-butylnitroxide moieties are considerably out of the *m*-phenylene plane (by 65.1 ° and 75.3 °) and in *syn* conformation. There is no interradical distance shorter than 7.04 A between the neighboring molecules; the formation of a dimer is not suggested. It is highly likely that the strong antiferromagnetic coupling is intramolecular. Lower π-spin polarization on the *m*-phenylene ring and a possible antiferromagnetic through-space interaction between the spins, rather localized on the nitroxide radicals at a distance of 5.74 Å (between the two middle points of the N-O bonds) may be responsible for the non-ferromagnetic coupling in 13. In toluene and PVC solid solutions, 13 is presumed to take another conformation, e.g., more planar or *anti* form in which such antiferromagnetic coupling is less effective.

In short, the 4,6-dimethoxy-*m*-phenylene unit in 13 is not a ferromagnetic coupling unit; the two nitroxide groups are coupled in an antiferromagnetic fashion strongly in crystals and weakly in solid solutions. Much care should be taken in the proper choice of a combination of spins and ferromagnetic couplers for designing high-spin polymers.

Conclusion

Classical magnetic susceptibility/magnetization measurements are instrumental in characterizing organic di-, tri-, tetra- and polyradicals and analyzing subtle interactions among the electron spins in them. These methods do not require samples with oriented principle axis of zero-field tensor. Non-Kekulé molecules have now been established to be classified further into the ones related in connectivity to TMM and with high-spin ground state, and the others related to TME and PMP with low-spin ground states. The former affords the ferromagnetic coupling units necessary for designing high-spin organic molecules. A number of unprecedentedly high-spin organic molecules have been produced as highlighted by a hydrocarbon with a $S = 9$ ground state. There are now several crystals found that exhibit ferromagnetic exchange coupling among the adjacent molecules at cryogenic temperature. Since the strategy for design of the two kinds of the ferromagnetic coupling within a molecule and among the molecules has been established, there is a bright future for obtaining organic ferromagnetic polyradicals with high Curie temperature.

Acknowledgment

The author acknowledges with great pleasure the invaluable contributions of our cowokers and graduate students. He is also thankful to Dr. Andrew S. Ichimura for his helpful and critical comments to the original manuscript. This work was supported by the Grant-in-Aid for Specially Promoted Research (No. 03102003) from the Ministry of Education, Science and Culture.

References

1. W. T. Borden, *Diradicals* (Wiley, New York , 1982) Chapter 1.
2. A. R. Lepley and G. L. Closs, *Chemically Induced Magnetic Polarization* (Wiley, New York, 1973).
3. M. S. Platz, in *Diradicals* , ed. W. T. Borden (Wiley, New York, 1982) Chapt 8.
4. N. Mataga, *Theor. Chim. Acta* **10** (1968) 372. H. Iwamura, *Pure Appl. Chem.* **58** (1986) 187; **59** (1987) 1595. H. Iwamura, *Adv. Phys. Org. Chem.* **26** (1990) 179. D. A. Dougherty, *Acc. Chem. Res.* **24** (1990) 88.
5. W. Schlenk and M. Brauns, *Ber. Dtsch. Chem. Ges.* **48** (1915) 661. G. Kothe, K.-H. Denkel and W. Summermann, *Angew. Chem., Int. Ed. Engl.* **9** (1970) 906.
6. E. Müller and I. Müller-Rodoloff, *Ann.* **517** (1935) 134.
7. A. Weiss and H. Witte, *Magnetochemie* (Verlag Chemie, Weinheim/ Bergstr., 1973). R. L. Carlin, *Magnetochemistry* (Springer, New York, 1986).
8. N. Nakamura, K. Inoue and H. Iwamura, *J. Am. Chem. Soc.* **114** (1992) 1484.
9. (a) K. Mukai, H. Nagai and K. Ishizu, *Bull. Chem. Soc. Jpn.* **48** (1975) 2381. (b) K. Ishida and H. Iwamura, *J. Am. Chem. Soc.* **113** (1991) 4238. (c) T. Watanabe and H. Iwamura, *Abstr. of papers, the 21st Symp. Structural Org. Chem.*, Tsukuba, October 1989.
10. N. Azuma, K. Ishizu and K. Mukai, *J. Chem. Phys.* **61** (1974) 2294.
11. E. F. Ullman, J. H. Osiecki, D. G. B.Boocock and R. J. Darcy, *J. Am. Chem. Soc.* **94** (1972) 7049.
12. J. Veciana, C. Rovira, O. Armet, V. H. Domingo, M. I. Crespo and F. Palacio, *Mol. Cryst. Liq. Cryst.* **176** (1989) 77. J.Veciana, C. Rovira, M. I. Crespo, O. Armet, V. M. Domingo and F. Palacio, *J. Am. Chem. Soc.* **113** (1991) 2552.
13. A. Rajca and S. Utamapanya and J. Xu, *J. Am. Chem. Soc.* **114** (1992) 1884.
14. K. Mukai and J. Sakamoto, *Chem. Phys.* **68** (1978) 1432.

324

15. T. Matsumoto, N. Koga and H. Iwamura, *J. Am. Chem. Soc.* **114** (1992) 5448. T. Matsumoto, T. Ishida, N. Koga and H. Iwamura, *J. Am. Chem. Soc.* **115** (1993) in press.

16. A. Fujii, T. Ishida, N. Koga, and H. Iwamura, *Macromol.* **24** (1991) 1077. N. Koga, K. Inoue, N. Sasagawa and H. Iwamura, *Mat. Res. Soc. Symp. Proc.* **173** (1990) 3949.

17. K. Inoue, N. Koga, and H. Iwamura, *J. Am. Chem. Soc.* **113** (1991) 9803.

18. Ref. 1. P. Dowd, *J. Am. Chem. Soc.* **92** (1970) 1066. P. Dowd, W. Chang and Y. H. Paik, *J. Am. Chem. Soc.* **108** (1986) 7416.

19. B. Bleaney and K. D. Bowers, *Proc. Royal Soc. London* **A214** (1952) 451.

20. L. Dulog and J. S. Kim, *J. Am. Chem. Soc.* **29** (1990) 415.

21. S. Kato, K. Morokuma, D. Feller, E. R Davidson, W. T. Borden, *J. Am. Chem. Soc.* **105** (1983) 1791. C. I. Ivanov, N. N. Tyutyulkov and S. Karabunarliev, *J. Magn. Magn. Mat.* **92** (1990) 171.

22. E. Wasserman, R. W. Murray, W. A. Yager, A. M. Trozzolo and G. Smolinsky, *J. Am. Chem. Soc.* **89** (1967) 5076.

23. H. Tukada, K. Mutai and H. Iwamura, *J. Chem. Soc., Chem. Commun.* (1987) 1159.

24. D. E. Seeger, P. M. Lahti, A. R. Rossi and J. A. Berson, *J. Am. Chem. Soc.* **108** (1986) 1251.

25. A. A. Ovchinnikov, *Theor. Chim. Acta* **47** (1978) 297.

26. W. T. Borden and E. R. Davidson, *J. Am. Chem. Soc.* **99** (1977) 4587.

27. J. Carilla, L. Julia, J. Riera, E. Brillas, J. A. Garrido, A. Labarta and R. Alcala, *J. Am. Chem. Soc.* **113** (1991) 8281.

28. D. F. Evans, *J. Chem. Soc.* (1959) 2003.

29. S. Murata, T. Sugawara and H. Iwamura, *J. Am. Chem. Soc.* **109** (1987) 1266. H. Iwamura and S. Murata, *Mol. Cryst. Liq. Cryst.* **176** (1989) 33. S. Murata and H. Iwamura, *J. Am. Chem. Soc.* **114** (1991) 5547.

30. S. Sasaki and H. Iwamura, *Chemistry Lett.* (1992) 1759.

31. D. A. Kaisaki, W. Chang and D. A. Dougherty, *J. Am. Chem. Soc.* **113** (1991) 2764.

32. T. Sugawara, S. Bandow, K. Kimura and H. Iwamura, *J. Am. Chem. Soc.* **106** (1984) 6449. T. Sugawara, S. Bandow, K. Kimura, H. Iwamura and K. Itoh, *J. Am. Chem. Soc.* **108** (1986) 368.

33. H. M. McConnell, *J. Chem. Phys.* **39** (1963) 1910. A. Izuoka, S. Murata, T. Sugawara and H. Iwamura, *J. Am. Chem. Soc.* **107** (1985) 1786. T. Sugawara, S. Murata, K. Kimura, H. Iwamura, Y. Sugawara and H. Iwasaki, *J. Am. Chem. Soc.* **107** (1985) 5293. A. Izuoka, S. Murata, T. Sugawara and H. Iwamura, *J. Am. Chem. Soc.* **109** (1987) 2631. K. Yamaguchi, Y. Toyoda and T, Fueno, *Chem. Phys. Lett.* **159** (1989) 459.

34. K. Inoue and H. Iwamura, *Adv. Mater.* **4** (1992) in press.
35. M. Kinohsita, P. Turek, M. Tamura, K. Nozawa, D. Shiomi, U. Nakazawa, M. Ishikawa, M. Takahashi, K. Awaga, T. Inabe and Y. Maruyama, *Chemistry Lett.* (1991) 1225. M. Tamura, Y. Nakazawa, D. Shiomi, K. Nozawa, Y. Hosokoshi, M. Ishikawa, M. Takahashi and M. Kinoshita, *Chem. Phys. Lett.* **186** (1991) 401.
36. A. Rajca, *J. Am. Chem. Soc.* **112** (1990) 5890. A. Rajca and S. Utamapanya and S. Thayumanavan, *J. Am. Chem. Soc.* **114** (1992) 1884.
37. F. Kanno, K. Inoue, N. Koga and H. Iwamura, *J. Am. Chem. Soc.* **115** (1993) in press.

PROGRESS TOWARD AN ALL-ORGANIC MAGNET.
HIGH SPIN MOLECULES AND POLYMERS

Dennis A. Dougherty
Division of Chemistry and Chemical Engineering
California Institute of Technology

1. Introduction

Of the many approaches to new magnetic materials, the all-organic strategy is the most challenging. By "all-organic" we mean a material composed primarily of C, H, O, and N, halogens, sulfur and other maingroup elements, but explicitly devoid of transition metals and rare earth elements. By this definition, the list of bone fide organic magnets is short indeed. The last decade has, however, seen significant advances in our understanding of the factors that control spin-spin interactions in organic structures.[1] Increasingly more rigorous design principles are emerging, and it seems likely that the next decade will see the preparation of a significant number of new organic materials that display interesting magnetic behaviors.

It may be fair to ask, "Why make an organic ferromagnet?". From our perspective, the primary motivation is an intellectual one. First, it should be appreciated that the quest for an "organic ferromagnet" is really a quest for new organic materials that display any of the large number of intriguing magnetic behaviors[2] that have been seen in inorganic materials, or that display totally new magnetic behaviors. An organic ferromagnet would be interesting, but so would an organic spin glass or ferrimagnet. We will use the term "organic magnet" to mean an organic material that displays any of the many kinds of cooperative magnetic behaviors.

It seems fair to say that a molecular-level understanding of the many forms of magnetism does not exist, and certainly at present we cannot rationally design new magnetic materials at will, whether they contain metals or not. The critical advantage of organic chemistry in this regard is that, more than any other field of science, organic chemistry has the ability to rationally and systematically modify structure through the power of organic synthesis. With organic magnetic materials, one could rationally discover what makes a spin glass a spin glass, and what it takes to convert it to a mictomagnet or superparamagnet. One could evaluate the features that are most important in establishing the critical temperature, or the coercivity, or the optical properties. The subtlety and breadth of modifications that can be made in organic structures far exceed those of metal-based systems. Thus, the study of organic magnets and related structures offers the possibility of developing a much deeper understanding of magnetism in general.

The development of organic magnets may also have technological implications. It is difficult to say with certainty what possible applications might be before one has such materials in hand. A few general observations are possible, however, based on the expected properties of organic magnets. First, it seems certain that organic magnets will not replace metal oxides and rare earth systems in most conventional applications of magnetism. The existing materials are inexpensive and powerfully magnetic, and the technology developed around them is extremely sophisticated.

In order to be technologically significant, an organic ferromagnet will have to offer new possibilities that are not available in inorganic systems. For example, organic ferromagnets should be lightweight, which may be useful in some applications. Improved processability, for example with regard to thin films or novel shapes, seems possible, especially for polymer-based organic magnets. One intriguing potential is the combination of magnetic properties with novel optical or electrical properties. For example, true magnetooptical switching – that is, the single photon interconversions of magnetic and non-magnetic states – seems possible with organics (see below).

The present work will review the progress to date in the development of organic magnetic materials. While this is certainly a "Research Frontier in Magnetochemistry", at this stage it is more "chemistry" than "magneto". We will divide the discussion into two broad classes of structures – molecular solids[3] and polymers. Much of our discussion will focus on high spin molecules, which serve a dual role in this field. First, they constitute important tests of spin coupling theories, and thus provide valuable design principles for solid state structures. In addition, solid samples of high spin molecules can themselves be magnetic, and such molecular magnets will be discussed. We will also describe polymer systems designed to be high spin and/or magnetic. These structures generally incorporate knowledge gained from the molecular studies, but may also contain design principles that are unique to the polymeric state.

2. Molecular Systems

Of course, to make a magnetic material one must have a large number of unpaired electrons, preferably in a form that is stable above room temperature. At first, this may seem like a major liability for magnetic organic materials. While it is certainly true that it is easier to put spin into a system by just adding iron, stable organic radicals have been known since 1900.[4] There is now a wide variety of organic radicals that are quite stable. More importantly, there is a generally good understanding of the structural features that enhance radical stability.[5] There are still challenges. Synthesis of samples with large numbers of stable radicals can be difficult. Also, the major factors that stabilize radicals – delocalization and steric protection – tend to increase the mass per spin, thereby diluting the potential spin density of a material. Still, conceptually, at least there are many ways to introduce stable spins.

The critical, and historically much less well understood, issue is the development of high spin and ultimately cooperative interactions among large numbers of organic radicals. Thus, the motivation for the study of high spin ($S >$ 1/2) molecules is apparent as a way to learn the rules of organic spin coupling. An additional motivation is the fact that simple mean field models[6] predict that for a molecular ferromagnet, the critical temperature scales as $S \bullet (S + 1)$, where S is the spin of the molecular building block. Thus, presumably the higher the spin state of the molecular building block, the higher the critical temperature.

2.1   Biradicals

To build an organic ferromagnet one must ensure that unpaired electrons (those not intimately involved in chemical bonds) experience high spin interactions (ferromagnetic coupling) rather than low spin (antiferromagnetic coupling). It should be appreciated that in most instances, two weakly interacting electrons will low spin couple. A significant manifestation of this is the fact that solid samples of stable organic radicals are almost always antiferromagnets at low temperatures. Thus, biradicals[7] – the two-electron system – have been studied extensively in the hope that such studies will produce general insights into the development of ferromagnetic coupling in organic systems. Much has been learned in recent years about this problem, and we will summarize the results here.[8] In terms of organic magnetochemistry, these can be viewed as the most fundamental studies, analogous to determining the factors that favor high spin vs. low spin Fe^{3+}.

In organic biradicals, there are several, apparently general ways to make the high spin, triplet (T) be the ground state. In each case, the underlying origin of the high spin preference is the same. As in all systems, the key to achieving a high spin ground state in an organic molecule is to maximize exchange interactions between the unpaired electrons (as manifest in the exchange integral K) while ensuring that the quantum mechanical overlap integral (S) between the orbitals that contain the spins (the magnetic orbitals) is zero, or very nearly so.[9]

All simple carbenes (Figure 1) have triplet (**T**) ground states. This is a direct extension of Hund's rule, as formulated for atomic systems. The carbene structures are well understood,[10] and directly analogous to high spin transition metal structures. Exceptions occur when strongly perturbing substituents are placed on the carbene, in ways that are easily anticipated.

FIGURE 1

T   S

There are also a number of planar, conjugated hydrocarbons that might be expected to have triplet ground states (Figure 2-4). Many of these are *non-Kekulé* molecules – structures for which one cannot write a classical Kekulé structure that has all the electrons paired into bonds.[11] At the Hückel level, these structures have a degenerate pair of non-bonding molecular orbitals (NBMO) that contain a total of two electrons. At higher levels of theory, the exact degeneracy of the NBMOs survives only in those (relatively rare) cases of molecules with 3-fold or higher rotation axes. Systems with fairly large deviations from exact degeneracy can still have triplet ground states (see for example **1**), but, of course there is a limit, beyond which spin pairing into the lower of the two "NBMOs" will occur.

FIGURE 2

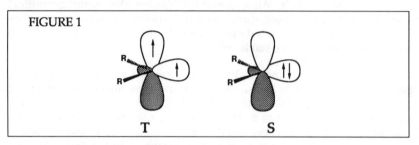

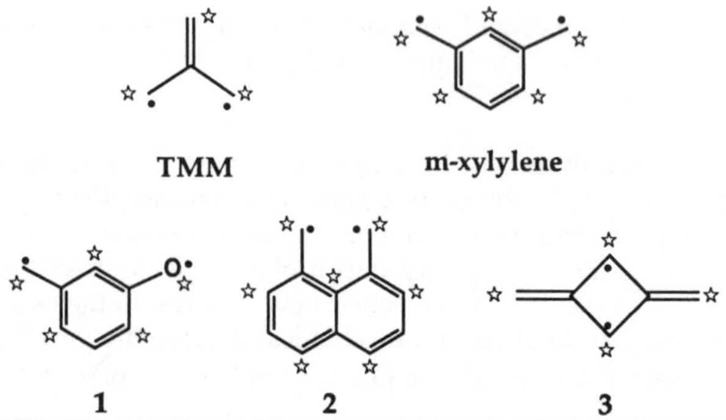

TMM   m-xylylene

1   2   3

For the most common π systems, a simple, topology-based rule allows one to anticipate when a structure will have a high spin ground state. Most systems studied to date are *alternant* hydrocarbons, systems for which all the atoms can be partitioned into two sets, termed starred and non-starred, such that no two atoms of the same set are adjacent (see Figure 2). If the number of starred atoms exceeds the number of non-starred by N, then the ground state multiplicity is generally expected to be $N + 1$.[12,13] The prototype is trimethylenemethane (TMM), which was first observed by Dowd in 1966[14] Several other typical examples are shown in Figure 2.[15-17] The rule is very simple, but it is firmly rooted in quantum mechanics – such topologies will produce magnetic orbitals that are orthogonal but are very much coextensive in space.[13] An attractive feature of this analysis is that it can easily be extended to very large, very high spin systems.[12]

There is also a group of potentially high spin molecules for which the numbers of starred and non-starred atoms are the same (Figure 3). For such systems there is no simple method to predict the ground spin state, but high levels of theory[13] often predict a low spin, singlet (S) ground state. There is, however, some controversy in this area. In fact, for the prototype system, TME, theory predicts a singlet ground state,[18] but experiment finds triplet ground states for the parent and several related structures.[19] Other structures such as **4** and **5** clearly show a singlet ground state,[20] but it is not clear how much the heteroatom perturbs the TME of **5**. In the present context it is perhaps best to say that these equal parity systems can be singlets or triplets. It also seems likely that in the systems with triplet ground states, the preference will not be large. Thus, they are perhaps not prime candidates for designing organic magnetic materials.

FIGURE 3

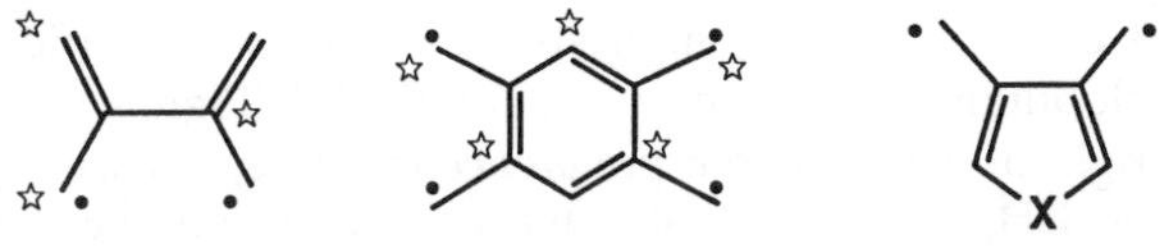

| TME | 4 | 5 (X = O,S) |

Another relevant class of structures is the [4n]annulenes (Figure 4). These also have (at Hückel level) a degenerate pair of NBMOs, and so can have a triplet ground state. The prototype – square cyclobutadiene (CBD) – has a *singlet* ground state.[21] In this structure there is an exact NBMO degeneracy even at high levels of theory. Thus, having a rigorously degenerate pair of NBMOs is neither a necessary (see above), nor a sufficient criterion for achieving a triplet ground state. The more relevant structures here, however, are derivatives of the benzene dication and dianion, and of cyclopentadienyl cation.[22] These fascinating systems have been developed extensively, especially by Breslow. As discussed elsewhere,[23] such structures can have either singlet or triplet ground states, and in many instances it appears that external influences (counterions or other environmental effects) can tilt the balance one way or the other. These structures are important in certain models of molecular ferromagnetism (see below), but a thorough understanding of the factors that determine ground state multiplicity is not available.

FIGURE 4

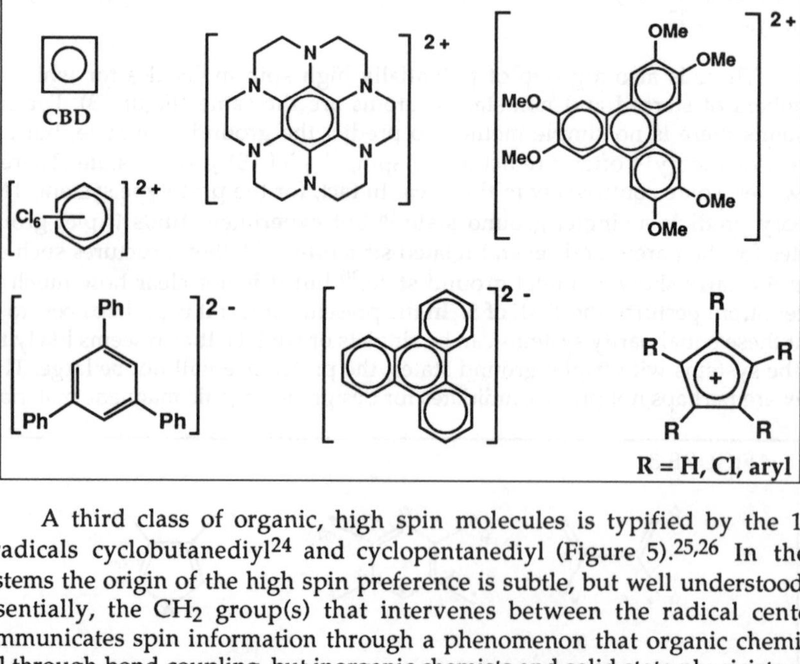

R = H, Cl, aryl

A third class of organic, high spin molecules is typified by the 1,3-biradicals cyclobutanediyl[24] and cyclopentanediyl (Figure 5).[25,26] In these systems the origin of the high spin preference is subtle, but well understood.[27] Essentially, the CH_2 group(s) that intervenes between the radical centers communicates spin information through a phenomenon that organic chemists call through-bond coupling, but inorganic chemists and solid state physicists will recognize as superexchange.[28] In a sense, the $C\bullet-CH_2-C\bullet$ system is directly analogous to the much-studied M–O–M (M = metal) system. In the 4- and 5-ring organic systems the balance between direct (through space) and superexchange (through bond) interactions produces a triplet ground state. This is not always the case, neither in inorganic structures, where low spin coupling is the norm, nor in the organic cases, where recent evidence suggests that the 6-ring structure (1,3-cyclohexanediyl) has a singlet ground state.[29]

FIGURE 5

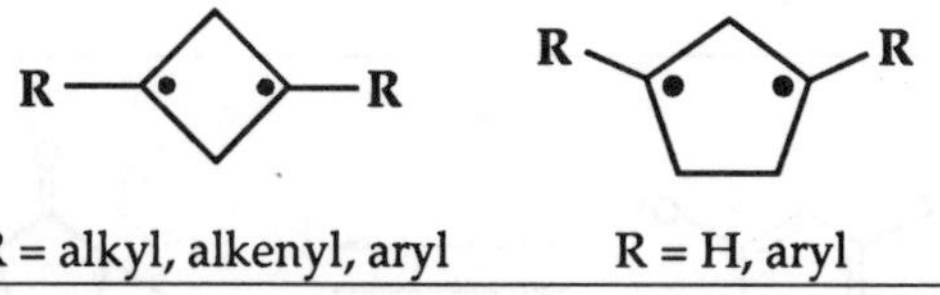

R = alkyl, alkenyl, aryl   R = H, aryl

There are, hopefully, other structural motifs that will produce general ferromagnetic coupling units. For example, recent results indicate that a biaryl system in which the rings are forced to be orthogonal (Figure 6) will produce high spin states when the rings are oxidized to radical cations[30] or reduced to radical anions[31]. It was not immediately obvious that such structures would be high spin, since 90° twisted ethylene – the prototype for these systems – has a singlet ground state.[21b] However, a theoretical justification of the high spin ground states of these more complicated structures has been presented.[32] Further work will be required to establish the magnitude of the high spin preference and the generality of this coupling mode.

FIGURE 6

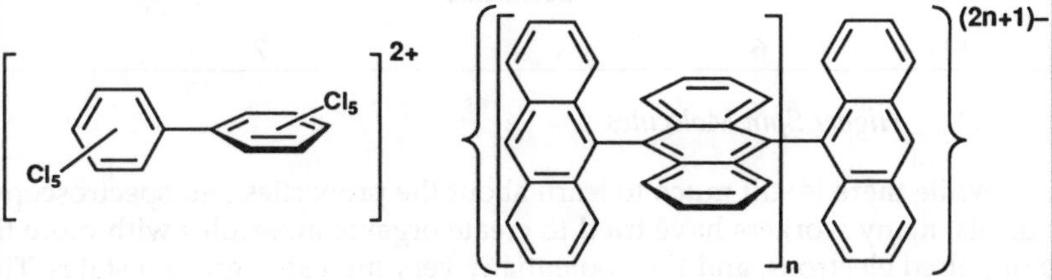

Apart from providing information on spin coupling mechanisms, biradicals can also suggest ways in which magnetic organic materials can have unique, perhaps useful properties. For example, every biradical has associated with it a closed shell, covalent isomer prepared by forming a bond with the two unpaired electrons. Of course, such a structure is necessarily diamagnetic. In several cases (Figure 7), both the biradical and the fully covalent forms of a structure are accessible, and are of comparable energy.[33] Conceptually, one can envision the interconversion of the two as a way to switch between "magnetic" and nonmagnetic structures. We have demonstrated that in one case (6 ⇌ 7), such an interconversion can be achieved photochemically.[34] This provides a prototype for true magnetooptical switching,[35] in a way that would be difficult to imagine with metallic systems. We emphasize that this is at best a prototype – the structures involved are quite exotic, and the switching is observable only at or below liquid nitrogen temperatures. Still this seems like an area where organic magnets could make a significant contribution.

FIGURE 7

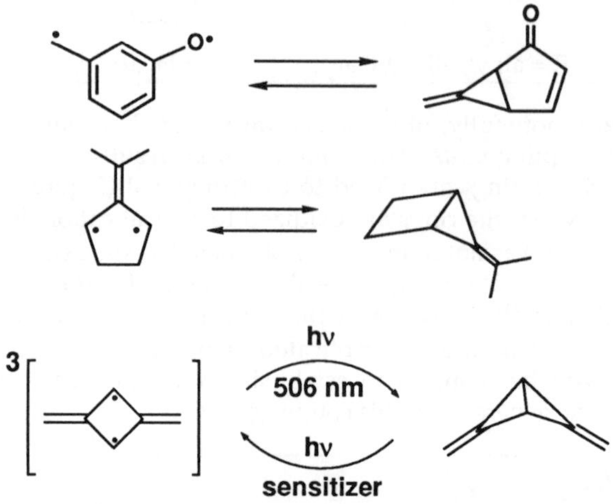

2.2   Higher Spin Molecules

While there is still much to learn about the properties and spectroscopy of biradicals, many workers have tried to create organic molecules with more than two unpaired electrons, and thus potentially very high spin ground states. These structures are intrinsically interesting in their own right, but also should provide valuable insights into the preparation of high spin and ultimately magnetic materials.

A useful notion in the design of high spin structures is the ferromagnetic coupling unit (FC).[8,37] This is a general structure that will ensure a high spin interaction between (among) two or more spin containing units that are linked through it. Certainly the longest studied and best documented FC is m-phenylene, and the majority of very high spin structures incorporate this motif. In fact, the unique situation of two radicals linked meta through a benzene has been a topic of theoretical and experimental interest since the earliest part of this century.[38]

The modern history begins with the observation of 8 (Figure 8), a hydrocarbon with a quintet ($S = 2$) ground state, reported simultaneously by Itoh[39] and Wasserman[40] in 1967. In this structure, two carbenes are linked by m-phenylene. The individual carbenes are triplets due to a Hund's rule type interaction, and their ferromagnetic coupling is ensured by the topology of the π system. For each carbene, one of the electrons is in a π-type orbital (Figure 1). Star/non-star analysis predicts a high spin coupling in the π system, so overall, a quintet is seen. It is an interesting historical fact that long before the ground spin state of the parent m-xylylene had been experimentally demonstrated, quintet **8** had established the validity of m-phenylene as a ferromagnetic coupling unit.

FIGURE 8

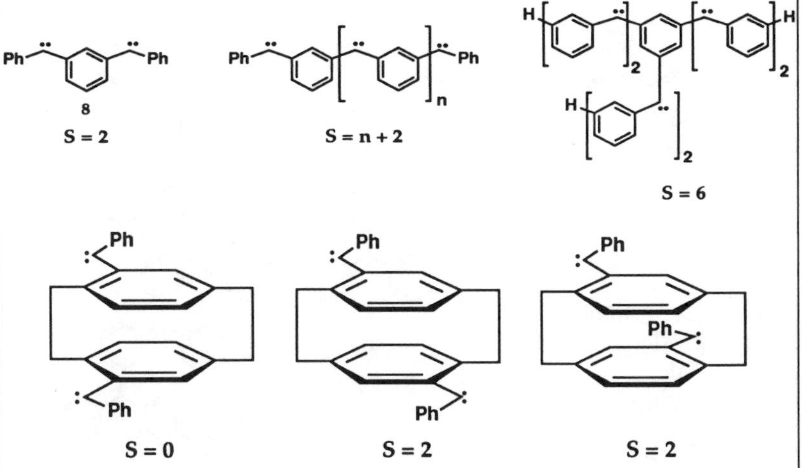

The initial observation of **8** was followed by a series of ground-breaking experiments by the Itoh and Iwamura groups on the now famous polycarbenes. This work has produced many historic firsts, including the "record" (achieved several times) for the highest molecular spin state, and the development of the novel spectroscopy of these unique structures. The work has been reviewed elsewhere,[41] and Figure 8 gives examples of the unique molecules prepared.[42-44]

The polycarbene work has gone a long way to establish m-phenylene as a ferromagnetic coupling unit, and others have extended the generality of this critical building block (Figure 9).[45] Along with the subsequently prepared m-xylylene type biradicals (Figure 2),[16,38] a number of systems with many radicals linked by m-phenylene have been prepared. It has been shown that the 1,3,5-substitution pattern is high spin, producing quartets if the substituents are radicals,[46] and septets if the substituents are carbenes or nitrenes.[41] Heptaradical **9** and related structures have been prepared by Rajca.[47] Also notable, although technically "just" biradicals, are the perchlorotriarylmethyl derivatives **10** of

Veciana and coworkers, which are stable *at room temperature in fluid media*.[48] Our group has shown that the spin containing unit need not be a one-center system, such as a carbene or radical. M-phenylene is also a ferromagnetic coupling unit for π-conjugated triplets of the TMM type (11)[29,37a] and for localized biradicals of the cyclobutanediyl type (12).[29,49] Berson has shown that more elaborate topologies (13), but still related to the m-xylylene system, can also produce high spin ground states, as anticipated by the star/non-star rule.[50]

FIGURE 9

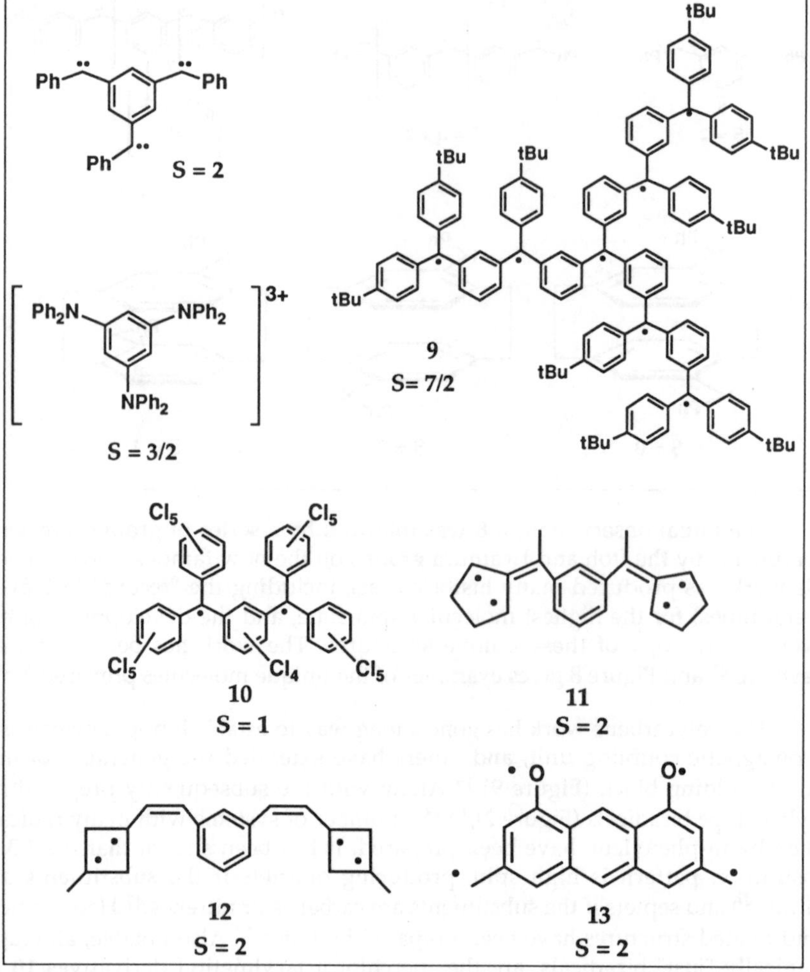

The biscarbene strategy has also been used to probe other potential ferromagnetic coupling units such as biphenyl, stilbene, and diaryldiacetylene derivatives, with results that are generally in accord with the topology-based rules.[51] An especially interesting study probes the role of heteroatoms in the coupling unit.[52]

It has also been established that the superexchange motif described above can be used to produce tetraradicals with quintet ground states (Figure 10). In particular, we have shown that, like m-phenylene, cyclobutanediyl can serve to high spin couple two triplets, in this case TMM derivatives, to make a quintet (**14**).[49] The cyclopentanediyl unit serves only as a weak ferromagnetic coupling unit, in that the quintet ground state of **15** is in thermal equilibrium with a tetraradical triplet state at temperatures as low as 40 K. In contrast, the six ring system (in a chair conformation) is an antiferromagnetic coupling unit, as evidenced by the fact that **16** has a singlet ground state.[29]

FIGURE 10

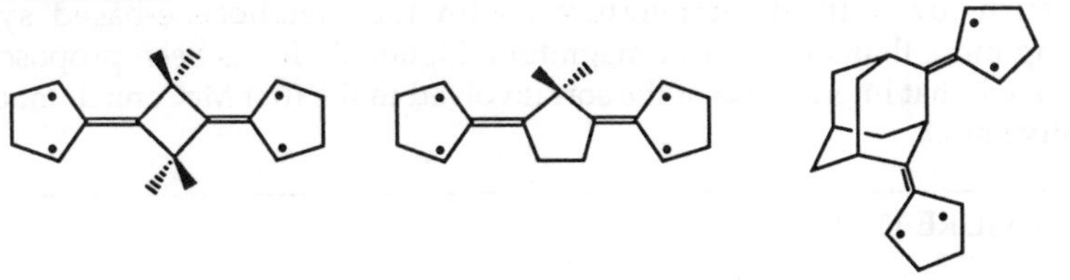

| 14 | 15 | 16 |
| S = 2 | S = 2 | S = 0 |

2.3 Molecular Solids

Several strategies have emerged for the development of bulk magnetic properties from samples of molecular, organic structures. Two models initially developed by McConnell have proven to be especially inspirational in considering intermolecular interactions that could be ferromagnetic. In the first,[53] ferromagnetic coupling between pairs of molecules is achieved if regions of positive spin density in one molecule interact most with regions of negative spin density in an adjacent molecule. Note that negative spin density is common in organic π radicals and polyradicals of the sort in Figure 2 and 3 due to spin polarization effects.[21] A brilliant test of the first McConnell model was made by Iwamura studying the cyclophane biscarbenes of Figure 8.[44] The ground states are exactly as expected based on the McConnell model and known spin densities in diphenyl carbenes.

338

The second McConnell model[54] involves charge transfer systems. Breslow was the first to explore possible experimental realizations of the model,[55] and the basic concept has been modified and expanded by several workers.[56] Fundamentally, the model involves mixed crystals of donors and acceptors, designed such that in one of the charge transfer configurations of the system, one of the partners exists as a high spin (usually triplet) state. Typically this is an excited state of the charge transfer system, but it influences the ground state ordering by configurational mixing, thereby producing ferromagnetic coupling among the molecular centers.

On the experimental side, the list of all-organic, molecular solids that have been shown unambiguously to exhibit three-dimensional, ferromagnetic order is quite short. The γ and β phases of the nitronyl nitroxide 17 (Figure 11) show bulk ferromagnetism, each with a critical temperature in the vicinity of 0.6 K.[57] This result is both encouraging, in that it establishes that organic magnetic molecular solids can be prepared, and discouraging, due to the extremely low critical temperature. As discussed elsewhere in this book, not all molecular magnets have such low critical temperatures, with the metallocene-based systems showing more than an order of magnitude higher T_c. It has been proposed, but not proven, that interactions of the sort involved in the first McConnell model are operative in 17.

FIGURE 11

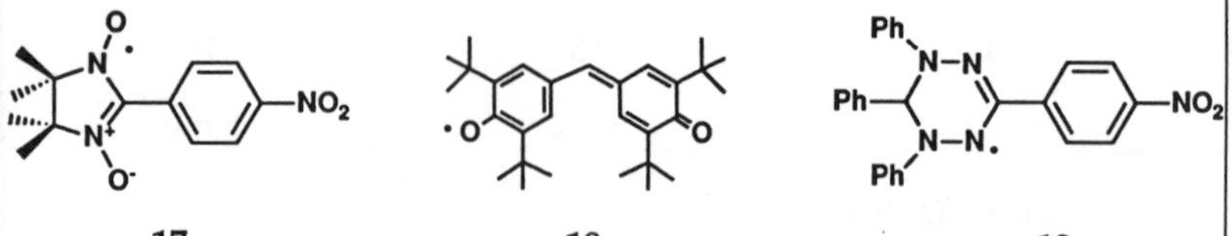

| 17 | 18 | 19 |

While 17 remains the only well-characterized molecular organic ferromagnet, there are a number of other organic, molecular solids with interesting magnetic properties. Galvinoxyl (18) is one of the best known of the stable organic radicals, and in the solid state it displays intriguing but ultimately frustrating properties.[58] Curie-Weiss data above 85 K give a quite large, positive Weiss constant of 19 K. Unfortunately, the molecule undergoes a crystal phase transition at 85 K which is accompanied by the development of antiferromagnetic interactions in the solid. Mixed crystals of galvinoxyl and galvinol do not display this phase transition, and the ferromagnetic couplings persist to low temperature, although no three-dimensional ordering is seen. Very recently, it has been shown that fine powders of pure galvinoxyl do not undergo the undersirable phase transition, allowing low temperature studies that reveal Θ equal to 4.4 K, but no long range ordering.[59]

Verdazyl derivative **19** (Figure 11) is closely related to **17** , and it has been cleverly designed to exploit the McConnell type I mechanism in the solid state. Although no three-dimensional ordering is seen, a positive Weiss constant of 1.6 K is reported.[60]

The pioneering work of Breslow[22a] has produced a number of interesting structures, including charge-transfer complexes intended to test the second McConnell mechanism. Key components of these structures are planar π systems that can be oxidized to give dications related to the benzene dications, of the sort shown in Figure 4. If these dications have triplet ground states, appropriate charge transfer interactions can lead to ferromagnetic coupling. An interesting variant of these structures involves chemical doping of the solid, charge transfer complexes.[61] Unfortunately, all these systems have thusfar shown predominantly antiferromagnetic interactions in the solid.

A charge-transfer salt of C_{60} and TDAE (tetrakis(dimethylamino)-ethylene) displays a sudden rise in the magnetization at 16 K, suggesting some sort of magnetic phase transition.[62] However, no hysteresis is seen, and no report of magnetization in zero field has been given. This material is still under study, but the most recent efforts suggest that it could be an itinerant ferromagnet.

3. High Spin Polymers

A fundamentally different approach to magnetic organic materials targets polymer-based structures. Of course, the goal is to combine magnetism with the highly useful materials properties usually associated with organic polymers. There has been a large amount of theoretical work on potentially ferromagnetic polymers. The topology based rules can be extended easily to infinite systems, and several clever implementations of this approach have been described.[12] There has been much recent effort in this area, leading to the production of organic materials with relatively large numbers of stable spins, including some with signs of short range order. However, no definitive evidence for three-dimensional magnetic ordering in a well characterized organic polymer has emerged.

We note at the outset that there is an ever-growing list of "materials" that are claimed to be organic ferromagnets. Many of these are simply the products of pyrolysis of organic molecules or polymers. Pyrolysis precursors include adamantane, 1,5,8,12-tetraazadodecane, polyacrylonitrile, polyvinylchloride, and others. Essentially nothing is known about the structures of these materials, and general concerns about the origin of the magnetic effects in these structures have been raised.[63] As such, these pyrolytic samples have produced no new insights into the magnetic organic materials problem.

There is another group of polymers that appear to have interesting magnetic properties. The structures are synthesized in a somewhat rational manner, but the syntheses ultimately produce intractable, ill-characterized materials of unknown composition. Such structures include: the I_2/triaminobenzene product of Torrance;[64] the polydiacetylenes of Ovchinnikov;[65] the COPNA resins of Ota;[66] and the Indigo-based polymers of Tanaka.[67] Again, the lack of characterization of these materials has prevented them from having a strong intellectual impact on the magnetic organic materials problem.

A recent review of the magnetic polymers field[63] has concluded that there are no adequately documented organic magnetic polymers, and suggests that in most, if not all, cases claimed to date, the observed magnetic properties are due to "extrinsic" factors, such as contamination by metals.

In contrast, there is a growing list of well-characterized polymers that contain many spins. In some cases, very high spin concentrations are obtained and the radical centers are quite stable. However, no three-dimensional ordering other than weak antiferromagnetism has been seen.

For the most part these high spin polymers involve a polyacetylene chain with pendant groups that contain the radical centers, although some other backbones have been explored.[68] Examples are shown in Figure 12. Since these polymers have conjugated backbones, one might expect that the π topology rules mentioned above would be helpful, and efforts to evaluate possible polymer topologies have been reported.[51,52] None of the polyacetylene-derived systems, however, show signs of long-range order.

An alternative strategy, in which the radical centers are an intrinsic part of the polymer chain, is represented by the so-called polaronic ferromagnet.[69] In this approach, spins are introduced by doping, producing radical cations or anions termed polarons. With a proper topology, ferromagnetic coupling among polarons along the chain is predicted.[69] We have found that doping to give the polymers in Figure 13 produces a significant number of stable spins, though not as many as in some of the pendant spin polymers described above. Magnetization studies reveal a net ferromagnetic coupling among the polarons, suggesting that at least in one dimension, the polaronic ferromagnet concept may be viable.[70]

FIGURE 12

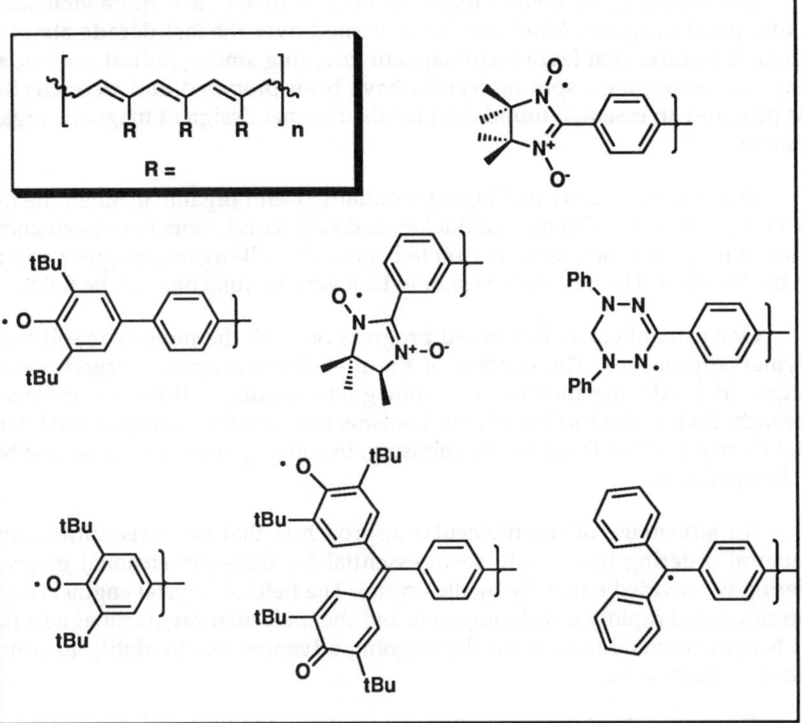

FIGURE 13

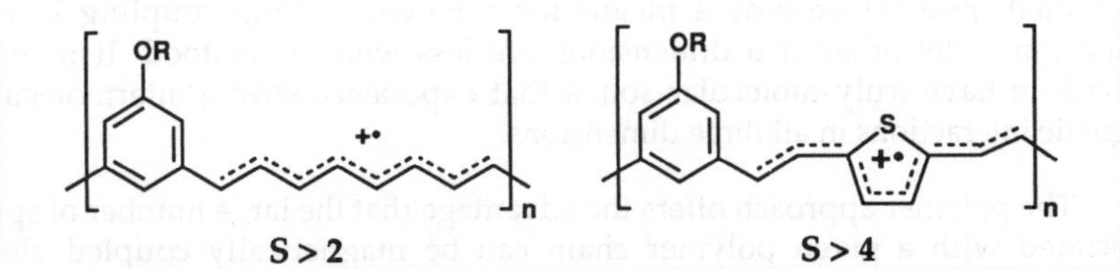

S > 2   S > 4

Essentially no progress has been reported on controlling magnetic interactions between polymer chains. Most of the high spin polymers display either no interchain interactions or weak antiferromagnetic interactions at very low temperatures.

4. Prospects

Intellectually, the field of magnetic organic materials is quite vital, and is making good progress. Much has been learned over the last decade about the structural features that favor ferromagnetic coupling among radical centers, and many interesting high spin molecules have been prepared and characterized. This provides an essential foundation for the rational design of magnetic organic materials.

In terms of actually making a potentially useful organic magnet, the field has a long way to go. Organic conductors and superconductors have been known for some time, and there is no reason to believe that all-organic magnets will not join the list soon. The next decade should be a very exciting one for the field.

One can anticipate continued progress on both the molecular solids and polymer approaches. The success of the metallocene-based charged transfer systems in producing molecular ferromagnets augurs well for the molecular approach. Both of the McConnell mechanisms have considerable potential, but it is not clear yet which is the most promising. In some systems it may be that both will be operative.

An advantage of the molecular approach is that the three-dimensional structural ordering that would seem essential for three-dimensional magnetic ordering is provided naturally by the crystal. The field of "crystal engineering" is an exciting and rapidly developing one, and the molecular organic magnets field will benefit tremendously from the ongoing advances our in ability to control solid state structures.

This strength of the molecular approach could also become a significant weakness. Generally, the intermolecular interactions in a crystal are relatively weak, energetically. Most of the models for molecular magnets (such as the McConnell models) provide a means for achieving strong coupling in one dimension – the other two dimensions are less well understood. It may be difficult to have truly molecular solids that experience strong intermolecular magnetic interactions in all three dimensions.

The polymer approach offers the advantage that the large number of spins associated with a given polymer chain can be magnetically coupled along covalent pathways. Such interactions are generally much stronger than noncovalent interactions. But the polymer approach has unique challenges of its own. Synthesis of polymers with a large number of spins is difficult, and though significant advances have been made, this will always be a challenge. The more difficult challenge, though, is the control of magnetic interactions between polymer chains. This is similar to the intermolecular coupling problem in the molecular magnets, but is more complicated because of the lack of an ordered structure. There are many techniques for aligning polymer chains, but it is not obvious that these will allow bulk magnetic behaviors to develop. An attractive alternative is to cross-link the polymer chains in a way that ensures ferromagnetic coupling between the chains. If done properly, cross-linking will ensure that all magnetic interactions are along covalent paths, and thus potentially quite strong. However, heavily cross-linked polymers tend to lack the desirable materials properties that provide the essential practical motivation for making magnetic organic materials. Some sort of balance between these two competing goals would have to be obtained.

We note that there are other ways to achieve the structural ordering that is required. For example, liquid crystalline materials have potentially interesting properties, and some efforts toward producing liquid crystalline magnetic materials have been reported.[71] Also, self assembling monolayers, Langmuir-Blodgett films and related "two-dimensional" systems produce significant alignment which could exploited. These and related techniques could by themselves or in combination with the molecular or polymer approaches produce interesting magnetic materials.

While substantial hurdles must be overcome before the rational development of a technologically useful organic magnet may be possible, one should not underestimate the potential intellectual impact organic materials can have on the field of magnetism. Low dimensional magnetic behaviors and new examples of novel magnetic behaviors will almost certainly emerge from studies of magnetic organic materials. In this way organic chemistry can have a significant impact on a vital field of condensed matter physics.

The field of magnetic organic materials has a long way to go, but all the pieces are in place for exciting new advances to made in the coming years. Creative and insightful application of the many design principles being developed, combined with a bit of good luck, will produce the breakthrough necessary to have organic magnets join the list of new materials with important intellectual and technological implications.

5. References

1. (a) Miller, J. S.; Dougherty, D. A., Eds. *Molecular Crystals and Liquid Crystals* 1989; Vol. 176., pp 1-562. (b) Gatteschi, D.; Kahn, O.; Miller, J. S.; Palacio, F. *Magnetic Molecular Materials*; Kluwer Academic Publishers: Boston, 1991; Vol. 198.

2. Hurd, C. M. *Contemp. Phys.* **1982**, *23*, 469-493.

3.   By a molecular solid we mean one that consists of discrete, molecular units that can redissolve in appropriate solvents. This contrasts metallic, ionic, or network solids.

4.   Gomberg, M. *J. Am. Chem. Soc.* **1900**, *22*, 757. Gomberg, M. *Chem. Ber.* **1900**, *33*, 3150.

5.   See, for example, Griller, D.; Ingold, K. U. *Acc. Chem. Res.* **1976**, *9*, 13-19.

6.   Kittel, C. *Introduction to Solid State Physics*; 6th ed.; John Wiley & Sons, Inc: New York, 1986; p. 426.

7.   For an overview of the biradicals field, see: Borden, W.T. *Diradicals*; John Wiley & Sons: New York, 1982.

8.   For a recent review, see: Dougherty, D. A. *Acc. Chem. Res.* **1991**, *23*, 88-94.

9.   See for example: Mattis, D.C. *The Theory of Magnetism I: Static and Dynamics*; Springer-Verlag: New York; 1988.

10.   See for example: Houk, K. N.; Rondan, N. G.; Mareda, J. *J. Am. Chem. Soc.* **1984**, *106*, 4291-4293.

11.   Dewar, M.J.S. in *The Molecular Orbital Theory of Organic Chemistry* ; McGraw-Hill: New York, 1969; pp. 232-233.

12.   Ovchinnikov, A. A. *Theoret. Chim. Acta (Berl.)* **1978**, *47*, 297-304. Klein, D. J.; Nelin, C. J.; Alexander, S.; Matsen, F. A. *J. Chem. Phys.* **1982**, *77*, 3101-3108.

13.   Borden, W. T.; Davidson, E. R. *J. Am. Chem. Soc.* **1977**, *99*, 4587-4594.

14.   Dowd, P. *Acc. Chem. Res.* **1972**, *5*, 242-248.

15.   Rule, M.; Matlin, A. R.; Hilinski, E. F.; Dougherty, D. A.; Berson, J. A. *J. Am. Chem. Soc.* **1979**, *101*, 5098-5099.

16.   Platz, M.S. in *Diradicals* ; Borden, W.T., Ed.; John Wiley & Sons, Inc.: New York: 1982; pp. 195-258.

17.   Snyder, G. J.; Dougherty, D. A. *J. Am. Chem. Soc.* **1985**, *107*, 1774-1775. Dowd, P.; Paik, Y. H. *J. Am. Chem. Soc.* **1986**, *108*, 2788-2790.

345

18. Du, P.; Borden, W. T. *J. Am. Chem. Soc.* **1987,** *109*, 930-931; Nachtigall, P.; Jordan, K.D. *J. Am. Chem. Soc.* **1992,** *114*, 4743-4747.

19. Dowd, P. *J. Am. Chem. Soc.* **1970,** *94*, 1066-1068. Dowd, P.; Chang, W.; Paik, Y. H. *J. Am. Chem. Soc.* **1986,** *108*, 7416-7417. Dowd, P.; Chang, W.; Paik, Y. H. *J. Am. Chem. Soc.* **1987,** *109*, 5284-5285.

20. Zilm, W. S.; Merrill, R. A.; Webb, G. G.; Greenberg, M. M.; Berson, J. A. *J. Am. Chem. Soc.* **1989,** *111*, 1533-1535. Stone, K. J.; Greenberg, M. M.; Goodman, J. L.; Peters, K. S.; Berson, J. A. *J. Am. Chem. Soc.* **1986,** *108*, 8088-8089. Reynolds, J. H.; Berson, J. A.; Kumashiro, K. K.; Duchamp, J. C.; Zilm, K. W.; Rubello, A.; Vogel, P. *J. Am. Chem. Soc.* **1992,** *114*, 763-764.

21. (a) Borden, W. T.; Davidson, E. R. *Acc. Chem. Res.* **1981,** *14*, 69-76. (b) Borden, W.T. in *Diradicals* ; Borden, W.T., Ed.; John Wiley & Sons: New York, 1982; pp. 1-72.

22. (a) Breslow, R. *Mol. Cryst. Liq. Cryst.* **1985,** *125*, 261-267. Thomaides, J.; Maslak, P.; Breslow, R. *J. Am. Chem. Soc.* **1988,** *110*, 3870-3979.LePage, T. J.; Breslow, R. *J. Am. Chem. Soc.* **1987,** *109*, 6412-6421. Breslow, R.; Maslak, P.; Thomaides, J. S. *J. Am. Chem. Soc.* **1984,** *106*, 6453-6454. Krusic, P.J.; Wasserman, E. *J. Am. Chem. Soc.* **1991 ,** *113* , 2322-2323. (b) Jesse, R.E.; Bileon, P.; Prins, R.; van Voorst, J.D.W.; Hoijtink, G.J. *Mol. Phys.* **1963,** *6*, 633. van Broekhoven, J.A.M.; Sommerdijk, J.L.; de Boer, E. *Mol. Phys.* **1971,** *20* , 993. van Willigen, H.; Broekhoven J.A.M.; de Boer, E. *Mol. Phys.* **1967,** *12*, 533. Glasbeek, M.; van Voorst, J.D.W.; Hoijtink, G.J. *J. Chem. Phys.* **1966,** *45*, 1852. (c) Saunders, M.; Berger, R.; Jaffe, A.; McBride, J.M.; O'Neill, J.; Breslow, R.; Hoffman, J.M., Jr.; Perchonock, C. Wasserman, E.; Hutton, R.S.; Kuck, V.J. *J. Am. Chem. Soc.* **1973,** *95*, 3017. Breslow, R.; Hill, R.; Wasserman, E. *J. Am. Chem. Soc.* **1964,** *86*, 5349. Breslow, R.; Chang, H.W.; Yager, W.A. *J. Am. Chem. Soc.* **1967,** *89*, 1112. Broser, W.; Kurreck, H., Siegle, P. *Chem. Ber.* **1967,** *100* , 788.

23. For a recent review, see: Dougherty, D. A. In *Kinetics and Spectroscopy of Carbenes and Biradicals.*; M. S. Platz, Ed.; Plenum Press: New York, 1990; pp 117-142.

24. Jain, R.; Snyder, G. J.; Dougherty, D. A. *J. Am. Chem. Soc..* **1984,** *106*, 7294-7295. Jain, R.; Sponsler, M. B.; Coms, F. D.; Dougherty, D. A. *J. Am. Chem. Soc.,* **1988,** *110*, 1356-1366. Sponsler, M. B.; Jain, R.; Coms, F. D.; Dougherty, D. A. *J. Am. Chem. Soc.* **1989,** *111*, 2240-2252.

25. Buchwalter, S. L.; Closs, G. L. *J. Am. Chem. Soc.* **1975,** *97*, 3857-3858. Buchwalter, S. L.; Closs, G. L. *J. Am. Chem. Soc.* **1978,** *101*, 4688-4694.

346

26.   Coms, F. D.; Dougherty, D. A. *Tetrahedron Lett.* **1988**, *29*, 3753-3756. Coms, F. D.; Dougherty, D. A. *J. Am. Chem. Soc.* **1989**, *111*, 6894-6896. Stewart, E. G. Ph.D. Thesis, California Institute of Technology, 1992.

27.   Goldberg, A. H.; Dougherty, D. A. *J. Am. Chem. Soc.* **1983**, *105*, 284-290. Doubleday, C., Jr.; McIver, J. W., Jr. *J. Am. Chem. Soc.* **1982**, *104*, 6533-6542. Pranata, J.; Dougherty, D. A. *J. Phys. Org. Chem.* **1989**, *2*, 161-176.

28.   The equivalence of these two concepts has been noted by Hoffman: Hay, P. J.; Thibeault, J. C.; Hoffmann, R. *J. Am. Chem. Soc.* **1975**, *97*, 4884-4899.

29.   Jacobs, S.J.; Shultz, D.A.; Jain, R.; Novak, J.; Dougherty, D.A., submitted.

30.   Veciana, J.; Vidal, J.; Jullian, N. *Mol. Cryst. Liq. Cryst.* **1989**, *176*, 443-450.

31.   Baumgarten, M.; Müller, U.; Bohnen, A.; Müller, K. *Angew. Chem. Int. Ed. Engl.* **1992**, *31*, 448-451.

32.   Müller, K.; Baumgarten, M.; Tyutyulkov, N.; Karabunarliev, S. *Syn. Metals* **1991**, *40*, 127-136.

33.   For most biradicals, such as TMM, TME, and the other structures of Figures 3 and 4, the covalent form is much more stable than the biradical, making viable switching systems unlikely.

34.   Snyder, G. J.; Dougherty, D. A. *J. Am. Chem. Soc.* **1989**, *111*, 3927-3942. Snyder, G. J.; Dougherty, D. A. *J. Am. Chem. Soc.* **1989**, *111*, 3942- 3954.

35.   Note that current "magnetooptical" systmes in fact use a laser essentially as a heat source to write data.[36]

36.   Greidanus, F.J.A.M.; Klahn, S. *Angew. Chem. Int. Ed. Engl.* **1989**, *28*, 235-241. Greidanus, F.J.A.M. *Philips J. Res.* **1990**, *45*, 19-34.

37.   (a) Dougherty, D. A. *Mol. Cryst. Liq. Cryst.* **1989**, *176*, 25-32. (b) Dougherty, D. A. *Pure and Appl. Chem.*, **1990**, *62*, 519-524. (c) Dougherty, D. A.; Kaisaki, D. A. *Mol. Cryst. Liq. Cryst.* **1990**, *183*, 71-79.

38.   Berson, J. A. in *The Chemistry of Quinonoid Compounds,*; Vol. II; Patai, S.; Rappoport, Z., Eds.; John Wiley & Sons, Ltd: New York, 1988; pp 455-536.

39.   Itoh, K. *Chem. Phys. Lett.* **1967**, *1*, 235.

40.   Wasserman, E.; Murray, R.W.; Yager, W.A.; Trozzolo, A.M.; Smolinsky, G. *J. Am. Chem. Soc.* **1967**, *89*, 5076-5078.

41.   Iwamura, H. *Pure & Appl. Chem.* **1986**, *58*, 187-196. Iwamura, H. *Pure & Appl. Chem.* **1987**, *59*, 1595-1604. Itoh, K. *Pure & Appl. Chem.*, **1978**, *50*, 1251-1259.

42.   Sugawara, T.; Bandow, S.; Kimura, K.; Iwamura, H.; Itoh, K. *J. Am. Chem. Soc.* **1986**, *108*, 368-371.

43.   Nakamura, N.; Inoue, K.; Iwamura, H. *J. Am. Chem. Soc.* **1992**, *114*, 1484-1485.

44.   Izuoka, A.; Murata, S.; Sugawara, T.; Iwamura, H. *J. Am. Chem. Soc.* **1985**, *107*, 1786-1787. Izuoka, A.; Murata, S.; Sugawara, T.; Iwamura, H. *J. Am. Chem. Soc.* **1987**, *109*, 2631-2639.

45.   There are exceptions if the radical groups are twisted out of conjugation. See: Dvolaitzky, M.; Chiarelli, R.; Rassat, A. *Angew. Chem. Int. Ed. Engl.* **1992**, *31*, 180-181.

46.   Yoshizawa, K.; Chano, A.; Ito, A.; Tanaka, K.; Yamabe, T.; Fujita, H.; Yamauchi, J.; Shiro, M. *J. Am. Chem. Soc.* **1992**, *114*, 5994-5998.

47.   Rajca, A.; Utamapanya, S.; Xu, J. *J. Am. Chem. Soc.* **1991**, 9235-9241. Rajca, A.; Utamapanya, S.; Thayumanavan, S. *J. Am. Chem. Soc.* **1992**, *114*, 1884-1885. Rajca, A.; Utamapanya, S. *J. Org. Chem.*. **1992**, *57*, 1760-1767.

48.   Veciana, J.; Rovira, C.; Crespo, M.I.; Armet, O.; Domingo, V.M.; Palacio, F. *J. Am. Chem. Soc.* **1991**, *113*, 2552-2561. See also, Carilla, J.; Juliá, L.; Riera, J.; Brillas, E.; Garrido, J.A.; Labarta, A.; Alcalá, R. *J. Am. Chem. Soc.* **1991**, *113*, 8281-8284.

49.   Novak, J. A.; Jain, R.; Dougherty, D. A. *J. Am. Chem. Soc.*, **1989**, *111*, 7618-7619.

50.   Seeger, D.E.; Lahti, P.M.; Rossi, A.R.; Berson, J.A. *J. Am. Chem. Soc.* **1986**, *108*, 1251-1265.

51.   Murata, S.; Iwamura, H. *J. Am. Chem. Soc.* **1991**, *113*, 5547-5556. Murata, S.; Sugawara, T.; Iwamura, H. *J. Am. Chem. Soc.* **1987**, *109*, 1266-1267. Okamoto, M.; Teki, Y.; Takui, T.; Kinoshita, T.; Itoh, K. *Chem. Phys. Lett.* **1990**, *173*, 265-270. Matsumoto, T.; Koga, N.; Iwamura, H. *J. Am. Chem. Soc.* **1992**, *114*, 5448-5449. Minato, M.; Lahti, P.M. *J. Phys. Org. Chem.* **1991**, *4*, 459-462.

348

52.   Itoh, K.; Takui, T.; Teki, Y.; Kinoshita, T. *Mol. Cryst. Liq. Cryst.* **1989**, *176*, 49-66.

53.   McConnell, H.M. *J. Chem. Phys.* **1963**, *39*, 1910.

54.   McConnell, H. *Proc. R.A. Welch Fdn* . **1967**, *11*, 144.

55.   Brelsow, R.; Jaun, B.; Kluttz, R.Q.; Xia, C.-Z. *Tetrahedron* **1982**, *38*, 863-867. Breslow, R. *Mol. Cryst. Liq. Cryst.* **1985**, *125*, 261-267.

56.   For an overview of the various extensions/interpretations of McConnell's second model, see: Wudl, F.; Closs, F.; Allemand, P.M.; Cox., S.; Hinkelmann, K.; Srdanov, G.; Fite, C. *Mol. Cryst. Liq. Cryst.* **1989**, *176*, 249-258. Miller, J. S.; Epstein, A. J. *J. Am. Chem. Soc.* **1987**, *109*, 3850-3855.

57.   Awaga, K.; Maruyama, Y. *Chem. Phys. Lett.* **1989**, *158*, 556-558. Turek, P.; Nozawa, K.; Shiomi, D.; Awaga, K.; Inabe, T., Maruyama, Y.; Kinoshita, M. *Chem. Phys. Lett.* **1991**, *109*, 327-331. Tamura, M.; Nakazawa, Y.; Shiomi, D.; Nozawa, K.; Hosokoshi, Y.; Ishikawa, M.; Takahashi, M.; Kinoshita, M. *Chem. Phys. Lett.* **1991**, *186*, 401-404. Takahashi, M.; Turek, P.; Nakazawa, Y.; Tamura, M.; Nozawa, K.; Shiomi, D.; Ishikawa, M.; Kinoshita, M. *Phys. Rev. Lett.* **1991**, *67*, 746-748. Kinoshita, M.; Turek, P.; Tamura, M.; Nozawa, K.; Shiomi, D.; Nakazawa, Y.; Ishikawa, M.; Takahashi, M.; Awaga, K.; Inabe, T. Maruyama, Y. *Chem. Lett.* **1991**, 1225-1228.

58.   Awaga, K.; Sugano, T.; Kinoshita, M. *J. Chem. Phys.* **1986**, *85*, 2211-2218.

59.   Chiang, L.Y.; Upasani, R.B.; Sheu, H.S.; Goshorn, D.P.; Lee, C.H. *J. Chem. Soc., Chem. Commun.* **1992** ,959-961.

60.   Allemand, P.-M.; Srdanov, G.; Wudl, F. *J. Am. Chem. Soc.* **1990**, *112*, 9391-9392.

61.   Chiang, L. Y.; Johnston, D. C.; Goshorn, D. P.; Block, A. N. *J. Am. Chem. Soc.* **1989**, *111*, 1925-1927.

62.   Allemand, P.-M.; Khemani, K.C.; Koch, A.; Wudl, F.; Holczer, K.; Donovan, S.; Grüner, G.; Thompson, J.D. *Science* **1991**, *253*, 301-303. Sparn, G.; Thompson, J.D.; Allemand, P.-M.; Li, Q.; Wudl, F.; Hjolczer, K.; Stephens, P.W. *Solid State Commun.* **1992**, *82*, 779-782.

63.   Miller, J.S. *Adv. Mater.* **1992**, *4*, 298-300. Miller, J.S. *Adv. Mater.* **1992**, *4*, 435-438.

64. Torrance, J.B.; Oostra, S.; Nazzal, A. *Synth. Metals* **1987**, *19*, 709-714. Torrance, J.B.; Bagus, P.S.; Johannsen, I.; Nazzal, A.I.; Parkin, S.S.P.; Batail, P. *J. Appl. Phys.* **1988**, *63*, 2962-2965.

65. Korshak, Y. V.; Medvedeva, T. V.; Ovchinnikov, A. A.; Spector, V. N. *Nature* **1987**, *326*, 370-372.

66. Ota, M.; Otani, S.; Igarashi, M. *Chem. Lett.* **1989**, 1183-1186.

67. Tanaka, H.; Tokuyama, K.; Sato, T.; Ota, T. *Chem. Lett.* **1990**, 1813-1816.

68. Yoshioka, N.; Nishide, H.; Tsuchida, T. *Mol. Cryst. Liq. Cryst.* **1990**, *190*, 45-53. Nishide, H.; Yoshioka, N.; Kaku, T.; Kaneko, T.; Yamazaki, M.; Tsuchida, E. *J. Macromol. Sci.* **1991**, *A28*, 1177-1187. Nishide, H.; Yoshioka, Kaneko, T.; Tsuchida, E. *Macromolecules* **1990**, *23*, 4487-4488. Rossitto, F.C.; Lahti, P.M. *J. Polym. Sci.* **1992**, *30*, 1335-1345. Fujii, A.; Ishda, T.; Koga, N.; Iwamura, H. *Macromolecules* **1991**, *24*, 1077-1082. Inoue, K.; Koga, N.; Iwamura, H. *J. Am. chem. Soc.* **1991**, *113*, 9803-9810. Alexander, C.; Feast, W.J. *Polym.. Bull.* **1991**, *26*, 245-252. Alexander, C.; Feast, W.J.; Friend, R.H.; Sutcliffe, L.H. *J. Mater. Chem..* **1992**, *2*, 459-465. Abdelkader, M.; Drenth, W.; Meijer, E.W. *Chem. Mater.* **1991**, *3*, 598-602. Miura, Y.; Inuui, K.; Yamaguchi, F.; Inoue, M.; Teki, Y.; Takui, T. Itoh, K. *J. Polym. Sci.* **1992**, *30*, 959-966.

69. Fukutome, H.; Takahashi, A.; Ozaki, M. *Chem. Phys. Lett.* **1987**, *133*, 34-38. Yamaguchi, K.; Toyoda, Y.; Fueno, T. *Synth. Met.* **1987**, *19*, 81-86.

70. Kaisaki, D. A.; Chang, W.; Dougherty, D. A. *J. Am. Chem. Soc.* **1991**, *113*, 2764-2766; Murray, M.; Dougherty, D.A., unpublished results.

71. Haase, W.; Borchers, B. in*Magnetic Molecular Materials*; Gatteschi, D.; Kahn, O.; Miller, J. S.; Palacio, F., Eds.; Kluwer Academic Publishers: Boston, **1991**; Vol. 198, pp. 245-253.

SPIN COUPLING CONCEPTS IN BIOINORGANIC CHEMISTRY

Christopher A. Reed and Robert D. Orosz

Department of Chemistry
University of Southern California
Los Angeles, California   90089-0744

1. Introduction

There is an inextricable relationship between spin coupling and molecular structure. This explains some of the recent interest in gaining an understanding of spin-coupling phenomena in bioinorganic chemistry: it is motivated by a desire to deduce the structure of active sites in metalloproteins which have interacting paramagnetic centers. On the other hand, recent synthetic work with spin-coupled metal complexes, some of which was motivated by a model compound approach to bioinorganic chemistry, has given the field of magnetic interactions a welcome new accessibility. There is now available a collection of exemplary metal complexes which illustrate many of the basic principles of spin coupling with an intuitive clarity heretofore unavailable. By developing a qualitative rationale for magnetic interactions in terms of the familiar concepts of orbital symmetry, orbital overlap and Hund's rule, the field of spin coupling is now taking advantage of precepts and a language already understood by the non-specialist. This review, selective rather than exhaustive in its approach, brings together the most useful examples of spin coupling from recent metalloporphyrin and copper chemistry to assess the current state of understanding of magnetic coupling as it relates to bioinorganic chemistry.

There is an impressive diversity of spin-coupled structures in nature. The field has been surveyed most recently by Solomon and Wilcox[1] and representative examples are given in Table I. Most often, the occurrence of interacting paramagnetic centers is associated with enzymes which carry out multi-electron chemistry. A well-known example is the active site of cytochrome oxidase where heme iron and copper are held in close proximity in order to catalyze the 4e⁻ reduction of dioxygen to water. The reverse reaction is catalyzed by interacting manganese centers in photosystem II. Binuclear copper and binuclear iron active sites are particularly widespread and are usually involved in 2- or 4-electron oxygen-carrying and oxygen-utilizing functions. In addition to multi-metal systems, there are cases of paramagnetic metals that are spin-coupled to organic radicals. The best known example is the so-called compound I intermediate of peroxidases where an $S = 1$ iron(IV) heme center is coupled to a porphyrin π-cation radical. One of the most interesting recent examples of spin-coupling is found in sulfite reductase where an Fe_4S_4 cluster appears as an axial ligand to siroheme. Outside of the realm of discrete

Table 1. Representative Spin-Coupled Centers in Metalloproteins

Protein	Spin-Coupled Entities	Function	Ref
cytochrome oxidase	Fe/Cu	$4 e^-$ O_2 reduction	2
laccase	Cu_2/Cu	$4 e^-$ O_2 reduction	3
ascorbate oxidase	Cu_2/Cu	O_2/substrate redox	3
sulfite reductase	Fe/Fe_4S_4	$6e^-$ SO_3^{2-} reduction	4
nitrogenase	Fe_7/Mo	$6e^-$ N_2 reduction	5
photosystem II	Mn/Mn	$4e^-$ O_2 oxidation	6
hemocyanin	Cu/Cu	O_2 carrier (2:1)	7
hemerythrin	Fe/Fe	O_2 carrier (2:1)	8-11
peroxidases	Fe/R•	$2e^-$ redox	12
galactose oxidase	Cu/R•	$2e^-$ redox	13
hydrogenase	Fe/Fe	$2e^-$ redox	14
iron sulfur proteins	Fe/Fe	$1e^-$ carriers	14,11
cytochrome c_3	Fe/Fe	$4e^-$ transfer	15
methane monooxygenase	Fe/Fe	CH_4/O_2 redox	8-11
ferritin	Fe_n	Fe storage	16
magnetotactic bacteria	magnetite	compass	17

molecular entities, magnetic solids have been identified in the form of magnetite in bacteria which apparently utilize the earth's magnetic field. More recently, magnetic particles have been identified in birds, bees and in the human brain.[18]

The connection between structure and observed signatures of spin coupling in metalloproteins has been established to different degrees in different systems. The fundamental understanding of these connections owes much to the study of model compounds where magnetostructural correlations can be established in a systematic way. Interest in bioinorganic chemistry has played an important role in the last two decades of stimulating a more general revival of interest in spin coupling phenomena. It has demanded a more chemically accessible comprehension of spin coupling and has contributed significantly to fundamental new insight into orbital interaction pathways.

2. Background and Concepts

The quotation "Few subjects in science are more difficult to understand than magnetism" certainly has validity[19] but the subject has become increasingly accessible in recent years. Spin coupling, magnetic interactions and exchange coupling† are synonymous terms for describing the interactions between unpaired electrons at nearby paramagnetic centers.[20] Typically, these interactions are weak compared to energies of typical chemical bonds and the paramagnetic centers have well-defined valencies which can be described by good spin quantum numbers ($S = 1/2, 1, 3/2$ ... etc. for one, two, three ... etc. unpaired electrons respectively). A well-defined valency usually means that unpaired spin is relatively localized in a describable atomic or molecular orbital, often referred to as a *magnetic orbital*.

Establishing discrete valencies is thus a prerequisite for exploring spin coupling. For example, the question of the formulation of compound I of peroxidases had to be rigorously decided in favor of an iron(IV) π-cation radical rather than iron(V)[21] before meaningful discussions of spin coupling could begin. Similarly, in model compounds, [FeCl(TPP)]$^+$ (TPP = tetraphenylporphinate) had to be reformulated as an iron(III) π-cation radical rather than an iron(IV) porphyrin before its magnetic properties could be understood.[22] The physical probes that have proved most useful in establishing discrete valence states include single crystal X-ray crystallography, X-ray absorption near edge spectroscopy (XANES), electronic spectroscopy,[23] vibrational spectroscopies, NMR, EPR[24] and ENDOR spectroscopies, polarized neutron diffraction (PND),[25] and in the case of iron, Mössbauer spectroscopy.[26] Magnetic susceptibility measurements, while critical for assessing overall spin content,[27] are a bulk property and can therefore be quite ambiguous as to the location of spin.

It is frequently useful to idealize the localization of an unpaired electron to a particular atomic or molecular orbital, e.g. a d orbital for a discrete transition metal

---

† The term exchange arises from the Heisenberg exchange effect. When considering the interaction of spins, quantum mechanics requires taking into account the degeneracy associated with the possibility of two electrons trading places.[31]

on, a molecular orbital on a delocalized cluster or a π orbital for an aromatic ligand radical. At the same time it is recognized that this spin is somehow delocalized onto adjacent centers, typically via orbitals on the connecting atoms. This mechanism of spin-spin interaction is referred to as *superexchange*. It is of major interest in bioinorganic chemistry because closed-shell diamagnetic ligands (oxide, peroxide, cyanide, imidazolate, carboxylate, thiolate etc.) mediate most if not all spin coupling in discrete molecular systems. Superexchange, or ligand-mediated exchange, is in principle distinguishable from *direct exchange* where direct overlap occurs between the magnetic orbitals of adjacent paramagnetic centers. Proven examples, however, are rare because it is often very difficult to distinguish through-space interactions from those mediated by the bridging atoms. Dimeric copper(II) acetate is the classic and long-debated case of this problem. A resolution in favor of contributions from both mechanisms comes from SCF calculations.[28] As will become evident throughout this review, the present understanding of the mechanisms of spin coupling relies heavily on the identification of the symmetry relationships between the magnetic orbitals (an extension of rules that originate with Goodenough[29] and with Kanamori[30])[20,24] and the nature of their overlap with the intervening atom(s).

To a good approximation, spin coupling between paramagnetic sites of specified local valency can be described by a spin Hamiltonian of the form

$$\mathcal{H} = -2J \, S_1 \cdot S_2$$

where S_1 and S_2 are the spin operators of the paramagnetic sites and J is the exchange coupling constant.[31] In practice, for moderately coupled systems, this usually reduces to a process of fitting variable temperature magnetic susceptibility data to readily available van Vleck expressions[32,33,24] which compute a value for an isotropic J in cm^{-1}. The magnitude of J is a measure of the strength of the magnetic coupling. The sign of J indicates whether the coupling is antiferromagnetic (-J, spin paired) or ferromagnetic (+J, parallel spin). In this review we are mainly concerned with understanding isotropic J values in the range 1 - 500 cm^{-1}. Smaller J values are typically the province of EPR spectroscopy and this field has recently been reviewed.[24]

In the simplest case of two interacting unpaired electrons, spin coupling leads to two overall spin states: a singlet ($S = 0$) and a triplet ($S = 1$). The energy separation is $2J$. In the case of a singlet ground state, the coupling is antiferromagnetic and J is negative. In the case of a triplet ground state, the coupling is ferromagnetic and J is positive. *Note:* For historical reasons there is no universal agreement on the form of the $-2JS_1 \cdot S_2$ spin Hamiltonian. A good body of literature uses spin Hamiltonians that omit the factor of 2. For this reason, care must be taken when comparing J values from different sources. In this review, all J values are in the $-2J S_1 \cdot S_2$ form as originally used by van Vleck.[31]

Another point to note in comparing the magnitude of J values in antiferromagnetically coupled dimers is that they decrease as the number of unpaired electrons increases. This is simply an inherent consequence of the spin Hamiltonian which separates the highest and lowest overall spin states by $2J$, $6J$, $12J$, $20J$ and $30J$ for dimers of $S = 1/2, 1, 3/2, 2, 5/2$ systems respectively. Thus, a J value of -150 cm^{-1} (and singlet/triplet splitting of 300 cm^{-1}) in an $S = 1/2, S = 1/2$ copper(II) dimer is equivalent to a J value of -10 cm^{-1} in an $S = 5/2, S = 5/2$ iron(III) dimer (where the $S = 0$ and $S = 5$ levels are also separated by 300 cm^{-1}). This suggests that comparing high/low spin energy gaps is a better way of comparing different spin systems than simply comparing J values.

Kahn[34] has shown that it is useful to decompose J into its two basic competing components: a negative antiferromagnetic component J_{AF}, and a positive ferromagnetic component J_F:

$$J = J_{AF} + J_F.$$

It can be shown that J_{AF} is proportional to the overlap density between the magnetic orbitals.[35] This is equivalent to the formation of a weak chemical bond, and in accord with simple molecular orbital theory and the Pauli exclusion principle, the electrons will be spin paired. Equating anti-ferromagnetic coupling with bond formation has proved to be an extremely powerful idea for rationalizing magnetic interactions in discrete molecular systems. Opposing this tendency for spin pairing is the so-called exchange energy stabilization that underlies Hund's rule of maximum spin multiplicity. Magnetic orbitals that are unable to overlap and form a bond can nevertheless interact and stabilize each other if the spins align in parallel fashion. In the language of spin coupling these orbitals are said to be *orthogonal* and the spin coupling is ferromagnetic. J_F is related to the familiar exchange integral of quantum mechanics. Equating ferromagnetic coupling with a Hund's rule-type exchange stabilization is the other most important concept that has helped to demystify magnetism. The mathematical expressions have been laid out succinctly by Kahn.[34]

There are three usefully differentiated ways to achieve orthogonality of magnetic orbitals and consequent ferromagnetic coupling. The first two, (a) *strict* orthogonality, and (b) *virtual* orthogonality, are closely related. They are readily viewed as a consequence of symmetry. Inspection of the symmetry relating the magnetic orbitals reveals overlap that is symmetry-forbidden. The third, (c) *accidental* orthogonality, is more readily viewed as a consequence of energy. A symmetry-allowed overlap gives rise to a fortuitous energetic degeneracy. Mathematically, of course, both symmetry-forbidden overlap and accidential degeneracy create algebraic orthogonality of wavefunctions. Hence, orthogonality is chosen as the noun common to all conditions of ferromagnetic coupling.

(a) Strict orthogonality arises when interacting magnetic orbitals are forbidden by symmetry to overlap. An excellent example is provided by the heterobinuclear complex $CuVO((fsa)_2en)(CH_3OH)$[34] shown in Figure 1.

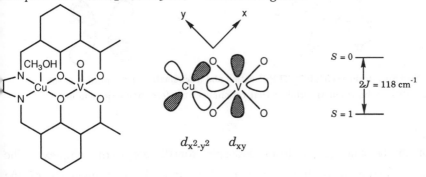

$$d_{x^2-y^2} \qquad d_{xy}$$

Figure 1. Structure of $CuVO((fsa)_2en)(CH_3OH)$ and sketch of orthogonal magnetic orbitals which give rise to ferromagnetic coupling and stabilization of the $S = 1$ state.[34]

358

The d^9 copper(II) ion has a $d_{x^2-y^2}$ magnetic orbital which will be σ-delocalized onto the bridging oxygen atoms. The d^1 vanadyl ion has a d_{xy} magnetic orbital which will be π-delocalized onto the bridging oxygen atoms. This σ/π symmetry mismatch means there is essentially zero overlap, no significant chemical bonding and insignificant antiferromagnetic coupling. The J value is +59 cm^{-1}. The spin coupling is ferromagnetic and the triplet-singlet energy gap is 118 cm^{-1}.

An easily visualized case of strict orthogonality in a more biologically relevant compound is found in the imidazolate-bridged trinuclear Cu-Fe-Cu species [Fe(CuIM)$_2$(TTP)]$^+$ illustrated in Figure 2.[36] The square planar copper(II) chelates

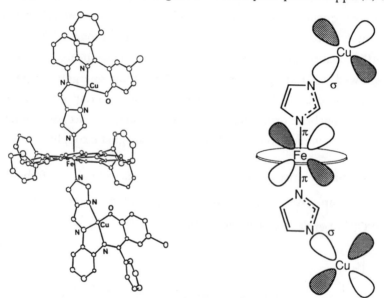

Figure 2. Structure of the [Fe(CuIM)$_2$(TPP)]$^+$ cation and a schematic representation of the σ/π orthogonality of the magnetic orbitals that rationalizes the ferromagnetic coupling.

supply bis-imidazolate ligation to the low-spin iron(III) porphyrin complex. The copper $d_{x^2-y^2}$ magnetic orbitals are delocalized into the σ framework of the imidazolates. The iron d_{xz} magnetic orbital is delocalized into the π framework of the imidazolates. The σ/π symmetry mismatch gives rise to ferromagnetic coupling.

The ground state is $S = 3/2$ and J is $+11$ cm^{-1}.

Another biologically relevant case of strict orthogonality is found in planar metalloporphyrin π-cation radicals. This is treated in section 3.

b) Virtual orthogonality is the name we propose for ferromagnetic coupling that arises when a magnetic orbital originating on one atom overlaps with an *empty* orbital on a second atom. The magnetic orbital on the second atom will then experience a Hund's-rule-like exchange interaction by virtue of this delocalization. The importance of this *"crossed interaction"* in rationalizing the magnetic interactions in μ–oxo dimers has recently been stressed by Girerd and Wieghardt.[37] It offers a convincing explanation for why Fe^{III}-O-Fe^{III} moieties are always antiferromagnetically coupled whereas structurally related Mn^{III}-O-Mn^{III} can be ferromagnetically coupled. The essential argument is that in the d^4 Mn(III) dimer, a d_{xz} magnetic orbital on one Mn atom can overlap via a $2p$ bridge orbital into the empty d_{z^2} orbital on the other (see Figure 3).

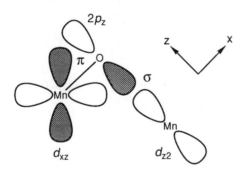

Figure 3. Overlap of the d_{xz} magnetic orbital on one Mn atom with the empty d_{z^2} on another via the filled bridging oxygen $2p_z$ orbital.

This virtual exchange gives rise to a relatively strong ferromagnetic interaction which can compete with the more familiar antiferromagnetic contributions arising from symmetry-allowed overlap of other magnetic orbitals. As always, it is the competition between J_F and J_{AF} pathways which determines the final outcome reflected in the measured J value.

360

(c) **Accidental orthogonality** arises from a fortuitous degeneracy of overlapping magnetic orbitals. It arises not so much from strict symmetry considerations in a molecule but from a particular set of structural parameters which tend to equalize overlap pathways. The classic case is found in dihydroxy-bridged copper(II) dimers where, for Cu-O-Cu angles $\theta < 97.6°$, ferromagnetic coupling dominates over antiferromagnetic coupling.[38] In fact, there is a linear relationship between J and θ in a series of gradually varying structures which maintain the planar $Cu_2(OH)_2$ core. The observation of this important magnetostructural correlation eventually led to an understanding of the mechanism of spin coupling[35,38] and remains a paradigm for present day investigations. The essence of the explanation for the angular dependence lies in the relative energies of the two molecular orbitals formed by overlap (i.e.. linear combination) of the magnetic orbitals.

From a simple inspection of the symmetries of the magnetic orbitals in a typical copper(II) dimer (see Figure 4), one might always expect overlap of the two half-filled d orbitals, bond formation and spin-pairing in the resulting lower energy molecular orbital i.e., antiferromagnetic coupling:

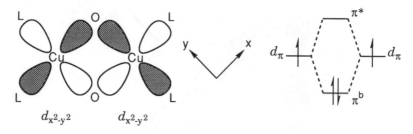

$d_{x^2-y^2}$ $d_{x^2-y^2}$

Figure 4. Magnetic orbitals of a typical planar di-μ-hydroxy dimer of copper (II)

However, this ignores superexchange i.e., the role of orbitals on the intervening atoms. The correct approach[35] is to consider how the symmetric and antisymmetric linear combinations (LC) of the magnetic orbitals are affected by specific bridging ligand orbitals. In di-hydroxy bridged copper(II) dimers, these are the $2p_x$, $2p_y$ and $2s$ atomic orbitals (AO) on the oxygen atoms. Their admixture with the magnetic orbitals is governed by the usual considerations of the LCAO approach, namely, correct symmetry and degree of overlap, which in turn, are critically dependent on the Cu-O-Cu bridge angles. In order to lay out the arguments clearly it is necessary to adopt coordinate axes that are rotated 45° from their traditional orientation (see Figure 5). Thus, the magnetic orbitals on copper which are usually denoted as $d_{x^2-y^2}$ become d_{xy}. The *antisymmetric* combination of these d_{xy} magnetic orbitals can mix with the $2p_x$ ligand orbitals to give φ_A as shown in Figure 5(a). The *symmetric* combination of the magnetic orbitals can mix with both the $2p_y$ and $2s$ ligand orbitals to give φ_S as shown in Figure 5(b).[†]

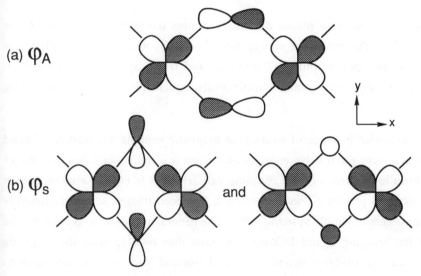

Figure 5. (a) Antisymmetric combination of copper(II) magnetic orbitals with $2p_x$ orbitals on the bridging O atoms. (b) Symmetric combination of copper(II) magnetic orbitals with $2p_y$ and $2s$ orbitals on the bridging O atoms.

At a particular Cu-O-Cu angle, in this case somewhere near 95°, the overlaps will fortuitously lead to φ_A and φ_S being equal in energy. This degeneracy is the condition of accidental orthogonality and ferromagnetic coupling arises because of Hund's rule:

$$\varphi_A \underline{\uparrow} \qquad \varphi_S \underline{\uparrow}$$

[†] Since the oxygen bridging ligand orbitals are filled, the overlaps in φ_A and φ_S are antibonding in character.

362

As the Cu-O-Cu angle increases, φ_A and φ_S become increasingly different in energy. At angles above 97.6°, the tendency towards spin pairing in the lower energy orbital is greater than the exchange energy and antiferromagnetic coupling prevails:[†]

At precisely 97.6° in these planar dihydroxy dimers the spliting of φ_A and φ_S is exactly equal to the exchange energy i.e., $J_{AF} = J_F$, and the dimer gives the appearance of no spin coupling. The singlet and triplet spin states are equal in energy and the situation is magnetically indistinguishable from non-interacting spins.

In any particular instance of overlap of magnetic orbitals via bridging ligand orbitals, it is always the relative energies of the linear combinations that determines whether fortuitous ferromagnetic coupling can arise. It is clearly a phenomenon that is difficult to predict a priori although it has been treated semiquantitatively with some success in simple systems.[35,38,39] The general approach is to consider the nature of the bridging ligand HOMOs and how they overlap with the magnetic orbitals.[35] Certain bridging ligands such as 1,1-bound azide and cyanate seem to show a particular propensity to facilitate degeneracy of φ_A and φ_S.[40] In addition, the concept of *spin polarization* has been invoked in such complexes to rationalize the large magnitude of the ferromagnetic coupling.[41] An instructive new case of accidental orthogonality was discovered recently in methemocyanin model compounds where the competition between dissimilar bridging ligands in copper(II) dimers promotes fortuitous degeneracy of the symmetric and antisymmetric linear combinations of the magnetic orbitals.[42] We treat this in section 4.

[†] It does not matter which orbital is raised or lowered in energy; it is the difference in energy between φ_A and φ_S that leads to antiferromagnetic coupling. Typically, it is φ_A which has the better overlap and is therefore higher in energy (because it has antibonding character).

In the competition between antiferromagnetic and ferromagnetic coupling pathways, antiferromagnetic coupling usually dominates. This is because the overlap energies of bond formation can easily overcome the relatively small quotients of energy involved in Hund's-rule-like exchange. In general, molecules of low symmetry will show antiferromagnetic coupling because ferromagnetic coupling arising from strictly orthogonal magnetic orbitals is absent. In section 3 we will show that the ruffling of a porphyrin core away from planarity is sufficient to convert a ferromagnetically coupled metal/ligand interaction into an antiferromagnetic one. Similarly, the large $-J$ value in an unsymmetrical Fe^{III}-O-Fe^{III} dimer compared to more symmetrical complexes[43] probably has its origin in the loss of strict pairwise orthogonal relationships between the d orbitals.

It is the dissection of spin coupling into its component antiferromagnetic and ferromagnetic parts and the identification of these with specific orbital interactions which represents the major advance of the last two decades. It has set the stage for understanding spin coupling in bioinorganic chemistry. It has also stimulated attempts at increasingly quantitative treatments of coupling mechanisms[44-46] and catalyzed a surge of interest in synthesizing molecular ferromagnets.[47-50]

3. Metalloporphyrin π-Cation Radicals

Iron(IV) porphyrin π-cation radicals are believed to be intermediates in various oxidative pathways catalyzed by hemoproteins. The best known examples are found in horseradish peroxidase (HRP) and catalase (CAT) where treatment of their iron(III) resting states with O-atom sources (peroxides, peracids etc) gives so-called compounds I as the first isolable intermediates:

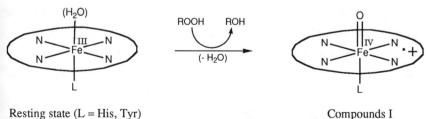

Resting state (L = His, Tyr)   Compounds I

The identity of HRP I and CAT I as iron(IV) porphyrin π-cation radicals with well-defined valencies is particulary evident from UV-VIS[21] and Mössbauer spectroscopies.[51] The iron(IV) center is paramagnetic with an $S = 1$ intermediate-spin state. Its immediate proximity to the $S = 1/2$ ligand radical sets up the opportunity for metal/ligand spin coupling. Our understanding of this coupling in hemoproteins is at an early stage but the conceptual groundwork has been laid by model compound studies with copper(II) and iron(III) metalloporphyrin π-cation radicals.

The essence of our present understanding of spin coupling in metalloporphyrin π-cation radicals is embodied in the comparison of $[Cu^{II}(TPP\cdot)]^+$ and $[Cu^{II}(TMP\cdot)]^+$ (where TPP = $meso$-tetraphenylporphinate and TMP = $meso$-tetramesityl-porphinate).[52] The spin coupling between the $S = 1/2$ copper(II) ion and the $S = 1/2$ porphyrin π-cation radical is dramatically different in these otherwise very closely related ions. The TPP derivative has a highly ruffled porphyrin core and is strongly $anti$ferromagnetically coupled.[52] Indeed, it is diamagnetic, making the singlet-triplet energy gap of the order of 1000 cm^{-1} or more and $|-J| \geq 500$ cm^{-1}. On the other hand, the TMP derivative has a nearly planar porphyrin core and is strongly $ferro$magnetically coupled.[53] The effective magnetic moment approaches that of a pure $S = 1$ system ($\mu = 2.9$ BM) over the entire 6 - 300 K temperature range, making the triplet-singlet energy gap at least 400 cm^{-1} and possibly > 1000 cm^{-1}. The J value is positive and is conservatively estimated to be ≥ 200 cm^{-1}. These features are illustrated in Figure 6 along with representations of the magnetic orbitals. The magnetic orbital on copper(II) is $d_{x^2-y^2}$. This designation is unambiguous and follows naturally from the d orbital splittings appropriate to a square planar crystal field and a d^9 configuration. In molecular orbital (or ligand field) terms it is the Cu-N σ-antibonding orbital of predominant $d_{x^2-y^2}$ character. On the other hand, the porphyrin magnetic orbital is of π-bonding character being made up of $2p_z$ contributions from the four nitrogen and twenty carbon atoms of the core. For tetraarylporphyrins with only $meso$-substitution this orbital is always of a_{2u} character whereas for pyrrole-substituted octaalkylporphyrins it is of a_{1u} character.[54,55] These are illustrated in Figure 7.

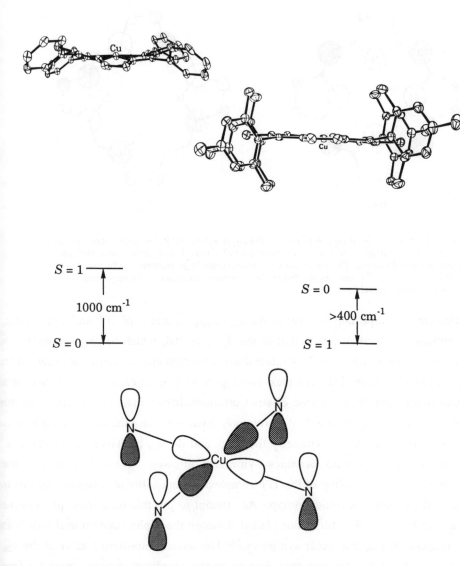

Figure 6. X-ray crystal structures of [Cu(TPP•)]⁺ and [Cu(TMP•)]⁺, spin state energy separations and a simplified representation of the σ/π orthogonality of the magnetic orbitals in D_{4h} symmetry.

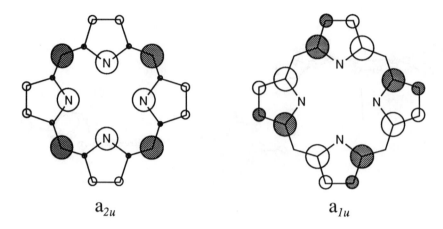

a_{2u}

a_{1u}

Figure 7. Atomic orbital representation of the a_{2u} magnetic orbital of *meso*-tetraarylporphyrin π-radical cations and the a_{1u} magnetic orbital of pyrrole-substituted octaethyl porphyrin π-radical cations. The circle sizes are approximately proportional to the atomic orbital coefficients. The open or shaded circles represent the signs of the uppermost lobes of the 2*p* atomic orbitals.

There are two important features of the a_{2u} magnetic orbital of TPP and TMP. First, its symmetry is orthogonal to that of the $d_{x^2-y^2}$ orbital, which has b_{1g} symmetry in D_{4h} molecular symmetry. This rationalizes the ferromagnetic coupling observed in the essentially planar TMP complex (see Figure 6). It is a nearly perfect case of σ/π symmetry mismatch, ie., a case of strict orthogonality of magnetic orbitals. In the highly ruffled core of the TPP derivative, however, the molecular symmetry is lowered to C_s. Under C_s symmetry the $d_{x^2-y^2}$ and the a_{2u} orbitals both reduce to a' symmetry, and overlap becomes symmetry allowed. This rationalizes the antiferromagnetic coupling of the TPP complex: orthogonality of magnetic orbitals is lost at the expense of some overlap. An attempt to put this on a semi-quantitative basis with INDO/SCF calculations failed although the triplet state ground state was reproduced in D_{4h} molecular symmetry.[56] The second important feature of the a_{2u} orbital is that it concentrates spin density on the coordinating nitrogen atoms (see Figure 7). Thus, a strong interaction with the $d_{x^2-y^2}$ orbital is likely regardless of whether it is ferromagnetic or antiferromagnetic in nature. This rationalizes the large magnitude of J in both the TPP and TMP complexes.

The concepts developed above for copper(II) are manifest with higher spin multiplicity in iron(III). In the six-coordinate complex $Fe(OClO_3)_2(TPP\cdot)$, the porphyrin core is essentially planar and the $S = 5/2$ iron is *ferro*magnetically coupled to the a_{2u} porphyrin radical.[22] The ground state is $S = 3$ and the $S = 2$ state is ca. 250 cm^{-1} higher in energy ($J \cong + 40$ cm^{-1}). This can be understood in terms of nearly strict orthogonality of the magnetic orbitals. The nearly planar porphyrin core and center of symmetry of this molecule impose effective D_{4h} symmetry upon the iron atom, and in this point group the five d magnetic orbitals (x^2-y^2, z^2, xz, yz, xy) have b_{1g}, a_{1g}, e_g, e_g and b_{2g} symmetry. The porphyrin π-cation radical, being that of a *meso*-tetraarylporphyrin, has a_{2u} symmetry. The six magnetic orbitals are all orthogonal to each other and Hund's rule-type exchange leads to an $S = 3$ ground state i.e., ferromagnetic coupling. If the molecular symmetry was rigorously D_{4h} the stabilization of the higher multiplicity state might be larger than the observed 250 cm^{-1}. Nevertheless, it is not much less than the 400 cm^{-1} or so observed in $[Cu(TMP)\cdot)]^+$ and the slight deviations from strict planarity of the porphyrin cores are similar.

By contrast, the closely related *five*-coordinate $[FeCl(TTP\cdot)]^+$ ion, is highly distorted from planarity. The iron atom is out of the porphyrin plane towards the axial chloride ligand and there is a saddle-shaped ruffling of the porphinato core similar to $[Cu(TPP\cdot)]^+$. The molecular symmetry is approximately C_{2v}. Upon such lowering of the site symmetry from D_{4h} the d orbital symmetries reduce to a_1, a_1, b_1, b_2 and a_2. The porphyrin a_{2u} orbital symmetry reduces to a_1, and we see that overlap between the ligand orbital and two of the metal orbitals becomes symmetry allowed. The overlap interactions apparently override any exchange interactions since the complex has an $S = 2$ ground state i.e., *antiferro*mgnetic coupling. The $S = 3$ state is not populated at room temperature placing it $\geq$ 1500 cm^{-1} above the ground state ($| -J | > 500$ cm^{-1}).

Once again, the large coupling constants observed in this pair of iron(III) complexes presumably arises from the high concentration of spin density at the coordinating nitrogen atoms of these a_{2u}-type porphyrin radicals. It suggests that structurally similar complexes with a_{1u}-type radicals might show much smaller coupling constants because a_{1u}-type radicals concentrate spin density at the β-pyrrole carbon atoms, remote from the metal (see Figure 7). Octaethylporphyrin (OEP) is known to form a_{1u}-type π-cation radicals;[55] the electron-donating ethyl groups at the β-pyrrole positions raise the a_{1u} HOMO above the a_{2u}. Thus, Morishima and coworkers prepared [FeCl(OEP·)]⁺ and Fe(OClO₃)₂(OEP·) in an attempt to prove the hypothesis that a_{1u} radicals are less strongly coupled than a_{2u} radicals.[57] Unfortunately, in the absence of Mössbauer and X-ray structural data necessary for the definitive characterization of these species, the interpretation of the magnetic susceptibility data have been shown to be in error, or at least ambiguous.[58] This illustrates the importance of adequately defining the valency and spin states of the components before considering spin coupling. For iron-containing systems, Mössbauer spectroscopy is uniquely powerful in this regard since the fundamental parameters of isomer shift (δ) and quadrupole splitting (ΔEq) are diagnostic of spin and oxidation state[59] and are essentially unaffected by spin-coupling.[22,27] The problem of a_{1u} versus a_{2u} spin coupling awaits a unique solution. The data on hemoprotein π-cation radicals that we discuss next suggest that a_{1u}-type radicals can couple only weakly with metal spins, but again, the results are by no means definitive.

Spin coupling in the iron(IV) π-cation radicals of horseradish peroxidase compound I and chloroperoxidase compound I (CPO I) have been estimated from EPR and Mössbauer measurements to have small J values, ca. -1 cm⁻¹ and -18 cm⁻¹ respectively.[51,59,60] HRP I has been treated anisotropically.[51] These appear to be the only biological systems measured to date. There is, however, a useful class of iron(IV) model compounds, represented by $Fe^{IV}(O)Cl(TMP·)$.[61] From Mössbauer measurements its ground spin state is determined to be S = 3/2, indicative of ferromagnetic coupling between the S = 1 iron(IV) and the S = 1/2 radical. No indication of a state of lower multiplicity is observed in data up to 60 K. This places a lower limit of about +30 cm⁻¹ on the value of J. The question then becomes one of understanding why the model compound is strongly ferromagnetically coupled while the hemoproteins are only weakly coupled. The model compound is a *meso*-tetrarylporphyrin of a_{2u} radical character and coupling is therefore expected to be strong. If it also has high symmetry, then orthogonality of the magnetic orbitals could rationalize the ferromagnetic nature of the coupling. In the reasonable assumption of at least C_{2v} molecular symmetry (it may be as high as C_{4v}), the d_{xz} and d_{yz} magnetic orbitals of iron(IV) have b_1 and b_2 symmetry while that of the porphyrin is a_1. This is sufficient to realize the condition of strict orthogonality. By comparison to the high spin iron(III) complexes discussed earlier, it is seen that the lack of d_{z^2} and $d_{x^2-y^2}$ occupation significantly reduces the opportunity for overlap that might lead to antiferromagnetic coupling. Nevertheless, HRP I and CPO I are not ferromagnetically coupled, and the coupling is weak. It is reasonable to ascribe this to two causes although it is presently difficult to assess their relative importance. Firstly, hemoproteins have no strict symmetry elements at the iron center. Axial ligation, varied heme peripheral substitution and the asymmetric protein wrapping all conspire to lower symmetry relative to a model complex. This would explain the negative sign of J: structural distortion tends to favor overlap at the expense of exchange i.e., antiferromagnetic coupling at the expense of ferromagnetic coupling. A fortuitous near equality of these opposing effects could lead to the condition of weak coupling. Secondly, there is the question of porphyrin radical type. If the radical is distinctly a_{1u} in character, the small magnitude of the J value could be rationalized by the relative remoteness of the ligand spin from the d orbitals. Given the β-substitution of naturally occurring porphyrins one might expect their π-cation radicals to have a_{1u} character like OEP. However, there is growing evidence that admixed a_{1u}/a_{2u} states rather than pure states must be considered as well as the possibility that π-bonding axial ligands can switch the ordering of these close-lying energy states.[55,62-64] Further understanding of this aspect in both proteins and model compounds is necessary before the spin coupling mechanisms can be fully understood.

Three further points deserve mention. The first is to note briefly a potentially important role for ^{1}H NMR spectroscopy in determining at least the sign, if not the magnitude of J in metalloporphyrin π-cation radicals. There are too few data available at present to be confident of generalities but there is explicit spin density information in both the magnitude and direction of NMR chemical shifts. The effects of particular d orbital occupations on the ^1H NMR shifts of metalloporphyrins are quite well understood.[65] An understanding of how they are modified by the additional effects of antiferro- or ferromagnetic coupling and a_{1u} or a_{2u} radical character has yet to be fully enunciated although some observations have been made.[22,62,66]

The second point of interest lies in gaining an understanding of how spin coupling varies as a function of the particular d orbital that is occupied. One might expect, for example, that a $(d_{xz})^1$ configuration at a metal would favor a ferromagnetic interaction with a porphyrin π-cation radical because orthogonality of the magnetic orbitals is difficult to break with small distortions. Similarly, the different d orbitals are expected to have different magnitudes of coupling arising from their different delocalizations or proximities to the ligand spin. Some progress has been made in this endeavor[52] but, as in the a_{1u} versus a_{2u} problem, the dimerization of most OEP and TPP π-cation radical complexes frequently complicates the interpretation of magnetic susceptibility data and may obscure the nature of the intramolecular coupling.

Thirdly, the question of why some metalloporphyrin π-cation radicals cores are essentially planar and some are highly ruffled has been addressed.[67] To date, all highly ruffled examples have been shown to arise from a face-to-face association in TPP-like complexes. The ruffling is a natural consequence of maximizing intermolecular face-to-face overlap and minimizing steric repulsion between interlocking phenyl substituents. Tetramesityl-substituted porphyrins (TMP) have sufficiently bulky side-groups that face-to-face core association is not possible and near planarity is observed, at least in the examples investigated to date. Thus, ruffling is not an intrinsic property of metalloporphyrin π-cation radicals. This does not mean that discrete monomeric complexes cannot be ruffled. There are other causes such as axial ligand effects or hole-size effects.[68] Indeed, since $[Fe^{III}Cl(TMP\cdot)]^+$ is an S = 2 system[69] we expect some intrinsic symmetry-lowering distortion must occur to break the strict orthogonality of the magnetic orbitals that might otherwise lead to an S = 3 ground state. (In the structurally characterized p-methyl-TPP derivative, a face-to-face association has been established by X-ray structure determination as the prime cause of ruffling).[22,67] Face-to-face intermolecular associations are always problematic in understanding spin-coupling because of the difficulty of separating *intra*- from *inter*-molecular effects. It has been argued that intermolecular coupling is weak in magnitude and antiferromagnetic in nature when TPP-type complexes dimerize (J = 0 to -54 cm^{-1}).[22,52] However, in sterically unencumbered OEP π-cation radicals, which form rather tight dimers in all four and five-coordinate complexes structurally characterized to date, the association is much stronger and most often leads to complete loss of spin via ligand/ligand antiferromagnetic coupling. Indeed, the tight π/π core dimerization can lead to an unexpectedly large alternation of C-C bond lengths in the OEP cores whose origin is not understood.[70]

Finally, we note that metal/ligand spin coupling is not restricted to metalloporphyrin π-cation radicals. It has been explored and, to a fairly satisfying extent, understood in metal nitroxyl complexes.[71-73] Semiquinone complexes have also been examined.[74,75] Particularly strong coupling of both a ferromagnetic and antiferromagnetic nature has been observed although there can be ambiguities of valence in the latter. In general, the principles used to understand these species are similar to those used with metalloporphyrins: determination of valencies and component spin states is followed by inspection of magnetic orbital symmetries. Frequently the magnitude of spin coupling can be rationalized from an intuitive appreciation of the extent of overlap of the magnetic orbitals. This may eventually lead to an understanding of spin coupling in galactose oxidase once the principles of metal/phenolate radical coupling are known in greater detail. The evidence to date is that a modified tyrosine ligand radical couples to a copper(II) center to give an ESR silent active site.[13] Strong coupling would be expected from this close approach of metal/ligand spins. The nature of the coupling, antiferro- or ferro-magnetic, will depend on the symmetry relationship of the magnetic orbitals. Other biological situations where metal/radical spin coupling has been proposed include the iron-substituted zinc enzyme HLADH, horse liver alcohol dehydrogenase. A very weakly coupled outer-sphere complex of high-spin iron(II) with triplet dioxygen is proposed ($|J| < 0.1$ cm^{-1}).[76] Coupling of an ubiquinone radical to high-spin iron(II)

372

(J = - 0.3 cm^{-1}) has been proposed in the photosystem II of bacteria.[77] EPR evidenc for metal/radical spin coupling has also been presented in vitamin B$_{12}$ studies[78] an in trimethylamine dehydrogenase.[79] Weak interactions such as these are probabl best viewed as through-space dipolar couplings without reference to specific orbit pathways. With certain assumptions, they can be used to estimate the separation of spins in the 5-15 Å region.[78]

4. Copper(II) Dimers:

The most important spin-coupling in copper proteins occurs in the binuclear site of EPR-silent Type III centers. The dioxygen-carrying protein hemocyanin is th simplest and best characterized example although the oxygenase tyrosinase, and th two oxidases, laccase and ascorbate oxidase, have also received much attention.[3]

Hemocyanin: Orbital Complementarity. The X-ray structure of deoxy-hemocyani by Hol[80] combined with the key model compound studies of Kitajima[7] lead to th structure for oxyhemocyanin represented in Figure 8.

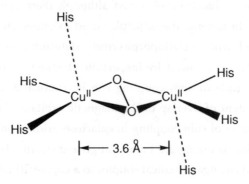

Figure 8. Representation of the probable structure of oxyhemocyanin (His = histidine). Dashed lines indicate longer, apical bonds.

The valencies are unambiguously established as copper(II) by many indicator chiefly the peroxidic nature of the coordinated dioxygen (ν_{O-O} = 750 cm^{-1}) and th electronic and EXAFS spectra of the copper.[3,81] The extent of magnetic coupling strong. Not only is oxyhemocyanin EPR-silent but SQUID magnetic susceptibili measurements indicate diamagnetism up to 260 K.[82] This places a lower limit o he magnitude of J i.e., $|-J| > 285$ cm^{-1}. Similarly, Kitajima's model complex, Cu(HB(3,5-i-Pr$_2$pz)$_3$)]$_2$O$_2$, is diamagnetic up to -10°C.[7] The essential tetragonality of he ligand geometry must generate a $d_{x^2-y^2}$ configuration for the magnetic orbitals on the d^9 copper centers. Given that there are two strongly bonded bridge atoms, his sets up an excellent opportunity for strong antiferromagnetic coupling provided, of course, that the bridging peroxide ligand does not lead to a situation of accidental orthogonality. By inspection, the dominant overlap will be the antisymmetric combination (φ_A) of the $d_{x^2-y^2}$ orbitals with the coplanar π^* HOMO of the peroxide as illustrated in Figure 9.

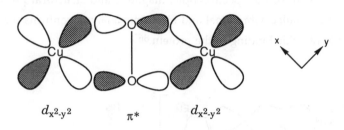

$$d_{x^2-y^2} \qquad \pi^* \qquad d_{x^2-y^2}$$

Figure 9. Representation of the antisymmetric combination of $d_{x^2-y^2}$ magnetic orbitals on copper(II) overlapping with the π^* HOMO on peroxide.

The strength of this interaction is borne out by extended Hückel[7] and SCF-X$_\alpha$-SW[83] calculations. The lack of peroxide orbitals near the HOMO level with the correct symmetry to overlap with the *symmetric* combination of $d_{x^2-y^2}$ orbitals (φ_S) means that the condition of accidental orthogonality cannot obtain. The dominant overlaps greatly destablize φ_A relative to φ_S and spin pairing in the lower energy orbital will occur. In other words, antiferromagnetic coupling with large $-J$ values is readily understood. Within the framework of the Hoffman model[35] used here to rationalize the diamagnetism of oxyhemocyanin, the energy gap of $-2J$ is equated with the HOMO/LUMO gap. An electronic Raman band at 1075 cm^{-1} has been tentatively assigned to this singlet to triplet transition.[84] This is consistent with the experimentally determined lower limit of ca. 550 cm^{-1} and with the ca. 1 ev (1000 cm^{-1}) estimates from recent calculations. [7,83]

Methemocyanin is the name given to dicopper(II) forms of hemocyanin that have bridging ligands other than peroxidic dioxygen. Even though they are of little biological significance, they were intensively investigated prior to the very recent elucidation of the novel $\mu-\eta^2$–peroxo structure of oxyhemocyanin because their properties were so similar to that of the oxy form. In addition, model compounds for methemocyanin were much more stable and easier to characterize than those for oxyhemocyanin.

The most informative studies have been with azide-methemocyanin, met(N_3)Hc and its models. The model compound [Cu_2(L-Et)(N_3)]$^{2+}$, illustrated in Figure 10 faithfully reproduces all of the spectroscopic, magnetic and structural properties of met(N_3)Hc[85] and a detailed vibrational and electronic analysis confirms the unusual 1,3-bridging mode of azide binding in the protein.[86]

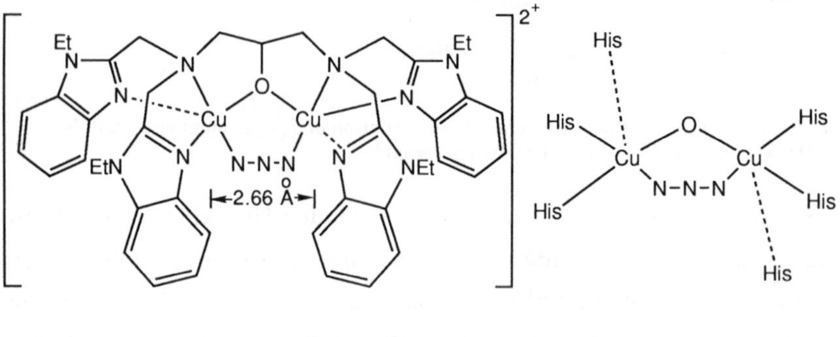

Figure 10. Structural representations of azide-methemocyanin (right) and the synthetic model compound [Cu_2(L-Et)(1,3-N_3)]$^{2+}$ (left).

Both the model compound and the protein are diamagnetic within the limits of detection, indicating large antiferromagnetic coupling. This is readily rationalized by inspection of how the bridging ligand HOMOs interact with the symmetric and antisymmetric combinations of the copper magnetic orbitals. As with oxyhemocyanin, the dominant interactions of both bridging ligands occur with the antisymmetric combination of $d_{x^2-y^2}$ orbitals. These are illustrated in Figure 11. The $d_{x^2-y^2}$ nature of the magnetic orbitals is again dictated by the essential tetragonality of the copper stereochemistry. This is known in the model compound

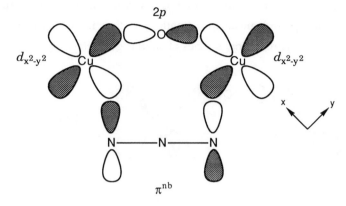

Figure 11. Representation of the complementary overlaps of the oxygen $2p$ and the azide π^{nb} HOMOs with the antisymmetric combination of Cu $d_{x^2-y^2}$ magnetic orbitals.

by X-ray crystallography[85] and is a safe assumption in the protein, given its electronic spectroscopy.[86] The dominant interacting HOMO of the 1,3-bridging azide is the in-plane π^{nb} orbital.[87] The relatively large Cu-O-Cu angle in the alkoxide bridge of the model complex (127°) means than only the $2p$ orbital on oxygen that is parallel to the Cu ... Cu vector will have a significant interaction with the copper $d_{x^2-y^2}$ orbitals. Thus, although the bridging ligands are dissimilar their dominant orbital phases are complementary such that they act in concert to destablize φ_A relative to φ_S and promote strong antiferromagnetic coupling. This reveals an important new concept in spin coupling that must always be considered when dissimilar bridging ligands are present. The phases of the bridging ligand HOMOs may be *complementary* or *anti-complementary*. In the latter case, it is possible that accidental orthogonality and ferromagnetic coupling occurs even though, on their own, both ligands would promote antiferromagnetic coupling. This was discovered in the acetate and nitrite analogues of the azide model compound.[42] Nishida and Kida[88] were first to use this rationale to explain why the addition of an acetate bridge to a mono-hydroxy copper(II) dimer dramatically decreased, rather than increased, the extent of antiferromagnetic coupling. The hydroxy ligand with its large obtuse Cu-O-Cu angle (~135°) interacts predominantly with φ_A via a $2p$ orbital whereas the

a_1 HOMO of acetate interacts predominantly with φ_S. This anticomplementarity is illustrated in Figure 12. The result is that φ_A and φ_S more nearly approach accidental orthogonality, and antiferromagnetic coupling in the hydroxy/acetate species is lower than in hydroxy-only species of similar geometry.

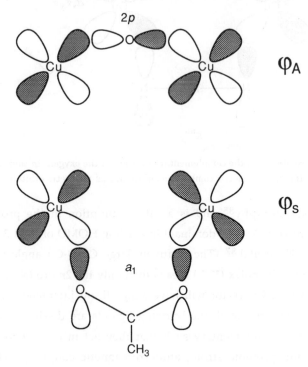

Figure 12. Antisymmetric and symmetric combinations of magnetic orbitals illustrating the anti-complementary nature of the hydroxy and acetate bridging HOMOs.

The strong spin coupling in the azide model compound $[Cu_2(N_3)(L\text{-}Et)]^{2+}$ discussed earlier arises from the combined complementary coupling pathways provided by the azide and the alkoxide bridging ligands. This was used to argue that met(N_3)Hc must also have two bridging ligands.[83] This is consistent with other titration and spectroscopic data[86,89] and, given the absence of an endogenous bridging ligand in the X-ray structure of hemocyanin, hydroxide is a logical exogenous candidate for the second bridge. This is also consistent with the tendency of copper(II) to be five-coordinate. However, implicit in the argument, is the assumption that a 1,3-azide bridge alone could not mediate such strong antiferromagnetic coupling. There are, as yet no appropriate mono-azido model compounds available to test this hypothesis. Even if there were, information on the dependence of J on the Cu-N-N angles would also be necessary to fully evaluate the possibilities. A relevant model compound is known, one with *two* 1,3-azide bridges acting in concert.[87] The complex is diamagnetic at room temperature indicating that antiferromagnetic coupling is very strong. Hückel calculations suggest a singlet-triplet energy gap of ca. 1100 cm^{-1}. Assuming that one azide can mediate half the coupling of two, then it does seem possible that a single 1,3-azide bridge could, in fact, promote near diamagnetism. This example illustrates both the promise and the difficulties of deducing precise structural information from magnetic data. A similar case arises in aquo-methemocyanin. Strong magnetic coupling could be rationalized by a single hydroxide bridge as long as the Cu-O-Cu angle is relatively obtuse. A number of examples are known in the model compound literature.[90-92]

Magnetic Orbital Identity and Orientation. Implicit in all of the above discussion has been that the bridging ligands lie in the tetragonal planes of copper(II) dimers having $d_{x^2-y^2}$ ground states. The previously discussed model compounds were selected to be consistent with this identity and relative orientation of the magnetic orbitals. However, given the stereochemical flexibility of copper(II), particularly when five-coordinate, there are many complexes where the ground state is different or even indeterminant. This can have a pronounced effect on magnetic coupling and is the most important of a number of other factors that must be considered when rationalizing J values.

Five coordinate copper(II) complexes can be tetragonal or trigonal bipyramidal, or almost anything in between along the Berry twist pathway of the C_{4v}/D_{3h} interconversion.[93] Tetragonal complexes with idealized square pyramidal geometry are readily assigned $d_{x^2-y^2}$ ground states. The ligands of the basal plane are not only identified by inspection of the geometry but more importantly by a significant extension of the unique metal-ligand bond that is designated as axial. Typically, this is 0.2 - 0.5 Å longer than a comparable bond in the basal plane. On the other hand, complexes whose geometries closely approach the trigonal bipyramidal ideal can usually be safely assigned a d_{z^2} ground state. Although some inequality of axial vs. equatorial bond length is found, the differences are not particularly diagnostic of a d_{z^2} ground state because opposing steric and electronic factors tend to equalize bond lengths. For structures whose stereochemistries are intermediate between the C_{4v} and D_{3h} ideals, considerable irregularity exists. There are geometrically-based criteria for deciding whether a particular complex is better described as tetragonal or trigonal bipyramindal[94,95] but when the decision is close, an admixture of $d_{x^2-y^2}$ and d_{z^2} ground states must be considered and this complicates a straightforward rationalization of the orbital pathway for spin coupling.[96]

The result of switching one or both of the individual ground states in a copper(II) dimer from $d_{x^2-y^2}$ to d_{z^2} is usually, but not always,[97] a diminution of spin-coupling. The underlying reason for this is a decrease in overlap of the magnetic orbitals. While the prolate lobes of the d_{z^2} orbital may be the equal of the $d_{x^2-y^2}$ orbital in delocalizing spin onto a bridging atom, the equatorial torus of the d_{z^2} orbital is not. This has been nicely illustrated in copper(II) cyano complexes where an axial cyanide bridge mediates stronger coupling than one that is equatorial,[98] in closely related monohydroxy-bridged species,[99] in oxalate-bridged complexes where spin-coupling can be progressively attenuated by d_{z^2} admixture,[100] and in trigonal bipyramidal nitrite and acetate analogues of the previously discussed azide methemocyanin model, $[Cu_2(N_3)(L-Et)]^{2+}$.[42]

Of course, it is not only important to determine the ground state(s) of the magnetic orbitals, it is necessary to inspect whether their relative orientations optimize or diminish σ overlap. Thus, apex-to-apex bridging of square pyrmidal copper complexes does not lead to any detectable spin coupling.[101] In a mono-bridged phenoxide complex where the bridging oxygen atom connects the apex of one tetragonal pyramid to the base of another, the overlap is negligible.[102] In fact, there is evidence of a weak ferromagnetic interaction ($J \cong +2$ cm^{-1}) presumably arising from the condition of σ/δ strict orthogonality. Similarly, ligands that bridge the apical positions of square pyramidal copper(II) structures can be safely assumed to have insignificant effects relative to ligands that bridge the basal plane.[91,103] In dibridged complexes that have mixed or indistinct ground states it is difficult to partition magnetic orbital orientation effects from orbital complementarity effects.[104,105]

In summary, the strength of antiferromagnetic coupling in copper(II) dimers is first and foremost a function of the coordinate geometry. This defines the relative orientation of the magnetic orbitals and determines which bridging ligand orbitals can mediate significant σ-overlap. Optimal conditions for σ-overlap can give rise to strong antiferromagnetic coupling, even at Cu ... Cu distances as great as 6 Å.[106] Indeed the cumulative success in rationalizing spin coupling over the past twenty years and the growing predictability of magnetic exchange endorses the basic validity of the σ mechanism. Antiferromagnetic coupling can be attenuated by a variety of factors which either (a) promote accidental orthogonality, and/or (b) diminish σ overlap. In the first category, we have shown that anticomplementarity of bridging ligand orbitals in unsymmetrical di-bridged systems is a common cause of weaker antiferromagnetic coupling. Bond angles to the bridging atoms(s) are also critical; typically, antiferromagnetic coupling diminishes as the bridgehead angle becomes more acute. In the second category, weaker antiferromagnetic coupling results from the tortional twist of tetragonal planes,[107] roof-like bending of tetragonal planes,[108] out-of-plane displacement of copper in tetragonal systems,[109] and pyramidalization of some,[110] but not all, [88,105] O-atom bridges. The possibility of some kind of π-contribution to coupling has been raised in some of these cases where there is distortion from ideal σ overlap.[110] We also note that the non-bridging, peripheral ligands should not be considered entirely innocent in spin-coupling. All other things being essentially equal, it has been shown that a nitrogen to sulfur change in a terminal donor ligand can effect a significant increase in antiferromagnetic coupling.[111] Similarly, bromide terminal ligands enhance antiferromagnetic exchange relative to chloride.[112]

Ranking of Ligands. Given the large body of data on copper(II) dimers, we can attempt to address the problem of ranking various bridging ligands in their ability to mediate antiferromagnetic coupling. This is a difficult task because it requires a systematic change in bridging ligand while at the same time holding everything else constant. Not only is this very difficult to realize, but when it is, the correlation is strictly valid only for the specific bridge dimensions that are present. Nevertheless, approximate constancy of geometry in certain series of copper(II) complexes allows some qualitative ranking of ligands. The order alkoxide ~ pyrazolate > acetate ~ hydroxide > chloride has been deduced from a series of mono- and dibridged copper(II) species.[110] It must be remembered, however, that alkoxide and hydroxide are particularly sensitive to bridge angle and the effects of ligand orbital complementarity usually cannot be ignored in dibridged systems. As discussed earlier, bridging η^2-peroxide is a very strong mediator of antiferromagnetic coupling. Bridging 1,2-peroxide is also capable of mediating strong antiferromagnetic coupling. Essential diamagnetism at room temperature is observed when 1,2-peroxide couples d_{z^2} ground states via the axial sites of trigonal bipyramidal copper(II) complexes.[113] The J value is conservatively estimated to be -300 cm-1. When bridging sulfur donor atoms replace comparable oxygen or nitrogen atoms, antiferromagnetic coupling typically increases.[114] The same conclusion has been reached in iron(III) dimers.[115] Within a series of carboxylate-bridged copper(II) dimers of interesting atypical structure, the order formate > pivalate > acetate has been established.[116] 1,3-Azide can be comparable to O-atom bridges (with large Cu-O-Cu angles) but potentially large antiferromagnetic coupling in 1,1-azide is countered by the ferromagnetic interaction arising from spin polarization.[41] Imidazolate-bridged dicopper(II) systems,

including cobalt- and copper-substituted bovine superoxide dismutase,[117] show moderate antiferromagnetic coupling when there is a σ orientation of the magnetic orbitals ($|-J| < 90$ cm-1).[118] The dependence of $-J$ on bridge bond angles[118] has been rationalized in terms of accidental degeneracy.[119] Moderate *ferro*magnetic coupling has been demonstrated in an imidazolate-bridged copper/iron system.[36] A *tri*nuclear imidazolate-bridged system of triangular structure has recently been reported.[120] Not only does this compound illustrate that the concepts developed in limers apply equally well to trimers, but the additional effect of *spin frustration* is manifest. If any two copper atoms are antiferromagnetically coupled, the third necessarily experiences simultaneous ferro- *and* antiferro-magnetic coupling i.e., its spin is said to be frustrated. An S = 1/2 ground state is the inescapable consequence of a triangular structure.

Other Copper Proteins. The preceding discussion of a triangular structure raises the question of understanding spin-coupling in the trinuclear center of the multicopper oxidases such as laccase and ascorbate oxidases. Originally deduced from EPR spectroscopy to contain three separate types of copper centers (Type I "blue", Type II "normal" and Type III "coupled binuclear"), it is now known from spectroscopic studies on laccase[121] and from the X-ray structure of ascorbate oxidase[122] that the EPR-"normal" Type II center is within ~ 3.9Å of the Type III coupled pair. A precise definition of the molecular structure is presently unavailable but the dimensions of the isosceles triangle of copper atoms suggest an O-atom-bridged binuclear pair like that proposed for methemocyanin. The Type II

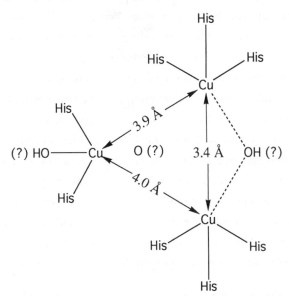

Figure 13. A schematic representation of the trinuclear site of Type II/III copper proteins.

copper may or may not be ligand-bridged to this pair. A representation of this structure is given in Figure 13.

At the present state of resolution it is not possible to conclusively determine whether this unit should be viewed magnetically as a strongly coupled Type III pair with a peripheral, weakly coupled Type II center or as a spin-frustrated trinuclear system. The problem has, however, been framed in considerable detail,[121] favoring the former description and model compounds for the isosceles triangle[123] and equilateral triangle[124] extremes are available.

5. Other Systems, Other Concepts

Given the complexity of interrelated factors that can influence magnetic coupling in an $S = \frac{1}{2} \big/ S = \frac{1}{2}$ copper(II) dimer, perhaps the conceptually most simple spin-coupled system, it is not surprising that the understanding of spin coupling in systems of higher spin multiplicity is less advanced. The well-defined magnetic orbitals of octahedral chromium(III) i.e., $\left(t_{2g}\right)^3$ have made spin-coupling in chromium(III) dimers approachable and understandable.[125] However, for the biologically relevant high-spin d^4, d^5 and d^6 configurations that are so widespread in iron-sulfur,[14] oxo-iron,[8-11] oxo-manganese[126] proteins and their model compounds, the sheer number and variety of possible interactions of the magnetic orbitals frequently precludes a ready understanding of the superexchange pathways. Nevertheless, some progress has been made. The approach follows the familiar sequence: (a) a rigorous determination of the valencies, the component spins, the molecular structure and the overall spin description, (b) a systematic search for magneto-structural correlations, (c) identification of the magnetic orbitals (d) deduction of dominant orbital interaction pathways and rationalization of J using the concepts outlined in part 2 of this review, and (e) an attempt to find support in Hückel, $X\alpha$ or *ab initio* calculations. We briefly review some of the highlights, focusing on features not found in metalloporphyrin or dicopper chemistry.

Magnetism in oxo-diiron proteins and models has been recently reviewed by Kurtz[8] so no detailed discussion is presented here. The general lack of linear relationships between J values and structural parameters points to a relatively complex interplay of competing orbital pathways for spin coupling and a sensitivity to multiple structural variables e.g. bond length, bond angle and spin distribution. What is clear is that μ-oxo linkages tend to dominate magnetic coupling and that like copper(II) dimers, antiferromagnetic coupling diminishes somewhat as the Fe-O-Fe angle becomes more acute. This has been identified with a π overlap pathway in the symmetric combination of magnetic orbitals (φ_S) with p_x on oxygen[127] as illustrated in Figure 14.

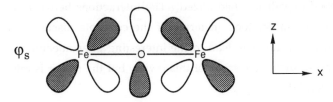

φ_S

Figure 14. π-Overlap of d_{xz} magnetic orbitals via $p_\pi(O)$ in linear or nearly linear μ-oxo iron(III) dimers.

This dominant overlap is attenuated as the Fe-O-Fe angle decreases from 180°. However, the angular dependence of J is not as marked as in copper(II) dimers because additional d_{xz}/d_{xz} interactions remain and a d_{xz}/d_{z^2} interaction increasingly comes into play as the angle becomes more acute.[37] One report of ferromagnetic coupling in an iron(III) dimer with very acute Fe-O-Fe angles to phenoxide bridges (92.5°) suggests that the condition of overall accidental orthogonality may be achievable even with five magnetic orbitals on each metal atom.[128]

While high-spin d^5 iron(III) dimers are almost invariable antiferromagnetically coupled comparable homo- and hetero-binuclear complexes with Ti, V, Cr and Mn show very revealing differences in the sign and magnitude of J. In an elegant recent study by Wieghardt and Girerd[37] the variation of magnetic coupling with d^n configuration in a series of μ-oxo di-μ-carboxylate species was used to reveal the nature of the dominant interactions: $d_{xz}/d_{xz}, d_{yz}/d_{yz}$ and d_{xz}/d_{z^2}. Importantly, the last of these is *antiferro*magnetic when both d orbitals are occupied e.g. iron(III) dimers, and *ferro*magnetic when only one is occupied e.g. manganese(III) dimers. The *ferro*magnetic interaction is the "crossed interaction" we refered to in part 2 of this review (i.e. Figure 3) and for which we suggest the potentially more consistent nomenclature "virtual orthogonality". It can successfully compete with the other antiferromagnetic interactions in Mn(III) dimers leading to an overall condition close to accidental orthogonality; J values are small and sometimes positive, A similar situation exists for Cr^{III}-O-Cr^{III} and Cr^{III}-O-Mn^{III} systems.[37] Dominant ferromagnetic coupling is also observed in certain mixed valence Fe(II/III) systems leading to ground or low-lying S = 9/2 states.[129,130] Such mixed valence compounds may be valence trapped or valence delocalized. The interactions between discrete valencies can probably be understood in terms of the concepts discussed in this review but valence delocalization can lead to the interesting phenomenon of *double exchange*. This concept is outside the scope of this review but the reader is refered to an excellent introduction.[131]

From the point of view of bioinorganic chemistry, magnetic coupling in high d^n systems has proved to be most useful in assessing congruence of structure between synthetic models and proteins. For example, although the active site structure of methemerythrin was known prior to the characterization of model compounds, the essential congruence of $-J$ for $\left[Fe_2^{II}(\mu-OH)(OAc)_2(MTACN)\right]^+$ and *deoxy*hemerythrin,[132] and the congruence for the many μ-oxo μ-dicarboxylato iron(III) dimers and *met*hemerythrin,[8-11] allowed the choice between μ-hydroxo and μ-oxo ligands to be made with confidence. In the mixed valence state of hemerythrin, however, the magnetism is not a conclusive guide to structure.[133] The growing classes of multi-iron and multi-manganese proteins[9] are likely to offer many more opportunities for magnetic properties to guide structure elucidation.[134]

Conclusion

In reviewing the concepts of spin coupling in bioinorganic chemistry and illustrating them chiefly with iron and copper chemistry, it has become clear that systematic studies with well-defined model compounds have been indispensible for gaining a measure of chemical understanding of the phenomenon of spin coupling. A good qualitative understanding of the more simple systems has emerged, particularly in $S = \frac{1}{2} \big/ S = \frac{1}{2}$ spin systems. Bimetallic systems of higher spin multiplicity are beginning to yield to analysis and there are prospects for understanding the complexity brought about by the clustering of more than two metals in iron-sulfur clusters and high nuclearity oxo-iron and manganese clusters. In many of these areas the overall spin states and structural classes are still at the stage of phenomenological definition. The sustained activity and continuing discovery in the area of iron sulfur clusters,[14,135] the growing interest in dinickel complexes as models for the active site of urease,[136-138] the intense exploration of multi-manganese chemistry as models for the oxygen-evolution site of Photosystem II[6,46,126,139,140] and the continuing interest in understanding the Fe/Cu site of cytochrome oxidase[141] are just a few areas where the application of spin coupling concepts, and perhaps the discovery of new ones, will be seen.

Spin coupling will always be important in defining molecular and electronic structure. It is the inevitable signature of bringing paramagnetic centers into close proximity and often, it is an exceedingly sensitive criterion of such structure. Eventually, however, it may be possible to relate spin coupling to the control of reactivity since the nature of a half occupied orbital is ultimately what influences electron transfer rates in redox proteins.

Acknowledgments

C. A. R. expresses his sincere appreciation to his students and collaborators for their major contributions. His interest in spin coupling began during a sabbatical leave in the laboratories of Dr. Jean-Claude Marchon in 1979 and has been sustained by a long term magnetostructural collaboration with Prof. W. Robert Scheidt. We thank Dr. Gerald Weunschell for preparing the Figures. This research is supported by the National Institutes of Health and the National Science Foundation.

References

1.   E. I. Solomon and D. E. Wilcox, in *Magneto-Structural Correlations in Exchange Coupled Systems*, ed. R. D. Willett, D. Gatteschi and O. Kahn (D. Reidel, Dordrecht, 1985) p. 463.

2.   G. Palmer, *Pure Appl. Chem.* **59** (1987) 749.

3. E. I. Solomon, K. W. Penfield and D. E. Wilcox, *Struct. Bonding (Berlin)* **53** (1983) 1.

4. J. A. Christner, E. Münck, T. A. Kent, P. A. Janick, J. C. Salerno and L. M. Siegel, *J. Am. Chem. Soc.* **106** (1984) 6786.

5. Burgess, B. K., *Chem. Rev.* **90** (1990) 1377; J. Kim and D. C. Rees, *Science*, **257** (1992) 1677.

6. G. W. Brudvig, in *Metal Clusters in Proteins*, ed. L. Que Jr., ACS Symposium Series **372** (1988) 221.

7. N. Kitajima, K. Fujisawa, C. Fujimoto, Y. Moro-oka, S. Hashimoto, T. Kitagawa, K. Toriumi, K. Tatsumi and A. Nakamura, *J. Am. Chem. Soc.* **114** (1992) 1277.

8. D. M. Kurtz Jr., *Chem. Rev.* **90** (1990) 585.

9. L. Que, Jr., and A. E. True, in *Progress in Inorganic Chemistry: Bioinorganic Chemistry. Vol. 38*, ed. S. J. Lippard (Wiley, 1990) p. 97.

10. J. B. Vincent, G. L. Olivier-Lilley and B. A. Averill, *Chem. Rev.* **9 0** (1990) 1447.

11. J. B. Howard and D. C. Rees, *Advances in Protein Chemistry* **42** (1991) 199.

12. T. L. Poulos, in *Heme Proteins. 7*, ed. G. L. Eichhorn and L. G. Marzilli (Elsevier, 1988) p. 1.

13. K. Clark, J. E. Penner-Hahn, M. M. Whittaker and J. W. Whittaker, *J. Am. Chem. Soc.* **112** (1990) 6433; N. Ito, S. E. V. Phillips, C. Stevens, Z. B. Orgel, M. J. McPherson, J. N. Keen, K. D. S. Yadav, and P. F. Knowles, *Nature*, **350** (1991) 87.

14. *Adv. Inorg. Chem: Iron-Sulfur Proteins*, ed. R. Cammack (Academic Press), **38** (1992) .

15. T. Yagi, H. Inokuchi and K. Kimura, *Acc. Chem. Res.* **16** (1983) 2.

16. T. G. Spiro and P. Saltman, *Struct. Bond.* **6** (1969) 116.

17. R. B. Frankel, R. P. Blakemore and R. S. Wolfe, *Science* **203** (1979) 1355.

18. J. L. Kirschvink, A. Kobayashi-Kirschvink and B. J. Woodford, *Proc. Nat. Acad. Sci. USA.* **89** (1992) 7683.

19. *Encyclopedia Britannica 15th Edition (1989)*, quoted by D. Jiles, in

Introduction to Magnetism and Magnetic Materials (Chapman and Hall, 1991).

20. A. P. Ginsberg, *Inorg. Chim. Acta Rev.* **5** (1971) 45.

21. D. Dolphin, A. Forman, D. C. Borg, J. Fajer and R. H. Felton, *Proc. Natl. Acad. Sci. USA.* **68** (1971) 614.

22. P. Gans, G. Buisson, E. Duée, J.-C. Marchon, B. S. Erler, W. F. Scholz and C. A. Reed, *J. Am. Chem. Soc.* **108** (1986) 1223.

23. E. I. Solomon, *Comments Inorg. Chem.* **5** (1984) 227.

24. A. Bencini and D. Gatteschi, in *EPR of Exchange Coupled Systems* (Springer- Verlag, Berlin, 1990).

25. B. N. Figgis, R. Mason, A. R. P. Smith, J. N. Varghese and G. A. Williams, *J. Chem. Soc. Dalton Trans.* (1983) 703.

26. D. P. E. Dickson, in *Mossbauer Spectroscopy Applied to Inorganic Chemistry. Vol. 1*, ed. G. J. Long (Plenum Press, New York, 1984) p. 339.

27. E. P. Day, T. A. Kent, P. A. Lindahl, E. Münck, W. H. Orme-Johnson, H. Roder and A. Roy, *Biophys. J.* **52** (1987) 837.

28. P. de Loth, P. Cassoux, J. P. Daudey and J. P. Malrieu, *J. Am. Chem. Soc.* **103** (1981) 4007.

29. J. B. Goodenough, *Magnetism and the Chemical Bond* (Interscience, New York, 1963).

30. J. Kanamori, *J. Phys. Chem. Solids* **10** (1959) 87.

31. J. H. Van Vleck, *The Theory of Electric and Magnetic Susceptibilities* (Oxford University Press, London, 1932).

32. R. L. Martin, in *Pathways in Inorganic Chemistry*, ed. E. A. V. Ebsworth, A. G. Maddock and A. G. Sharpe (University Press, Cambridge, 1968) chapt 9.

33. E. Sinn, *Coord. Chem. Rev.* **5** (1970) 313.

34. O. Kahn, *Inorg. Chim. Acta* **62** (1982) 3.

35. P. J. Hay, J. C. Thibeault and R. Hoffman, *J. Am. Chem. Soc.* **97** (1975) 4884.

36. C. A. Koch, C. A. Reed, G. A. Brewer, N. P. Rath, W. R. Scheidt, G. Gupta and G. Lang, *J. Am. Chem. Soc.* **111** (1989) 7645.

388

37. R. Hotzelmann, K. Wieghardt, U. Flörke, H.-J. Haupt, D. C. Weatherburn, J. Bonvoisin, G. Blondin and J.-J. Girerd, *J. Am. Chem. Soc.* **114** (1992) 1681.

38. D. J. Hodgson, *Prog. Inorg. Chem.* **19** (1975) 173.

39. O. Kahn and M. F. Charlot, *Nouv. J. Chim.* **4** (1980) 567.

40. J. Comarmond, P. Plumeré, J.-M. Lehn, Y. Agnus, R. Louis, R. Weiss, O. Kahn and I. Morgenstern-Badarau, *J. Am. Chem. Soc.* **104** (1982) 6330.

41. O. Kahn, S. Sikorav, J. Gouteron, S. Jeannin and Y. Jeannin, *Inorg. Chem.* **22** (1983) 2877.

42. V. McKee, M. Zvagulis and C. A. Reed, *Inorg. Chem.* **24** (1985) 2914.

43. P. Gómez-Romero, E. H. Witten, W. M. Reiff, G. Backes, J. Sanders-Loehr and G. B. Jameson, *J. Am. Chem. Soc.* **111** (1989) 9039.

44. H. Astheimer and W. Haase, *J. Chem. Phys.* **85** (1986) 1427.

45. K. Yamaguchi, T. Fueno, N. Ueyama, A. Nakamura and M. Ozaki, *Chem. Phys. Lett.* **164** (1989) 210.

46. E. A. Schmitt, L. Noodleman, E. J. Baerends and D. N. Hendrickson, *J. Am. Chem. Soc.* **114** (1992) 6109.

47. O. Kahn, *Struct. Bonding (Berlin)* **68** (1987) 89.

48. D. A. Dougherty, *Acc. Chem. Res.* **24** (1991) 88.

49. J. S. Miller, A. J. Epstein and W. M. Reiff, *Acc. Chem. Res.* **21** (1988) 114.

50. C. J. Cairns and D. H. Busch, *Coord. Chem. Rev.* **69** (1986) 1.

51. C. E. Schulz, R. Rutter, J. T. Sage, P. G. Debrunner and L. P. Hager, *Biochemistry* **23** (1984) 4743.

52. B. S. Erler, W. F. Scholz, Y. J. Lee, W. R. Scheidt and C. A. Reed, *J. Am. Chem. Soc.* **109** (1987) 2644.

53. H. Song, C. A. Reed and W. R. Scheidt, *J. Am. Chem. Soc.* **111** (1989) 6865.

54. M. Gouterman, *J. Mol. Spectr.* **6** (1961) 138.

55. P. O. Sandusky, A. Salehi, C. K. Chang and G. T. Babcock, *J. Am. Chem. Soc.* **111** (1989) 6437.

56. W. D. Edwards, G. H. F. Diercksen and M. C. Zerner *J. Mol. Struct. (Theochem)* **199** (1989) 137.

57. S. Nakashima, H. Ohya-Nishiguchi, N. Hirota, H. Fujii and I. Morishima, *Inorg. Chem.* **29** (1990) 5207.

58. W. R. Scheidt, H. Song, K. J. Haller, M. K. Safo, R. D. Orosz, C. A. Reed, P. G. Debrunner and C. E. Schulz, *Inorg. Chem.* **31** (1992) 939.

59. L. P. Hager, D. L. Doubek, R. M. Silverstein, J. H. Hargis and J. C. Martin, *J. Am. Chem. Soc.* **94** (1972) 4364.

60. S. P. Cramer, J. H. Dawson, K. O. Hodgson and L. P. Hager *J. Am. Chem. Soc.* **100** (1978) 7282.

61. B. Boso, G. Lang, T. J. McMurry and J. T. Groves, *J. Chem. Phys.* **79** (1983) 1122.

62. I. Morishima, Y. Takamuki and Y. Shiro, *J. Am. Chem. Soc.* **106** (1984) 7666.

63. R. S. Czernuszewicz, K. A. Macor, X-Y. Li, J. R. Kincaid and T. G. Spiro, *J. Am. Chem. Soc.* **111** (1989) 3860.

64. W.-J. Chuang and H. E. Van Wart, *J. Biol. Chem.* **267** (1992) 13293.

65. G. N. La Mar and F. A. Walker in *The Porphyrins*, Vol. IV, ed D. Dolphin (Academic Press, 1979) pp. 61-157.

66. G. M. Godziela and H. M. Goff, *J. Am. Chem. Soc.* **108** (1986) 2236; A. Nanthakumar and H. M. Goff, *Inorg. Chem.* **30** (1991) 4460.

67. W. R. Scheidt and Y. J. Lee *Struct Bonding (Berlin)* **64** (1987) 1.

68. W. R. Scheidt in *The Porphyrins*, Vol. III, ed D. Dolphin (Academic Press, 1978) pp. 463-511.

69. K. Hatano, personal communication.

70. H. Song, R. D. Orosz, C. A. Reed and W. R. Scheidt, *Inorg. Chem.* **29** (1990) 4274.

71. C. G. Pierpont and R. M. Buchanan, *Coord. Chem. Rev.* **38** (1981) 45.

72. A. Caneschi, D. Gatteschi and P. Rey, in *Progress in Inorganic Chemistry, Vol. 39*, ed. S. J. Lippard (John Wiley and Sons, 1991).

73. A. Caneschi, D. Gatteschi, R. Sessoli and P. Rey, *Acc. Chem. Res.* **22** (1989) 392.

74. O. Kahn, R. Prins, J. Reedijk and J. S. Thompson, *Inorg. Chem.* **26** (1987) 3557.

75. C. Benelli, A. Dei, D. Gatteschi and L. Pardi, *Inorg. Chem.* **27** (1988) 2831.

76. E. Bill, C. Haas, X-Q Ding, W. Maret, H. Winkler, A. X. Trautwein and M. Zeppezauer, *Eur. J. Biochem.* **180** (1989) 111.

77. W. F. Butler, D. C. Johnston, H. B. Shore, D. R. Fredkin, M. Y. Okamura and G. Feher, *Biophys. J.* **323** (1980) 967.

78. J. F. Boas, P. R. Hicks, J. R. Pilbrow and T. D. Smith *J. Chem. Soc. Faraday Trans. II* **74** (1978) 417.

79. D. J. Steenkamp, T. P. Singer and H. Beinert, *Biochem J.* **169** (1978) 361.

80. A. Volbeda and W. G. Hol *J. Mol. Biol.* **209** (1989) 249.

81. M. S. Co, K. O. Hodgson, T. K. Eccles and R. Lontie, *J. Am. Chem. Soc.* **103** (1981) 984.

82. D. M. Dooley, R. A. Scott, E. Ellinghaus, E. I. Solomon and H. B. Gray, *Proc. Nat. Acad. Sci. USA.* **75** (1978) 3019.

83. P. K. Ross and E. I. Solomon, *J. Am. Chem. Soc.* **113** (1991) 3246.

84. J. A. Larrabee and T. G. Spiro, *J. Am. Chem. Soc.* **102** (1980) 4217.

85. V. McKee, M. Zvagulis, J. V. Dagdigian, M. G. Patch and C. A. Reed, *J. Am. Chem. Soc.* **106** (1984) 4765.

86. J. E. Pate, P. K. Ross, T. J. Thamann, C. A. Reed, K. D. Karlin, T. N. Sorrell and E. I. Solomon, *J. Am. Chem. Soc.* **111** (1989) 5198.

87. J. Comarmond, P. Plumeré, J.-M. Lehn, Y. Agnus, R. Louis, R. Weiss, O. Kahn and I. Morgenstern-Badarau *J. Am. Chem. Soc.* **104** (1982) 6330.

88. Y. Nishida and S. Kida, *J. Chem. Soc. Dalton Trans.* (1986) 2633.

89. D. E. Wilcox, J. R. Long and E. I. Solomon, *J. Am. Chem. Soc.* **106** (1984) 2186.

90. M. G. B. Drew, M. McCann and S. M. Nelson, *J. Chem. Soc. Dalton Trans.* (1981) 1868.

91. P. K. Coughlin and S. J. Lippard, *J. Am. Chem. Soc.* **103** (1981) 3228.

92. P. L. Burk, J. A. Osborn and M.-T. Youinou, *J. Am. Chem. Soc.* **103** (1981) 1273.

93. B. J. Hathaway in *Comprehensive Coordination Chemistry* Vol 5, ed. G. Wilkinson, Pergamon (1987) p. 607.

94. E. L. Muetterties and L. J. Guggenberger, *J. Am. Chem. Soc.* **96** (1974)

1748.

95. A. W. Addison, T. Nageswara Rao, J. Reedijk, J. van Rijn and G. C. Verschoor, *J. Chem. Soc. Dalton Trans.* (1980) 1272.

96. T. N. Sorrell, D. L. Jameson, C. J. O'Connor, *Inorg. Chem.* **23** (1984) 190.

97. T. R. Felthouse, E. J. Laskowski and D. N. Hendrickson, *Inorg. Chem.* **16** (1977) 1077.

98. D. S. Bieksza and D. N. Hendrickson, *Inorg. Chim. Acta* **82** (1984) 13.

99. M. G. B. Drew, J. Nelson, F. Esho, V. McKee and S. M. Nelson, *J. Chem. Soc. Dalton Trans.* (1982) 1837.

100. M. Julve, M. Verdaguer, A. Gleizeo, M. Philoiche-Levisalles and O. Kahn, *Inorg. Chem.* **23** (1984) 3808.

101. M. G. B. Drew, M. McCann and S. M. Nelson *J. Chem. Soc. Chem. Commun.* (1979) 481.

102. H. P. Berends and D. W. Stephan, *Inorg. Chem.* **26** (1987) 749.

103. P. Chaudhuri, K. Oder, K. Wieghardt, B. Nuber and J. Weiss, *Inorg. Chem.* **25** (1986) 2818.

104. T. N. Sorrell, C. J. O'Connor, O. P. Anderson and J. H. Reibenspies, *J. Am. Chem. Soc.* **107** (1985) 4199.

105. N. A. Bailey, D. E. Fenton, J. Lay, P. B. Roberts, J.-M. Latour and D. Limosin, *J. Chem. Soc. Dalton Trans.* (1986) 2681.

106. C. Chauvel, J. J. Girerd, Y. Jeannin, O. Kahn and G. Lavigne, *Inorg. Chem.* **18** (1979) 3015.

107. E. Sinn, *Inorg. Chem.* **13** (1974) 2013.

108. M. F. Charlot, S. Jeannin, Y. Jeannin, O. Kahn, J. Lucrece-Abaul and J. Martin-Frere, *Inorg. Chem.* **18** (1979) 1675.

109. B. Chiari, O. Piovesana, T. Tarantelli and P. F. Zanazzi, *Inorg. Chem.* **27** (1988) 4149.

110. W. Mazurek, B. J. Kennedy, K. S. Murray, M. J. O'Conner, J. R. Rodgers, M. R. Snow, A. G. Weed and P. R. Zwack, *Inorg. Chem.* **26** (1985) 3258.

111. O. Kahn, T. Mallah, J. Gouteron, S. Jeanmin and Yves Jeannin, *J. Chem. Soc. Dalton Trans.* (1989) 1117.

112. L. K. Thompson, F. L. Lee and E. J. Gabe, *Inorg. Chem.* **27** (1988) 39.

392

113. K. D. Karlin, Z. Tyeklar, A. Farooq, R. R. Jacobson, E. Sinn, D. W. Lee, J. E. Bradshaw, L. J. Wilson *Inorg. Chim. Acta* **182** (1991) 1.

114. O. Kahn, *Angew. Chem. Int. Ed. Engl.* **24** (1985) 850.

115. J. R. Dorfman, J.-J. Girerd, E. D. Simhon, T. D. P. Stack and R. H. Holm, *Inorg. Chem.* **23** (1984) 4407.

116. T. Tokii, N. Watanabe, M. Nakashima, Y. Muto, M. Morooka, S. Ohba and Y. Saito, *Bull Chem. Soc. Jpn.*, **63** (1990) 364.

117. I. Morgenstern-Badarau, D. Cocco, A. Desideri, G. Rotilio, J. Jordanov and N. Dupre *J. Am. Chem. Soc.* **108** (1986) 300.

118. K. G. Strothkamp and S. J. Lippard, *Acc. Chem. Res.* **15** (1982) 318.

119. A. Bencini, C. Benelli, D. Gatteschi and C. Zanchini, *Inorg. Chem.* **25** (1986) 398.

120. P. Chaudhuri, I. Karpenstein, M. Winter, C. Butzlaff, E. Bill, A. X. Trautwien, U. Flörke, H.-J. Haupt, *J. Chem. Soc. Chem. Commun.* (1992) 289.

121. J. L. Cole, P. A. Clark and E. I. Solomon, *J. Am. Chem. Soc.* **112** (1990) 9534.

122. A. Messerschmidt, A. Rossi, R. Ladenstein, R. Huber, M. Bolognesi, G. Gatti, A. Marchesini, R. Petruzzelli and A. Finazzi-Agro, *J. Mol. Biol.* **206** (1989) 513; A. Messerschmidt and R. Huber, *Eur. J. Biochem.* **187** (1990) 341.

123. H. Adams, N. A. Bailey, M. J. S. Dwyer, D. E. Fenton, P. C. Hellier and P. D. Hempstead, *J. Chem. Soc. Chem. Commun. (1991)* 1297.

124. R. J. Butcher, C. J. O'Connor and E. Sinn, *Inorg. Chem.* **20** (1981) 537.

125. D. J. Hodgson in *Magneto-Structural Correlations in Exchange Coupled Systems*, ed. R. D. Willett, D. Gatteschi and O. Kahn (D. Reidel, Dordrecht, 1985) p. 497.

126. W. H. Armstrong in *Bioinorganic Chemistry of Manganese* ed. V. L. Pecoraro (VCH, New York, 1992) p. 261.

127. R. N. Mukherjee, T. D. P. Stack and R. H. Holm, *J. Am. Chem. Soc.* **110** (1988) 1850.

128. M. Mikuriya, Y. Kakuta, K. Kawano and T. Tokii, *Chem. Lett. (Japan)*

(1991) 2031.

129. K. K. Surerus, E. Münck, B. S. Snyder and R. H. Holm, *J. Am. Chem. Soc.* **111** (1989) 5501 and references therein.

130. M. S. Mashuta, R. J. Webb, J. K. McCusker, E. A. Schmitt, K. J. Oberhausen, J. F. Richardson, R. M. Buchanan and D. N. Hendrikson *J. Am. Chem. Soc.* **114** (1992) 3815 and references therein.

131. E. Münck, V. Papaefthymiou, K. K. Surerus, J.-J. Girerd in *Metal Clusters in Proteins* ed. L. Que Jr. (ACS Symposium Series **372**, Washington DC, 1988) p.302.

132. J. R. Hartman, R. L. Rardin, P. Chaudhuri, K. Pohl, K. Wieghardt, B. Nuber, J. Weiss, G. C. Papaefthymiou, R. B. Frankel, S. J. Lippard, *J. Am. Chem. Soc.* **109** (1987) 7387.

133. M. J. Maroney, D. M. Kurtz Jr., J. M. Nocek, L. Pearce and L. Que *J. Am. Chem. Soc.* **108** (1986) 6871.

134. See for example E. I. Solomon and Y. Zhang, *Accounts Chem. Res.* **25** (1992) 343.

135. R. H. Holm, S. Ciurli and J. A. Weigel *Progr. Inorg. Chem.* **38** (1990) 1.

136. P. A. Clark and D. E. Wilcox *Inorg. Chem.* **28** (1989) 1326.

137. C. A. Salata, M.-T. Youinou and C. J. Burrows *Inorg. Chem.* **30** (1991) 3454.

138. T. R. Holman, M. P. Hendrich and L. Que Jr. *Inorg. Chem.* **31** (1992) 937 and references therein.

139. D. N. Hendrikson, G. Christou, E. A. Schmitt, E. Libby, J. S. Bashkin, S. Wang, H.-L Tsai, J. B. Vincent, P. D. W. Boyd, J. C. Huffman, K. Folting, Q. Li and W. E. Streib, *J. Am. Chem. Soc.* **114** (1992) 2455 and references therein.

140. S. Pal, M. K. Chan and W. H. Armstrong *J. Am. Chem. Soc.* **114** (1992) 6398.

141. A. Nanthakumar, M. S. Nasir, K. D. Karlin, N. Ravi and B. H. Huynh, *J. Am. Chem. Soc.* **114** (1992) 6564.

MULTIFIELD SATURATION MAGNETIZATION OF
METALLOPROTEINS

Edmund P. Day and Mariana S. Sendova

Emory Physics Department, Emory University, Atlanta, GA 30233

Multifield saturation magnetization (MSM) measurements of metalloproteins have led to discoveries of new EPR signals and the re-interpretation of other EPR signals and of several Mössbauer spectral features. These new magnetization measurements[1,2] are a third technique along with Mössbauer and EPR spectroscopies required in any thorough study of the magnetic properties of Fe-containing sites. For other transition metals where Mössbauer spectroscopy does not apply, EPR and magnetic circular dichroism (MCD) spectroscopies need be applied in combination with the MSM technique in order to arrive at a comparably complete picture of the magnetic state of the metal ion. This strategy is gaining currency of combining a number of measurements including both resonance spectroscopies and thermal equilibrium techniques to arrive at a single description of the magnetic properties of an active site.[3,4,5]

The MSM technique is an extension of a magnetic susceptibility measurement from the Curie region to low temperatures and high magnetic fields where saturation occurs. It is a thermal equilibrium measurement which complements resonance techniques such as Mössbauer, MCD, and EPR spectroscopies. In previous susceptibility measurements the Curie slope was combined with metal analysis to determine the spin of the paramagnetic metal site. In an MSM study a nested set of saturation magnetization curves is fit to determine the spin (S), spin concentration ({S}), g value, and zero-field splitting parameters (D and E/D) without using the metal content. The new methodology is clearly superior in its ability to measure five or six properties of the sample (S, {S}, g, D, E/D, and the exchange coupling J) from the magnetic data alone whereas susceptibility data has to be combined with metal analysis to arrive at a one or two quantities (the spin S and exchange coupling J). Both techniques require a well-characterized, magnetically pure sample although the MSM technique is superior in its ability to detect the presence of small amounts of unexpected impurities.

The surprising results from early multifield saturation magnetization studies of metalloproteins came as the result of unsuspected limitations in what had been state-of-the-art EPR studies. The discovery of the broad EPR signal of the phosphate complex of reduced uteroferrin[4] and the discovery of the spin S = 3/2 state of the Fe protein of nitrogenase[6] are two such cases. This review of MSM studies of metalloproteins will highlight the limitations of resonance techniques when used alone. Examples will be given of the misinterpretation of the apparent absence of EPR signals or of the misinterpretation of the loss of an EPR signal with increasing temperature. The complementarity of MSM and resonance techniques will be emphasized. MSM detects all of the paramagnetism in a metalloprotein sample including any impurities which are present. Therefore, resonance spectroscopies such as EPR, Mössbauer, and MCD are required in any MSM study to characterize each MSM sample, measure its impurities, and detect the presence of unwanted redox states of the metalloprotein to avoid misinterpretation of the MSM data. On the other hand, MSM data are required for each resonance study to detect EPR-silent or MCD-silent or non-Fe paramagnetism, resolve ambiguities in the interpretation of resonance signals, or to quantitate MCD or EPR signals. Indeed, it is now possible to quantitate diamagnetic (spin S = 0) states of metalloproteins by using MSM in combination with metal analysis.

This review of MSM measurements of metalloproteins will cover a number of proteins and will focus on the magnetic properties of a specific redox state of the active site in each case. Reviews (the first citations within each section) should be consulted for further information on the structure and function of each protein.

1. Previously Undetected EPR Signals of Spin S = 1/2, 3/2, ... States

Kramers theorem guarantees a two-fold degeneracy in half-odd integer spin states in the absence of a magnetic field. Application of a magnetic field lifts this degeneracy in a predictable way leading to routine detection of these Kramers states by EPR spectroscopy. Only in rare circumstances are half-odd integer spin states not detected by EPR. Several of the first MSM studies of metalloproteins were motivated by the apparent absence of the EPR signal of a Kramers system and, in two cases, led to discoveries of the missing EPR signals.

1.1. Phosphate Complex of Reduced Uteroferrin:[7,8,9] Discovery of a New Class of Spin S = 1/2 EPR Signals

The struggle to understand the magnetic properties of the phosphate complex of reduced uteroferrin furnishes an example of the danger of misinterpreting the apparent loss of an EPR signal. In this study the MSM technique led to the discovery of a broad EPR signal that had been detected previously but rejected as an artifact.[10]

Oxidized uteroferrin is purple (λ_m = 550 nm), catalytically inactive, and EPR-silent. The EPR-silence of this state is due to the strong antiferromagnetic coupling between the two high spin ferric sites at the active site which produces a diamagnetic spin S = 0 ground state. Reduced uteroferrin is pink (λ_m = 510 nm), catalytically active, and exhibits an EPR signal with g_{av} = 1.75 (Figure 2A). This EPR signal with a g value substantially below 2 is the definitive signature of the class of metalloproteins containing a μ-oxo bridged iron dimer at the active site.[11] In the mixed valent state the antiferromagnetically coupled iron dimer is in a spin S = 1/2 state. The low g value of this spin S = 1/2 state is due to the antiferromagnetic coupling of the g = 2 spin S = 5/2 ferric ion with the spin S = 2 ferrous state having a g value greater than 2.

The spin Hamiltonian for an exchange coupled binuclear iron center is

$$H = \Sigma H_i + H_{exc} \qquad (1)$$

with

$$H_i = D_i \{(S_{zi}^2 - S_i (S_i + 1)/3) + E_i/D_i (S_{xi}^2 - S_{yi}^2)\} + \beta S_i \bullet g_i \bullet H \qquad (2)$$

and

$$H_{exc} = J S_1 \bullet S_2 \qquad (3)$$

Antiferromagnetic exchange coupling is described by positive J. D_i and E_i are the zero-field splitting parameters of the ferrous (i = 1) and ferric (i = 2) sites respectively.

When phosphate is added to reduced uteroferrin there is an immediate shift in the optical absorption to higher wavelengths, immediate loss of the g = 1.75 EPR signal, and slow loss of catalytic activity. On the one hand, the loss of the EPR signal and shift in

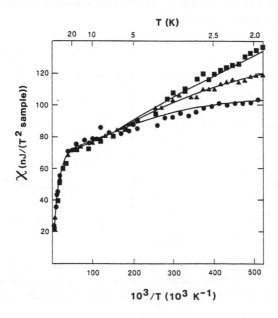

Figure 1. The susceptibility (M/H) for the phosphate complex of reduced uteroferrin[4] at three fixed fields (● 5 T; ▲ 2.5 T; and ■ 0.3125 T). The theoretical curves were calculated from the spin Hamiltonian ($J_{red:PO4}$=6.0(5) cm^{-1}; S_1=2; D_1=8 cm^{-1}; E_1/D_1=0.08; g_1=(2.18, 2.22, 2.00); S_2=5/2; D_2=1 cm^{-1}; E_2/D_2=0; g_2=(2.04, 2.07, 2.04)).

398

optical absorption were interpreted by Aisen et al.[12] to indicate oxidation of the enzyme. On the other hand, Zerner et al.[13] and Que et al.[14] emphasized the slow loss of activity as evidence against an immediate oxidation. Mössbauer data[14] indicated the presence of a paramagnetic state following phosphate addition giving further evidence against the interpretation of the loss of the EPR signal as loss of paramagnetism. MSM data[4] verified the paramagnetism of the phosphate complex of reduced uteroferrin first seen by Mössbauer spectroscopy. Moreover, it became apparent in fitting the MSM data that the spin S = 1/2 paramagnetic state following phosphate addition was dramatically different than that of reduced uteroferrin. The exchange coupling was three times smaller (J_{red} = 19.8(2) cm^{-1}; $J_{red:PO4}$ = 6.0(5) cm^{-1}); the g strain was nine times larger; and the anisotropy in g was three times as great. These changes resulted in a dramatically altered EPR signal which was broader and much more difficult to detect (Figure 2B,C). Richard Dunham was able to find the EPR signal at 1.5 K of the phosphate complex of reduced uteroferrin once these characteristics had been found from fitting the MSM data. This new class of EPR signals is now routinely used to characterize different complexes of uteroferrin and purple acid phosphatase.[15,16]

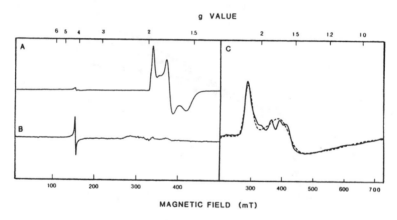

Figure 2. (A) The X-band EPR spectra at 8 K of reduced uteroferrin,[4] and (B,C) of its phosphate complex. (A,B) were recorded at 10 G modulation amplitude and 0.2 mW microwave power and (C) at 40 G modulation amplitude and 200 mW microwave power.

The MSM study of the reduced enzyme also cleared up the discrepancy between the Faraday balance[17] and superconducting susceptometer[18] results on the one hand (J_{red} > 35 to 100 cm^{-1}) and room temperature NMR (J_{red} = 20 cm^{-1})[19] and EPR loss of signal (J_{red} > 11, 14 cm^{-1})[20,21] on the other. Fitting the MSM data indicated an exchange coupling (J_{red} = 19.8(2) cm^{-1}) consistent with the room temperature NMR measurement.

Earlier Faraday balance and superconducting susceptometer studies had not dealt properly with sample magnetization noise and did not have the requisite sensitivity to measure exchange coupling in a metalloprotein.

1.2. Reduced Fe Protein of Nitrogenase:[22,23,24] Discovery of a Spin S = 3/2 EPR Signal

At the time of the MSM study of the Fe protein of nitrogenase from *Azotobacter vinelandii*[6] the active site was believed to be a single {4Fe-4S} cluster.[25] Metal analysis indicated four Fe atoms per protein molecule.[26,27,28,29,30] EPR studies of the reduced protein gave the g = 1.94 signal characteristic of a {4Fe-4S} cluster.[31] However, quantitation of the g = 1.94 EPR signal gave a result ({0.3 electrons})[31,32,33,34,35] well short of the one electron observed in redox titration studies. Attempts to explain this discrepancy included proposals for a neighboring spin to broaden the EPR signal by relaxation effects.[36] It is worth noting that by the time the MSM study was performed in the mid-1980's more than ten years had passed since the g = 1.94 EPR signal of the reduced Fe protein had been published. The discrepancy in the quantitation of the signal had been apparent for more than ten years. Every resonant technique including EPR, MCD, and Mössbauer spectroscopy had been applied to this problem. It remained for MSM to resolve the difficulty.

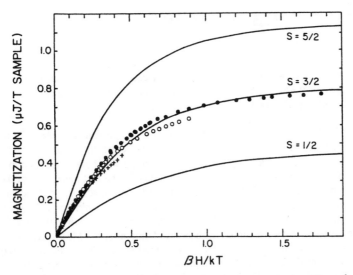

Figure 3. Magnetization of the reduced Fe protein of *Azotobacter vinelandii*[6] at three fixed fields (● 5.0 T; ○ 2.5 T; and + 1.25 T) over the temperature range from 1.8 to 70 K. The Brillouin curves of S=5/2, S=3/2, and S=1/2 are shown.

400

The nested MSM data[6] (shown in Figure 3) was dominated by a higher spin than the expected spin S = 1/2 state detected by EPR which would have yielded a single superimposed Brillouin curve (as in Figure 10 for the spin S = 1/2 state of oxidized ferredoxin II). Comparison of this data with half-odd integer Brillouin curves (solid lines in Figure 3) indicated the presence of a spin S = 3/2 center. At this point the EPR signal was re-investigated and the broad signals at g = 5.8 and 5.2 were discovered. It is

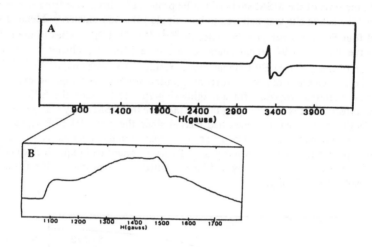

Figure 4. (A) X-band EPR spectra at 8.8 K of the reduced Fe protein of *Azotobacter vinelandii*[6] at 0.2 mW and 10 G modulation amplitude. (B) EPR spectra in the g=5 region at 5 mW and 10 G modulation amplitude.

interesting to note that this region of the EPR spectrum was published in the 1972 study[31] to indicate the absence of the spin S = 3/2 state of the MoFe protein. Here is a case where the difference in the EPR signatures of two different spin S = 3/2 states resulted in one of these states remaining hidden for fifteen years! This, despite the fact this region of the EPR spectrum was investigated in detail by highly skilled EPR spectroscopists aware of a discrepancy in the EPR quantitation of the g = 1.94 signal. The spin S = 3/2 signature of the native FeMo-co factor was readily observed in the early 1970's while the spin S = 3/2 EPR signal of the reduced Fe protein lay undiscovered until after the MSM study found this paramagnetic state in the mid-1980's. A second point worth stressing is that quantitation of the broad spin S = 3/2 EPR signal of the Fe protein is very difficult.[37] One general use of the new MSM technique is reliable quantitation of spin states whose EPR signals are particularly difficult to quantitate.

1.3. Sulfite Complex of Oxidized Sulfite Reductase:[38,39,40] *EPR-Silent Spin S = 1/2 State*

At the time of the MSM study of the sulfite complex of oxidized sulfite reductase, there was an apparent discrepancy in the optical and Mössbauer data on the one hand, and EPR spectroscopy on the other. The optical[41] and Mössbauer data[42] indicated a spin S = 1/2 state arising from the low spin ferric heme complexed with sulfite coupled to the spin S = 0 diamagnetic {4Fe-4S} cluster. Yet there was no spin S = 1/2 EPR signal.[43] The only feature in the EPR spectrum was a broad feature at g = 2.7 which quantitated to less than 10% of an electron. Similar observations were made for the nitrite complex of oxidized nitrite reductase except that no EPR feature was observed in this case.[44] The MSM study of the sulfite complex of oxidized sulfite reductase[42] detected the spin S = 1/2 state verifying the interpretation of the optical and Mössbauer data. The form of the saturation magnetization curves ruled out extreme anisotropy in the g values as an explanation for the lack of an EPR signal. The remaining explanation for the absence of the spin S = 1/2 EPR signal is extreme g-strain. Sulfite and nitrite rarely bind to a hemeprotein.[45] In this study of two exceptions to this general rule, it appears that the resulting adducts have exceptional heterogeneity in the g values of the spin S = 1/2 bound complex. It is intriguing that only the substrates of each enzyme show this extreme heterogeneity in g values.

2. Measurement of Zero: Quantitation of the Spin S = 0 State

It is surprisingly difficult to quantitate the spin S = 0 state. Mössbauer spectroscopy can reliably measure the fraction of Fe in the diamagnetic (S = 0) state and the spin state markers of optical techniques may give broad hints of the presence of diamagnetism at a potentially paramagnetic site. But reliable quantitation of the diamagnetic state of a potentially paramagnetic site has been difficult. Because the MSM technique measures all of the paramagnetism present in a sample it is capable (when combined with careful metal analysis) of quantitating the amount of spin S = 0 diamagnetism present at a metal site.

2.1. The S-Site of the MoFe Protein of Nitrogenase:[22,46,24] *Re-Interpretation of Mössbauer Spectra*

The earliest Mössbauer studies of the MoFe protein of nitrogenase indicated a quadrupole doublet feature attributed to the S-site involving two Fe.[47] A careful study of an [57]Fe-enriched sample enabled both Mössbauer and MSM data to be collected on the self-same sample of native MoFe protein from *Azotobacter vinelandii*[1] The sample's magnetization was found to be comprised of only two parts: the known spin S = 3/2 state of the two FeMo-co factors in the protein and a spin S = 2 ferrous iron impurity detected by Mössbauer. The spin of the S-site was therefore measured to be spin S = 0. It was this measurement which re-enforced the possibility that the S-site was not a separate entity. This, in turn, led to a re-assessment of the quadrupole doublet originally attributed to the separate S-site as part of the P cluster Mössbauer spectra.[48,49] The MSM measurement of a spin S = 0 state of a metalloprotein in the presence of a substantial signal from an impurity required full use of the new methodology of fitting MSM data of a sample thoroughly characterized by both Mössbauer and EPR spectroscopies. These resonant spectroscopies were crucial in identifying and quantitating the impurity and in establishing the absence of any oxidized protein in the magnetization sample.

2.2. {NiFeSe} Hydrogenase:[50]MSM Used to 'Detect' Spin S = 0 Diamagnetism

The MSM study of the Ni(II) site of EPR silent {NiFeSe} hydrogenase[51] was made difficult by the presence of two {4Fe-4S} clusters in this metalloprotein in addition to the Ni(II) site of interest. This resulted in small amounts of high spin ferric (Fe(III)) (detected by EPR spectroscopy) and high spin ferrous (Fe(II)) (detected by Mössbauer spectroscopy) impurities which had to be accounted for properly if the spin of the Ni(II)

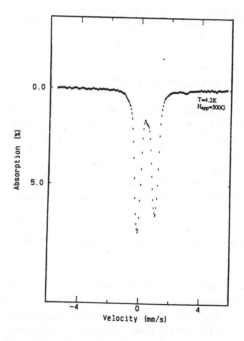

Figure 5. The Mössbauer spectrum of {NiFeSe} hydrogenase at 4.2 K in 500 G. The small absorption peak near 3.0 mm/s corresponds to 2.1(4)% of the total iron in the sample.

site were to be accurately measured. For this reason, the metalloprotein was isolated from bacteria grown on an [57]Fe-enriched medium and the self-same magnetization sample was studied successively by both the Mössbauer and magnetization techniques. The Mössbauer spectra of the magnetization sample verified that the two {4Fe-4S} clusters were diamagnetic as expected for the 'EPR-silent' state under study and measured the small amount of Fe(II) impurity. EPR studies of a parallel sample were performed in order to verify the EPR-silent state of the sample and to detect the Fe(III) impurity. The Ni(II) site was found to be diamagnetic. In combination with separate EXAFS studies,[52] this indicated that the Ni(II) site is five-coordinate.

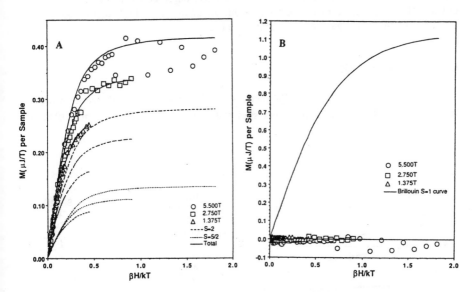

Figure 6. (A) Multifield saturation magnetization data for native *D. baculatus* hydrogenase[51] at three fixed fields over the temperature range from 2 to 100 K. The dashed curves are from the 2.1% Fe(II) impurity. The dotted curves arise from 0.8% Fe(III) impurity. (B) The same data as those shown in (A) after subtracting the impurity signals. The solid line indicates the signal expected for 1 mole of spin S=1 (Brillouin curve). This is the expected signal if the Ni(II) site were high spin.

2.3. Urease:[53,54] No Evidence for Exchange Coupling Between the Two Ni(II) Ions at the Active Site in the Native State

At the time of the recent MSM study of urease, a previous magnetic susceptibility study[55] had concluded that the two Ni(II) ions at the active site were exchange coupled in

404

the native state of the enzyme. The MSM study found no evidence for exchange coupling.[56] The MSM study did find that the fraction of Ni(II) in the high spin S = 1 state decreased with increasing pH.

There are many examples in the inorganic literature of Ni(II) systems which have a fraction of the sites with high spin S = 1 and a fraction of the sites with low spin S = 0.[57,58] This heterogeneous population distribution of spin states will often change with solvent properties including pH. Under these circumstances, the MSM technique combined with careful metal analysis is essential in order to measure both states (the high spin S = 1 and the low spin S = 0) of the Ni(II) site. This general characteristic of the magnetochemistry of Ni(II) contrasts with that of ions such as Fe(III) and Fe(II) or Mn(III) and Mn(II) which more often than not adopt an unchanging high or low spin state depending on the local ligand environment at the ion.

3. Integer Spin EPR

An important new area of EPR spectroscopy has been developed recently with the detection of EPR signals from integer spin states.[59,60,61,62,63,64] A number of these

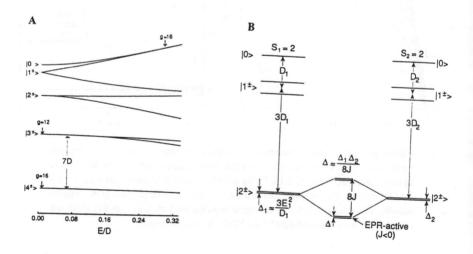

Figure 7. Two alternative energy level schemes to explain the integer EPR spectra of the azide complex of reduced hemerythrin. (A) Energy levels of an S=4 manifold (strong coupling regime) for H=0 and D < 0. (B) Effect of a weak exchange interaction on the $|2^{\pm}\rangle$ states of two S=2 sites (weak coupling regime).[65]

signals have been quantitated.[66,65,67,68] The fundamental dilemma in quantitating integer spin EPR signals is that many known integer spin states have yielded no signal despite extensive searches. The reason for the lack of signal is well understood as due to zero-field splitting larger than the microwave quantum of existing EPR spectrometers. In these integer spin, non-Kramers systems there is no physical principle requiring degeneracy in the absence of a magnetic field (as there is for half-odd integer, Kramers systems). The result is often a splitting in zero field which prevents the detection of an EPR signal in the presence of a field. In other cases, when the zero-field splitting is small enough to yield an EPR signal, there often is more than one energy level scheme to explain all of the integer spin EPR spectra (Figure 7). In light of these circumstances, an MSM study combined with integer spin EPR spectroscopy is called for to detect EPR-silent integer spin states, to verify the quantitation of integer spin EPR signals when they are detected, and to identify and resolve ambiguities when they arise in the interpretation of integer spin EPR spectra.

3.1. Azide Complex of Reduced Hemerythrin:[7,8,9,69] Resolving the Ambiguities of an Integer Spin EPR Signal

Study of the magnetic properties of the azide complex of fully reduced

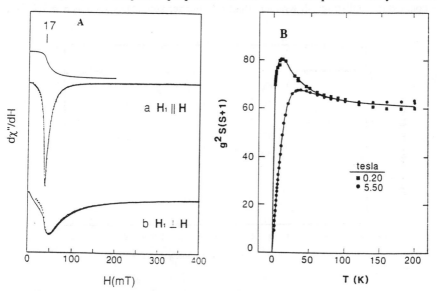

Figure 8. (A) X-band EPR spectra (-) and simulations (⋯) of deoxyhemerythrin azide at 4 K with microwave field H_1 parallel and perpendicular to the static field H. (B) Magnetization of the same sample at the extreme fields plotted as μ_{eff} versus temperature. The simulations in both figures are based on the weak coupling model assuming two identical ferrous sites and $J = -3.cm^{-1}$.[65]

(Fe(II)Fe(II)) hemerythrin[65] highlighted the ambiguities inherent in integer spin EPR signals. It was possible to fit all of the integer spin EPR spectra (Figure 8A) (parallel and perpendicular mode X-band as well as Q-band) equally well with either of two schemes (Figure 7). In each scheme the EPR spectra came from transitions between two closely spaced low-lying energy levels. In Scheme I (the strong exchange coupling regime) the two low lying levels arise from the ± 4 levels of the ferromagnetically coupled diferrous dimer. In Scheme II (the weak exchange coupling regime) the two low lying levels arose from the quartet of states arising from weak exchange coupling of the ± 2 levels of the ferrous ions. Fits to the MSM data resolved this ambiguity in the integer spin EPR data (Figure 8B).[65] Parameters derived from either fit to the EPR data were capable of fitting all of the low temperature MSM data. However, only the Scheme II (weak exchange coupling) EPR parameters were also capable of fitting the high temperature MSM data.

3.2. The Oxidized P Clusters of the MoFe Protein of Nitrogenase:[22,46,24] A Truly Exceptional Mössbauer Spectrum

The story of the spin of the oxidized P-clusters of the MoFe protein of nitrogenase furnishes one very dramatic exception to the general rule that Mössbauer spectroscopy can be counted on to distinguish Kramers (half-odd integer spin states such as S = 1/2, 3/2,

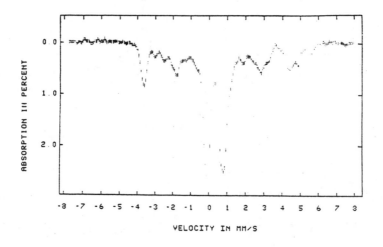

Figure 9. Mössbauer spectrum of thionine-oxidized MoFe protein from *C. pasteurianum* taken at 4.2 K in a magnetic field of 600 G.[70]

5/2, ...) from non-Kramers (integer spin states such as S = 0, 1, 2, 3, ...) states. The data in Figure 9[70] indicate the stark contrast between the low temperature, low field Mössbauer spectrum of a non-Kramers system (the two large peaks associated with a single quadrupole doublet at the center of the spectrum) and that of a Kramers system (the many low features extending from - 4 mm/s to + 6 mm/s on either side of the central quadrupole doublet). These differences are due to the two-fold degeneracy in the absence of a magnetic field required by Kramers theorem for all half-odd integer spins in contrast with the general non-degeneracy for even spin systems. In this case, however, the magnetic spectrum of Figure 9 is now known[68] to arise from an accidently degenerate integer spin paramagnetic center. For almost twenty years this misleading Mössbauer spectrum resulted in difficulties in understanding the magnetic properties of the oxidized P clusters. There were hints that the oxidized P clusters might be integer spin from EPR spectroscopy.[71] Resolution of this dilemma awaited the X-ray crystallographic data[46,72] which convincingly demonstrated that there are two, eight-iron P-clusters per MoFe protein rather than four, four-iron P clusters as had been thought previously.

A MSM study of the oxidized MoFe protein of *Azotobacter vinelandii* was carried out in the mid 1980's but not published until recently because no fit could be found to the data under the assumption there were four oxidized P-clusters. Today the data is readily understood as arising from two, spin S = 4, eight-Fe P-clusters.[73]

3.3. Cytochrome c Oxidase:[74,75,76,77] MSM Used to Distinguish 'Fast' from 'Slow'.

The oxidized state of cytochrome *c* oxidase gives dramatic evidence for the accidental nature of integer spin EPR signals. The fast form of oxidized cytochrome *c* oxidase reacts rapidly with cyanide while the slow form reacts 300 times more slowly with the same irreversible inhibitor.[78] The slow form has an integer spin EPR signal at 'g = 12' while the fast form lacks this signal. An MSM study of both 'fast' and 'slow' forms of the oxidized enzyme[79] indicates that the coupled Fe(III)-Cu(II) site is integer spin S = 2 (or, possibly S = 1 for a fraction of the fast preparation) for both. Although the average magnetic properties determined by MSM differ for the fast and slow forms of the enzyme, these average magnetic properties do not explain why the slow form shows an integer spin g = 12 signal while the fast form does not.

4. Unexpected Susceptibility Results at High Temperatures

The MSM results for the {3Fe-4S} cluster in both oxidized and reduced ferredoxin II of *Desulfovibrio gigas* were unexpected.

4.1. Oxidized {3Fe-4S} Cluster: Re-Interpretation of the Loss of the EPR Signal

In the case of the oxidized {3Fe-4S} center, the loss upon warming of the EPR signal from the spin S = 1/2 states observed at low temperatures had been interpreted as due to population of a low-lying spin S = 3/2 excited state.[80] This interpretation of the

408

EPR data predicted a small value for the exchange coupling ($J_{ox} \approx 20$ cm^{-1}; $JS_1 \cdot S_2$) between pairs of the three coupled Fe sites of the center. This small J value was also used to explain the long tail in the EPR signal as arising from perturbations due to small J/D terms in the Hamiltonian.[81] Both of these interpretations were found to be inconsistent with the magnetization data which showed J to be very large ($J_{ox} > 200$ cm^{-1}; $JS_1 \cdot S_2$).[82]

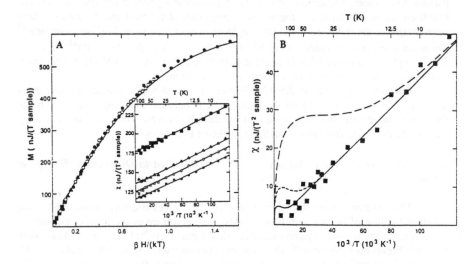

Figure 10. (A) Magnetization of oxidized Fd II[82] at four fixed fields (● 4.5 T; ○ 2.5 T; ▲ 1.25T; ■ 0.3 T). The fit to the Curie region for each field and all of the high temperature data are shown in the inset. (B) Sample susceptibility of the same sample. The upper dashed line was calculated from the spin Hamiltonian with J=40 cm^{-1}; the lower dashed line with J=100 cm^{-1}; and the solid line with J=200 cm^{-1}.

The loss of EPR signal is more likely due to mutual relaxation effects between the two spin S = 1/2 states known to be present at low temperatures.

4.2. Reduced {3Fe-4S} Cluster: Re-Interpretation of the High Temperature Mössbauer Data

The Mössbauer spectrum of the reduced {3Fe-4S} cluster furnishes clear evidence for a Fe(III)Fe(II) ion pair which share a single delocalized electron on the time scale of the Mössbauer effect (10^{-7} s).[83,61] One quadrupole doublet has the quadrupole splitting and isomer shift reflecting averages of a Fe(III) site and a Fe(II) site. This quadrupole doublet has twice the area as the second quadrupole doublet. The second quadrupole doublet observed at low temperatures in low field has the quadrupole splitting and isomer shift of a localized Fe(III) site. At 20 K the Mössbauer spectrum begins to change indicating population of a state which is fully delocalized over the three iron sites. The quadrupole splitting and isomer shift of the high temperature quadrupole doublet are averages of two Fe(III) ions and one Fe(II).[61] The fully delocalized excited state was expected to have a very different spin than the spin S = 2 ground state.[84,61] However, the magnetization data obeyed the Curie law to 200 K (the highest measured temperature).[82] This indicated that any excited states populated at 20 K had to have the same spin (S = 2) as the ground state. This, in turn, raises questions with there being any fundamental change in delocalization beginning at 20 K since this would be expected to alter the spin state of the system. Delocalization favors a ferromagnetically coupled high spin system state over the antiferromagnetically coupled low spin system state encountered in most cases where localized spins are exchange coupled.

5. Other MSM Studies of Metalloproteins

5.1. Manganese Superoxide Reductase: MSM Used to Quantitate the Metal Content

Jim Peterson, working as a postdoctoral fellow with E. P Day at the Gray Freshwater Biological Institute of the University of Minnesota, was the first to emphasize the capability of MSM to determine the spin quantitation ($\{S\}$) without metal analysis. He carried out a study of manganese superoxide reductase which exploited this new capability.[85] The ground state magnetic properties of the native and reduced enzyme were found to differ from those of known manganese model compounds.

5.2. Nitrous Oxide Reductase:[86,87] MSM Used to Quantitate the Paramagnetism

MSM data of nitrous oxide reductase[88] were consistent with recent EPR studies of this enzyme[89,90] in indicating 25% of the four copper ions per active site were spin S = 1/2 Cu(II). No evidence was found for population of a spin ladder of excited states in the temperature range from 2 to 200 K. Therefore, if the EPR silence of the remaining three copper sites is to be explained, in part, by antiferromagnetic exchange coupling between a pair of spin S = 1/2 Cu(II) sites, the exchange coupling must be relatively strong ($J > 200$ cm^{-1}; $JS_1 \bullet S_2$).

6. Shortcomings of the MSM Technique

Apart from the technical difficulties involved in the MSM technique which requires magnetically pure and well characterized samples, the fundamental limitations of MSM arise either when antiferromagnetic coupling reduces the signal of interest below the sensitivity of the instrument or when the magnetic properties of alternative models for the data cannot be resolved. We have run into the problem of a crucial lack of resolution in studies which attempt to resolve a spin S = 2 state from a spin S = 1 state. Normally this is straightforward whenever both alternatives have g values near 2. Unfortunately, there are several situations where a spin S = 1 state can have an unusually large g value. When this is the case it is not possible to distinguish the shape of the multifield saturation magnetization data for the spin S = 2 state with g = 2 from the spin S = 1 state having a large g value.

7. The Impact of the MSM Technique on Magnetochemistry

Until very recently there has been little reason to adopt the new MSM technique when measuring the magnetic properties of model compounds. Magnetochemists have had extensive experience with magnetic susceptibility measurements. It is a routine tool used to characterize all new cluster compounds. "If it ain't broke, why fix it?" would be reason enough to avoid the more elaborate MSM technique in favor of the familiar and simpler magnetic susceptibility measurement. There are other apparently good reasons not to adopt the new methodology:

1. Multifield measurements are not needed until you go below 20 K because all the data at all fields above this temperature are the same;

2. Going to low temperatures is not needed in order to measure exchange coupling when J is larger than 20 cm^{-1} ($JS_1 \cdot S_2$) because data above 20 K can be used to measure J precisely;

3. Going to low temperatures costs a lot of effort because impurities, even small amounts of impurities having high spin, will affect the data greatly. Sorting out these magnetic impurities does not seem worth the effort, even if it is possible;

4. Going to low temperatures for concentrated powder samples introduces the need to deal with lattice exchange coupling effects which are not of interest in cluster studies;

5. It has been time-consuming to collect several fields of data over the temperature range from 2 to 200 K even with the first commercial superconducting susceptometer and impossible with the best Faraday instruments.

A crucial reason to make the effort now to adopt the new MSM methodology is that models of μ-oxo bridged proteins are needed. The J values for many of the interesting redox states of these proteins are small (J < 20 cm^{-1}). These small J values can only be measured properly using the new MSM methodology. Great care in synthesizing fully characterized model clusters of bridged metal dimers will be wasted if the exchange coupling is measured using data collected at a single field - or using multifield data fit assuming D can be ignored. The only way J can be determined precisely when J < 20 cm^{-1} is through the use of the full MSM methodology which yields precise information on all the relevant parameters (J, S, {S}, g_{av}, D, E/D) even in the presence of small amounts of impurities.

The new MSM methodology can now be carried out far more conveniently than even a year or two ago due to several developments:

1. The commercial superconducting susceptometer available from Quantum Design collects six fields of magnetization data over the temperature range from 2 to 200 K (300 data points) (or five fields over the temperature range from 2 to 300 K) in less than 24 hours of fully automated, unattended operation. This speed and convenience of data collection exceeds that of the previous commercial superconducting susceptometer built by SHE (now BTi) by more than a factor of three;

2. Techniques to overcome noise introduced by the sample have been worked out and can be routinely employed when needed;[1,2]

3. Commercial software to fit MSM data is now available from WEB, Inc. This software can fit MSM data assuming either a monomer, a dimer, or any combination of the two types of paramagnetic center. This makes it possible to fit data containing small amounts of impurities. Combined with the availability of inexpensive desktop computers of increasing power, the WEB software quickly performs elaborate fits to MSM data.

It is likely that inorganic magnetochemistry will be advanced by the routine application of the MSM methodology. Synthetic methods to eliminate magnetic impurities cannot be developed until the magnetic impurities can be routinely detected and characterized. Meaningful models of exchange coupled metalloprotein sites cannot be developed unless the exchange coupling of both the metalloprotein and the models are each measured precisely. It is probable that the MSM technique which was originally developed for the purpose of gaining reliable information on the magnetic properties of metalloproteins will eventually have an even larger impact on inorganic magnetochemistry.

References

(1)   Day, E. P.; Kent, T. A.; Lindahl, P. A.; Münck, E.; Orme-Johnson, W. H.; Roder, H.; Roy, A. *Biophys. J.* **1987**, *52*, 837-853.

(2)   Day, E. P. In *Metallobiochemistry, Part C. Spectroscopic and Physical Methods for Determination of Metal Ion Environments in Metalloenzymes and Metalloproteins*; J. F. Riordan and B. L. Vallee, Ed.; Academic Press, Inc.: Orlando, FL, 1992; in press.

(3)   Gupta, G. P.; Lang, G.; Scheidt, W. R.; Geiger, D. K.; Reed, C. A. *J. Chem. Phys.* **1985**, *83*, 5945-5952.

(4)   Day, E. P.; David, S. S.; Peterson, J.; Dunham, W. R.; Bonvoisin, J. J.; Sands, R. H.; Que, L., Jr *J. Biol. Chem.* **1988**, *263*, 15561-15567.

(5)   **Trautwein, A. X.; Bill, E.; Bominaar, E. L.;** Winkler, H. *Struct. Bonding (Berlin)* **1991**, *78*, 1-95.

(6)   Lindahl, P. A.; **Day, E. P.; Kent,** T. A.; Orme-Johnson, W. H.; Münck, E. *J. Biol. Chem.* **1985**, *260*, 11160-11173.

412

(7)   Que, L., Jr.; True, A. E. In *Progress in Inorganic Chemistry: Bioinorganic Chemistry*; S. J. Lippard, Ed.; Wiley (Interscience): New York, 1990; Vol. 38; pp 97-200.

(8)   Kurtz, D. M., Jr. *Chem. Rev.* **1990**, *90*, 585-606.

(9)   Vincent, J. B.; Olivier-Lilley, G. L.; Averill, B. A. *Chem. Rev.* **1990**, *90*, 1447-1467.

(10)   Personnel communication from Dr. Michael P. Hendrich.

(11)   Bertrand, P.; Guigliarelli, B.; More, C. *New J. Chem.* **1991**, *15*, 445-454.

(12)   Antanaitis, B. C.; Aisen, P. *J. Biol. Chem.* **1985**, *260*, 751-756.

(13)   Keough, D. T.; Beck, J. L.; de Jersey, J.; Zerner, B. *Biochem. Biophys. Res. Commun.* **1982**, *108*, 1643-1648.

(14)   Pyrz, J. W.; Sage, J. T.; Debrunner, P. G.; Que, L., Jr. *J. Biol. Chem.* **1986**, *261*, 11015-11020.

(15)   David, S. S.; Que, L., Jr *J. Am. Chem. Soc.* **1990**, *112*, 6455-6463.

(16)   Crowder, M. W.; Vincent, J. B.; Averill, B. A. *Biochemistry* **1992**, *31*, 9603-9608.

(17)   Antanaitis, B. C.; Aisen, P.; Lilienthal, H. R.; Roberts, R. M.; Bazer, F. W. *J. Biol. Chem.* **1980**, *255*, 11204-11209.

(18)   Davis, J. C.; Averill, B. A. *Proc. Natl. Acad. Sci. USA* **1982**, *79*, 4623-4627.

(19)   Lauffer, R. B.; Antanaitis, B. C.; Aisen, P.; Que, L., Jr. *J. Biol. Chem.* **1983**, *258*, 14212-14218.

(20)   Antanaitis, B. C.; Aisen, P.; Lilienthal, H. R. *J. Biol. Chem.* **1983**, *258*, 3166-3172.

(21)   Averill, B. A.; Davis, J. C.; Burman, S.; Zirino, T.; Sanders-Loehr, J.; Loehr, T. M.; Sage, J. T.; Debrunner, P. G. *J. Am. Chem. Soc.* **1987**, *109*, 3760-3767.

(22)   Smith, B. E.; Eady; R, R. *Eur. J. Biochem.* **1992**, *205*, 1-15.

(23)   Georgiadis, M. M.; Komiya, H.; Chakrabarti, P.; Woo, D.; Kornuc, J. J.; Rees, D. C. *Science* **1992**, *257*, 1653-1659.

(24)   Burris, R. H. *J. Biol. Chem.* **1991**, *266*, 9339-9342.

(25)   Documented by P. A. Lindahl in his Ph.D. thesis, Massachusetts Institute of Technology, Cambridge, MA.

(26)   Gillum, W. O.; Mortenson, L. E.; Chen, J. S.; Holm, R. H. *J. Am. Chem. Soc.* **1977**, *99*, 584-594.

(27)   Burgess, B. K.; Jacobs, D. B.; Steifel, E. I. *Biochim. Biophys. Acta* **1980**, *614*, 196-209.

(28)   Nelson, M. J.; Levy, M. A.; Orme-Johnson, W. H. *Proc. Natl. Acad. Sci. USA* **1983**, *80*, 147-150.

(29)   Hausinger, R. P.; Howard, J. B. *J. Biol. Chem.* **1983**, *258*, 13486-13492.

(30)   Wang, Z. C.; Burns, A.; Watt, G. D. *Biochemistry* **1985**, *24*, 214-221.

(31)   Palmer, G.; Multani, J. S.; Cretney, W. C.; Zumft, W. G.; Mortenson, L. E. *Arch. Biochem. Biophys.* **1972**, *153*, 325-332.

(32)   Zumft, W. G.; Palmer, G.; Mortenson, L. E. *Biochim. Biophys. Acta* **1973**, *292*, 413-421.

(33) Orme-Johnson, W. H.; Davis, L. C.; Henzl, M. T.; Averill, B. A.; Orme-Johnson, N. R.; Munck, E.; Zimmermann, R. In *Recent Developments in Nitrogen Fixation*; W. Newton and J. Postgate, Ed.; Academic Press: New York, NY, 1977; pp 133-178.

(34) Anderson, G. L.; Howard, J. B. *Biochemistry* **1984**, *23*, 2118-2122.

(35) Haaker, H.; Braaksma, A.; Cordewener, J.; Klugkist, J.; Wassink, H.; Grande, H.; Eady, R. R.; Veeger, C. In *Advances in Nitrogen Fixation Research*; C. Veeger and W. E. Newton, Ed.; Nijhoff/Junk: The Hague, Netherlands, 1984; pp 123-131.

(36) Lowe, D. J. *Biochem. J.* **1978**, *175*, 955-957.

(37) Aasa, R.; Vänngård, T. *J. Magn. Reson.* **1975**, *19*, 308-315.

(38) Siegel, L. M.; Davis, P. S. *J. Biol. Chem.* **1974**, *249*, 1587-1598.

(39) Christner, J. A.; Münck, E.; Janick, P. A.; Siegel, L. M. *J. Biol. Chem.* **1981**, *256*, 2098-2101.

(40) McRee, D. E.; Richardson, D. C.; Richardson, J. S.; Siegel, L. M. *J. Biol. Chem.* **1986**, *261*, 10277-10281.

(41) Janick, P. A.; Rueger, D. C.; Krueger, R. J.; Barber, M. J.; Siegel, L. M. *Biochemistry* **1983**, *22*, 396-408.

(42) Day, E. P.; Peterson, J.; Bonvoisin, J. J.; Young, L. J.; Wilkerson, J. O.; Siegel, L. *Biochemistry* **1988**, *27*, 2126-2132.

(43) Siegel, L. M.; Rueger, D. C.; Barber, M. J.; Krueger, R. J.; Orme-Johnson, N. R.; Orme-Johnson, W. H. *J. Biol. Chem.* **1982**, *257*, 6343-6350.

(44) Vega, J. M.; Kamin, H. *J. Biol. Chem.* **1977**, *252*, 896-909.

(45) Young, L. J.; Siegel, L. M. *Biochemistry* **1988**, *27*, 2790-2800.

(46) Kim, J.; Rees, D. C. *Science* **1992**, *257*, 1677-1681.

(47) Zimmermann, R.; Münck, E.; Brill, W. J.; Shah, V. K.; Henzl, M. T.; Rawlings, J.; Orme-Johnson, W. H. *Biochim. Biophys. Acta* **1978**, *537*, 185-207.

(48) McLean, P. A.; Papaefthymiou, V.; Orme-Johnson, W. H.; Münck, E. *J. Biol. Chem.* **1987**, *262*, 12900-12903.

(49) McLean, P. A.; Papaefthymiou, V.; Munck, E.; Orme Johnson, W. H. In *Nitrogen Fixation: Hundred Years After*; Bothe, d. Bruijn and Newton, Ed.; Gustav Fischer: Stuttgart New York, 1988; pp 101-106.

(50) Fauque, G.; Peck, H. D., Jr; Moura, J. J. G.; Huynh, B. H.; Berlier, Y.; DerVartanian, D. V.; Teixeira, M.; Przybyla, A. E.; Lespinat, P. A.; Moura, I.; LeGall, J. *FEMS Microbiol. Rev.* **1988**, *54*, 299-344.

(51) Wang, C.-P.; Franco, R.; Moura, J. J. G.; Moura, I.; Day, E. P. *J. Biol. Chem.* **1992**, *267*, 7378-7380.

(52) Maroney, M. J.; Colpas, G. J.; Bagyinka, C.; Baidya, N.; Mascharak, P. K. *J. Am. Chem. Soc.* **1991**, *113*, 3962-3972.

(53) Andrews, R. K.; Blakeley, R. L.; Zerner, B. In *The Bioinorganic Chemistry of Nickel*; J. R. Lancaster Jr, Ed.; VCH Publishers, Inc: New York, 1988; pp 141-165.

(54) Mobley, H. L. T.; Hausinger, R. P. *Microbiol. Rev.* **1989**, *53*, 85-108.

(55) Clark, P. A.; Wilcox, D. E. *Inorg. Chem.* **1989**, *28*, 1326-1333.

(56) Day, E. P.; Peterson, J.; Sendova, M. S.; Todd, M. J.; Hausinger, R. P. *Inorg. Chem.* **1993**, in press.

(57) Cotton, F. A.; Wilkinson, G. *Advanced Inorganic Chemistry;* 5 ed.; John Wiley & Sons: New York, 1988.

(58) Sacconi, L. In *Transition Metal Chemistry. A Series of Advances*; R. L. Carlin, Ed.; Marcel Dekker, Inc.: New York, 1968; Vol. 4; pp 199-298.

(59) Hagen, W. R. *Biochim. Biophys. Acta* **1982**, *708*, 82-98.

(60) Dunham, W. R.; Sands, R. H.; Shaw, R. W.; Beinert, H. *Biochim. Biophys. Acta* **1983**, *748*, 73-85.

(61) Papaefthymiou, V.; Girerd, J. J.; Moura, I.; Moura, J. J. G.; Münck, E. *J. Am. Chem. Soc.* **1987**, *109*, 4703-4710.

(62) Hendrich, M. P.; Debrunner, P. G. *Biophys. J.* **1989**, *56*, 489-506.

(63) Dexheimer, S. C.; Gohdes, J. W.; Chan, M. K.; Hagen, K. S.; Armstrong, W. H.; Klein, M. P. *J. Am. Chem. Soc.* **1989**, *111*, 8923-8925.

(64) Conover, R. C.; Kowal, A. T.; Fu, W.; Park, J.-B.; Aono, S.; Adams, M. W. W.; Johnson, M. K. *J. Biol. Chem.* **1990**, *265*, 8533-8541.

(65) Hendrich, M. P.; Pearce, L. L.; Que, L., Jr; Chasteen, N. D.; Day, E. P. *J. Am. Chem. Soc.* **1991**, *113*, 3039-3044.

(66) Hendrich, M. P.; Münck, E.; Fox, B. G.; Lipscomb, J. D. *J. Am. Chem. Soc.* **1990**, *112*, 5861-5865.

(67) Juarez-Garcia, C.; Hendrich, M. P.; Holman, T. R.; Que, L.; Münck, E. *J. Am. Chem. Soc.* **1991**, *113*, 518-525.

(68) Surerus, K. K.; Hendrich, M. P.; Christie, P. D.; Rottgardt, D.; Orme-Johnson, W. H.; Münck, E. *J. Am. Chem. Soc.* **1992**, *114*, 8579-8590.

(69) Reem, R. C.; Solomon, E. I. *J. Am. Chem. Soc.* **1987**, *109*, 1216-1226.

(70) Zimmermann, R.; Münck, E.; Brill, W. J.; Shah, V. K.; Henzl, M. T.; Rawlings, J.; Orme-Johnson, W. H. *Biochim. Biophys. Acta* **1978**, *537*, 185-207.

(71) Hagen, W. R.; Wassink, H.; Eady, R. R.; Smith, B. E.; Haaker, H. *Eur. J. Biochem.* **1987**, *169*, 457-465.

(72) Bolin, J. T.; Ronco, A. E.; Mortenson, L. E.; Morgan, T. V.; Williamson, M.; Xuong, N. H. In *Nitrogen Fixation: Achievements and Objectives*; P. M. Gresshoff, L. E. Roth, G. Stacey and W. E. Newton, Ed.; Chapman and Hall: New York, 1990; pp 117-122.

(73) Day, E. P.; Lindahl, P. A.; Orme-Johnson, W. H. *J. Biol. Chem.* **1993**, manuscript in preparation.

(74) Wikström, M.; Krab, K.; Saraste, M. *Cytochrome Oxidase: A Synthesis*; Academic: New York, 1981.

(75) Palmer, G. *Pure Appl Chem* **1987**, *59*, 749-758.

(76) Malmström, B. G. *Chem. Rev.* **1990**, *90*, 1247-1260.

(77) Babcock, G. T.; Wikström, M. *Nature* **1992**, *356*, 301-309.

(78) Baker, G. M.; Noguchi, M.; Palmer, G. *J. Biol. Chem.* **1987**, *262*, 595-604.

(79)   Day, E. P.; Peterson, J.; Sendova, M. S.; Schoonover, J.; Palmer, G.
        Biochemistry **1993**, submitted.
(80)   Gayda, J. P.; Bertrand, P.; Theodule, F. X.; Moura, J. J. G. *J. Chem. Phys.*
        1982, *77*, 3387-3391.
(81)   Guigliarelli, B.; More, C.; Bertrand, P.; Gayda, J. P. *J. Chem. Phys.* **1986**, *85*,
        2774-2778.
(82)   Day, E. P.; Peterson, J.; Bonvoisin, J. J.; Moura, I.; Moura, J. J. G. *J. Biol.
        Chem.* **1988**, *263*, 3684-3689.
(83)   Huynh, B. H.; Moura, J. J. G.; Moura, I.; Kent, T. A.; LeGall, J.; Xavier, A. V.;
        Münck, E. *J. Biol. Chem.* **1980**, *255*, 3242-3244.
(84)   Girerd, J. J.; Papaefthymiou, V.; Surerus, K. K.; Münck, E. *Pure Appl Chem*
        1989, *61*, 805-816.
(85)   Peterson, J.; Fee, J. A.; Day, E. P. *Biochim. Biophys. Acta* **1991**, *1079*, 161-
        168.
(86)   Payne, W. J. *Denitrification*; Wiley: New York, 1981.
(87)   Knowles, R. *Microbiol. Rev.* **1982**, *46*, 43-70.
(88)   Dooley, D. M.; Landin, J. A.; Rosenzweig, A. C.; Zumft, W. G.; Day, E. P. *J.
        Am. Chem. Soc.* **1991**, *113*, 8978-8980.
(89)   SooHoo, C. K.; Hollocher, T. C.; Kolodziej, A. F.; Orme-Johnson, W. H.;
        Bunker, G. *J. Biol. Chem.* **1991**, *266*, 2210-2218.
(90)   Zhang, C.-s.; Hollocher, T. C.; Kolodziej, A. F.; Orme-Johnson, W. H. *J. Biol.
        Chem.* **1991**, *266*, 2199-2202.